CAMBRIDGE TRACTS IN MATHEMATICS

General editors

Hopf algebras

EIICHI ABE
Professor of Mathematics, University of Tsukuba

Hopf algebras

translated by Hisae Kinoshita and Hiroko Tanaka

CAMBRIDGE UNIVERSITY PRESS

CAMBRIDGE
LONDON NEW YORK NEW ROCHELLE
MELBOURNE SYDNEY

PUBLISHED BY THE PRESS SYNDICATE OF THE UNIVERSITY OF CAMBRIDGE
The Pitt Building, Trumpington Street, Cambridge, United Kingdom

CAMBRIDGE UNIVERSITY PRESS
The Edinburgh Building, Cambridge CB2 2RU, UK
40 West 20th Street, New York NY 10011–4211, USA
477 Williamstown Road, Port Melbourne, VIC 3207, Australia
Ruiz de Alarcón 13, 28014 Madrid, Spain
Dock House, The Waterfront, Cape Town 8001, South Africa

http://www.cambridge.org

Translated by Hiroko Tanaka and Hisae Kinoshita

Originally published in Japanese by Iwanami Shoten, Publishers, Tokyo, 1977
English edition first published by Cambridge University Press, 1980
First paperback edition 2004

A catalogue record for this book is available from the British Library

Library of Congress cataloguing in publication data

Abe, Eiichi, 1927-
Hopf algebras.
(Cambridge tracts in mathematics)
Includes index.
1. Hopf algebras. I. Title. II. Series.
QA613.8.A213 512′.55 79-50912

ISBN 0 521 22240 0 hardback
ISBN 0 521 60489 3 paperback

Contents

Preface

Let G be a finite group and k a field. The set $A = \text{Map}\,(G, k)$ of all functions defined on G with values in k becomes a k-algebra when we define the scalar product and the sum and product of functions by

$$(\alpha f)(x) = \alpha f(x), \qquad (f + g)(x) = f(x) + g(x),$$
$$(fg)(x) = f(x)g(x), \ f, g \in A, \alpha \in k, x \in G.$$

In general, a k-algebra A can be characterized as a k-linear space A together with two k-linear maps

$$\mu : A \otimes_k A \to A, \quad \eta : k \to A,$$

which satisfy axioms corresponding to the associative law and the unitary property respectively. If we identify $A \otimes_k A$ with Map $(G \times G, k)$ where $A = \text{Map}\,(G, k)$, and if the operations of G are employed in defining the k-linear maps

$$\Delta : A \to A \otimes_k A, \quad \varepsilon : A \to k$$

respectively by $\Delta f(x \otimes y) = f(xy)$, $\varepsilon f = f(e)$ for x, $y \in G$ and where e is the identify element of G, then Δ and ε become homomorphisms of k-algebras having properties which are dual to μ and η respectively. In general, a k-linear space A with k-linear maps μ, η, Δ, ε defined as above is called a k-bialgebra. Furthermore, we can define a k-linear endomorphism of $A = \text{Map}\,(G, k)$

$$S : A \to A, \quad (Sf)(x) = f(x^{-1}), \quad f \in A, \quad x \in G$$

such that the equalities

$$\mu(1 \otimes S)\Delta = \mu(S \otimes 1)\Delta = \eta \circ \varepsilon$$

hold. A k-bialgebra on which we can define a k-linear map S as above is called a k-Hopf algebra. Thus, a k-Hopf algebra is an algebraic system which simultaneously admits structures of a k-algebra as well as its dual, where these two structures are related by a certain specific law. For a finite group G, its group ring kG over a field k is the dual space of the k-linear space $A = \mathrm{Map}(G, k)$ where its k-algebra structure is given by the dual k-linear maps of Δ and ε. Moreover, kG admits a k-Hopf algebra structure when we take the dual k-linear maps of μ, η, and S. In other words, kG is the dual k-Hopf algebra of $\mathrm{Map}(G, k)$.

If, for instance, we replace the finite group G in the above argument by a topological group and k by the field of real numbers or the field of complex numbers, or if we take G to be an algebraic group over an algebraically closed field k and A is replaced by the k-algebra of all continuous representative functions or of all regular functions over G, then A turns out to be a k-Hopf algebra in exactly the same manner. These algebraic systems play an important role when studying the structure of G. Similarly, a k-Hopf algebra structure can be defined naturally on the universal enveloping algebra of a k-Lie algebra. The universal enveloping algebra of the Lie algebra of a semi-simple algebraic group turns out to be (in a sense) the dual of the Hopf algebra defined above. These constitute some of the most natural examples of Hopf algebras. The general structure of such algebraic systems has recently become a focus of interest in conjunction with its applications to the theory of algebraic groups or the Galois theory of purely inseparable extensions, and a great deal of research is currently being conducted in this area.

It has only been since the late 1960s that Hopf algebras, as algebraic systems, became objects of study from an algebraic standpoint. However, beginning with the research on representation theory through the use of the representative rings of Lie groups by Hochschild–Mostow (*Ann. of Math.* **66** (1957), 495–542, **68** (1958), 295–313) and in subsequent studies (cf. references [4], [7], [8] for Chapter 3), Hopf algebras have been taken up extensively as algebraic systems and also used in applications. On the other hand, in algebraic topology, the concept of graded Hopf algebras was derived at an even earlier date from an axiomatization of the works of H. Hopf relating to topological properties of Lie groups (cf. *Ann. of Math.* **42** (1941), 22–52). Hence the name, 'Hopf algebra'. For instance, if G is a connected Lie group, the cohomology group $H^*(G)$ or the homology group $H_*(G)$ of G with coefficients in a field k has a multiplication or a dual multiplication induced by the diagonal map $d: G \to G \times G$ ($d(x) = (x, x)$, $x \in G$) and is moreover a commutative k-algebra. In addition, the map $m: G \times G \to G$ defined via the Lie group multiplication induces the maps

$$\Delta: H^*(G) \to H^*(G) \otimes H^*(G) \quad \text{or} \quad \Delta: H_*(G) \to H_*(G) \otimes H_*(G)$$

which respectively make $H^*(G)$ or $H_*(G)$ a k-Hopf algebra. These are graded Hopf algebras and such structures can be defined also for H-spaces, and have been generalized by A. Borel, J. Leary, and others (cf. A. Borel: *Ann. of Math.* **57** (1953), 115–207). For details on the algebraic properties of Hopf algebras, the reader is referred to J. Milnor–J. C. Moore: *Ann of Math.* **81** (1965), 211–64.

This book has as its main objective algebraic applications of (non-graded) Hopf algebras, and an attempt has been made to acquaint the reader with the elementary properties of Hopf algebras with a minimal amount of preliminary knowledge. There is an excellent work on the subject, *Hopf algebras* by Sweedler (Benjamin, 1969), to which this book owes a great deal. But here the central theme will revolve around applications of Hopf algebras to the representative rings of topological groups and to algebraic groups. Some recent developments on the subject have also been incorporated.

This book consists of five chapters and an appendix. Chapter 1 is preparation for the central subject of the book and deals with some basic properties of modules and algebras which become necessary in the sequel. Some simple properties of groups, fields, and topological spaces have been used without proofs. With regard to the solutions to some of the exercises and in the treatment of finitely generated commutative algebras, where I have either omitted or condensed a number of the proofs of well-known theorems, the reader is asked to refer to other texts. In Chapter 2, coalgebras, bialgebras, and Hopf algebras are defined, and their fundamental properties are outlined. Chapter 3 takes up the structure theorem of bimodules and properties of Hopf algebras similar to those of the representative rings of topological groups. The proofs of the existence and uniqueness of the integral of commutative Hopf algebras are due to J. B. Sullivan. In Chapter 4, fundamental properties of affine algebraic groups are proven through an application of the theory of Hopf algebras. The construction of factor groups, the proof of the decomposition theorem of solvable groups, and the proofs of theorems related to completely reducible groups are respectively due to Mitsuhiro Takeuchi, J. B. Sullivan, and M. E. Sweedler. Although properties pertaining to the representation of affine algebraic groups can be described satisfactorily by such an approach, there are drawbacks as well as merits in pursuing a general theory in this context. For instance, it may distort the overall view of the development of the subject matter. It should be interesting, for instance, to attempt an exposition of the theory of Borel subgroups, which is an important topic in the theory of linear algebraic groups, not touched upon in this book. Chapter 5 contains a brief glance at the Galois theory of purely inseparable extensions. Although there are numerous texts dealing with Galois theory, here I have presented a treatment of the subject by D. Winter. The appendix contains a sketch of the theory of categories, where notions such as categories, functors and, for a category $\mathscr{C}$, $\mathscr{C}$-groups and $\mathscr{C}$-cogroups, which are used throughout this book, are defined.

Since commutative Hopf algebras are precisely the commutative algebras which represent affine group schemes, applications of Hopf algebras also naturally develop along these lines. The reader will do well to refer to other texts on this matter.

The writing of this book was suggested to me by Professor Nagayoshi Iwahori of the Faculty of Science of the University of Tokyo, to whom I would like to express my most sincere gratitude. I am also very grateful to Professors Yukio Doi of Fukui University and Mitsuhiro Takeuchi of the University of Tsukuba for many valuable words of advice which they have given me in preparing the manuscript.

Eiichi Abe
Tokyo, Japan

Notations

Sets

$a \in M$ or $M \ni a$	a is an element of the set M
$M \subset N$ or $N \supset M$	$a \in M$ implies $a \in N$, namely, M is contained in N
$M \cup N$; $\bigcup_{\lambda \in \Lambda} M_\lambda$	the union of M and N; the union of M_λ ($\lambda \in \Lambda$)
$M \cap N$; $\bigcap_{\lambda \in \Lambda} M_\lambda$	the intersection of M and N; the intersection of M_λ ($\lambda \in \Lambda$)
$\emptyset$	the empty set
$\mathbb{Z}$	the set of all integers
$\mathbb{N}$	the set of all natural numbers

Maps

$f: M \to N$	a map from the set M to the set N
$x \mapsto y$	the image $f(x)$ of $x \in M$ under the map f is $y \in N$

When $f(M) = N$, f is called a surjection.
If $x, x' \in M$, $x \neq x'$ implies $f(x) \neq f(x')$, then f is called an injection.
For $M' \subseteq M$, we denote the restriction of f to M' by $f|_{M'}$.
If $f: M \to N, g: N \to P$ are two maps, the composition of f and g is written $g \circ f$, $x \mapsto g(f(x))$.

Logical symbols

$A \Rightarrow B$	If proposition A holds, then proposition B holds
$A \Leftrightarrow B$	$A \Rightarrow B$ and $B \Rightarrow A$
$\forall x \in M$	for any element x of M

1
Modules and algebras

1 Modules

This section deals with direct sums, direct products, tensor products, and the projective and inductive limits of modules. Proofs of some fundamental properties of such constructions have been omitted and left as exercises. For these, the reader is asked to refer to texts such as [3] or [5].

1.1 Modules

A set A with two operations – addition and multiplication – which satisfies properties (1) to (3) below is said to be a **ring with identity**. Since this book deals exclusively with this type of ring, we will call them simply **rings**.

(1) The addition $+$ makes A an abelian group.

(2) The multiplication $\cdot$ makes A a semigroup with identity element 1.

(3) The distributive law holds. Namely, for $a, b, c \in A$, we have

$$a \cdot (b + c) = a \cdot b + a \cdot c, \quad (a + b) \cdot c = a \cdot c + b \cdot c.$$

A ring with a commutative multiplication is called a **commutative ring**. Henceforth, the product $a \cdot b$ of $a, b \in A$ will be written ab. Let A and B be rings. If a map $u : A \to B$ satisfies the properties

$$u(a + b) = u(a) + u(b), \quad u(ab) = u(a)u(b), \quad u(1) = 1, \quad a, b \in A,$$

then u is said to be a **ring morphism** from A to B. The category of rings (resp. commutative rings) will be denoted **Alg** (resp. **M**) and the set of all ring morphisms from A to B will be written **Alg** (A, B) (resp. $\mathbf{M}(A, B)$ when A, B are commutative rings).

For a ring A and an abelian group M, suppose we are given a map $\varphi : A \times M \to M$ (resp. $\psi : M \times A \to M$). We signify the group operation on M by addition $+$, and for $a \in A$, $x \in M$, we write $\varphi(a, x) = ax$ (resp. $\psi(x, a) = xa$). When conditions (1) to (4) (resp. (1′) to (4′)) below hold for $a, b \in A$ and $x, y \in M$, M is called a **left A-module** (resp. **right A-module**), and φ (resp. ψ) is said to be the **structure map** of the left (resp. right) A-module.

(1)	$a(x+y) = ax + ay$,	resp.	(1′)	$(x+y)a = xa + ya$,
(2)	$(a+b)x = ax + bx$,		(2′)	$x(a+b) = xa + xb$,
(3)	$(ab)x = a(bx)$,		(3′)	$x(ab) = (xa)b$,
(4)	$1x = x$.		(4′)	$x1 = x$.

Moreover, given rings A, B, if M is both a left A-module and a right B-module satisfying the condition

$$(ax)b = a(xb), \quad a \in A, \quad b \in B, \quad x \in M,$$

then M is called a **two-sided (A, B)-module**. A two-sided (A, A)-module is called simply a **two-sided A-module**. If A is a commutative ring, a left A-module can be regarded as a right A-module, and is often simply called an A-module. For instance, an abelian group is a $\mathbb{Z}$-module. In the case of a ring A, when the map defining the multiplication $\mu : A \times A \to A$ given by $\mu(a, b) = ab$, $a, b \in A$ is taken to be the structure map, A becomes a left A-module as well as a right A-module. Moreover, A is a two-sided A-module. For a field k, a k-module is also called a **k-linear space** or a **k-vector space**.

Let M, N be left A-modules. A map $f : M \to N$ such that

$$f(ax + by) = af(x) + bf(y), \quad a, b \in A, \quad x, y \in M,$$

is called a **left A-module morphism** from M to N. If k is a field, a k-module morphism is sometimes called a **k-linear map**. The category of left A-modules is denoted ${}_A\mathbf{Mod}$, and the set of all left A-module morphisms from M to N is denoted ${}_A\mathbf{Mod}(M, N)$. Similarly, given right A-modules M, N, we can define **right A-module morphisms** and the set of all right A-module morphisms from M to N, which we write $\mathbf{Mod}_A(M, N)$. In particular, $\mathbf{Mod}_A(M, M)$ is written $\mathrm{End}_A(M)$. If $f \in {}_A\mathbf{Mod}(M, N)$ or $f \in \mathbf{Mod}_A(M, N)$ is bijective, f is said to be an **isomorphism**. The identity map from M to M is an isomorphism, denoted by 1_M or simply by 1.

Now let $f, g \in {}_A\mathbf{Mod}(M, N)$. Defining

$$(f \pm g)(x) = f(x) \pm g(x), \quad x \in M,$$

$f \pm g$ becomes a left A-module morphism from M to N. Under this operation, ${}_A\mathbf{Mod}(M, N)$ becomes an abelian group. If N is also a two-sided A-module, then by defining

$$(fa)(x) = f(x)a, \quad a \in A, \quad x \in M,$$

we have $fa \in {}_A\mathbf{Mod}(M, N)$, and hence ${}_A\mathbf{Mod}(M, N)$ becomes a right A-module. When in particular $N = A$, N is a two-sided A-module, and here, the right A-module ${}_A\mathbf{Mod}(M, A)$ is called the **dual right A-module** of the left A-module M, which is denoted by M^*. If A is commutative, ${}_A\mathbf{Mod}(M, N)$ can be regarded as a left A-module.

EXERCISE 1.1 Given a left A-module morphism $f : M \to N$, f is an isomorphism $\Leftrightarrow$ there exists a left A-module morphism $g : N \to M$ such that $f \circ g = 1_N$ and $g \circ f = 1_M$.

EXERCISE 1.2 Let A be a commutative ring. Given left A-modules M, M', N, N' and left A-module morphisms $g : M \to M'$, $h : N \to N'$, the maps

$$g^* : {}_A\mathbf{Mod}(M', N) \to {}_A\mathbf{Mod}(M, N), \quad f \mapsto f \circ g,$$
$$h_* : {}_A\mathbf{Mod}(M, N) \to {}_A\mathbf{Mod}(M, N'), \quad f \mapsto h \circ f,$$

are left A-module morphisms.

We observe that if a subgroup N of a left A-module M satisfies the condition

$$x \in N, \quad a \in A \Rightarrow ax \in N,$$

then N is a left A-module. Such an N is called a **left A-submodule** of M. The factor group M/N also inherits a left A-module structure, and M/N is called a **factor left A-module**. Regarding a ring A as a left A-module (resp. right A-module; two-sided A-module), then an A-submodule of A is simply a **left ideal** (resp. **right ideal**; **two-sided ideal**).

Suppose now that the only left A-submodules of a left A-module M

are $\{0\}$ and M. In this situation, we call M a **simple** (or **irreducible**) **left A-module**. Given a left A-module morphism $f: M \to N$, the sets

$$\operatorname{Ker} f = \{x \in M;\ \ f(x) = 0\},$$
$$\operatorname{Im} f = \{f(x) \in N;\ \ x \in M\}$$

are left A-submodules of M and N respectively and are called the **kernel** of f and the **image** of f. The smallest left A-submodule which contains a subset S of a left A-module M is written $\langle S \rangle$ and called the left A-submodule generated by S.

Let Λ be a finite or infinite sequence of consecutive integers and let M_i $(i \in \Lambda)$ be left A-modules. Suppose we are given left A-module morphisms $f_i: M_i \to M_{i+1}$ $(i, i+1 \in \Lambda)$. When $\operatorname{Ker} f_{i+1} = \operatorname{Im} f_i$ $(i, i+1 \in \Lambda)$ for the sequence of left A-module morphisms

$$\cdots \to M_i \to M_{i+1} \to M_{i+2} \to \cdots, \tag{1.1}$$

then (1.1) is said to be an **exact sequence**. For instance, when $0 \to M \xrightarrow{f} N$ (resp. $M \xrightarrow{f} N \to 0$; $0 \to M \xrightarrow{f} N \to 0$) is an exact sequence, then f is injective (resp. surjective; bijective), and the converse also holds.

EXERCISE 1.3 Let A be a commutative ring. For a sequence $M' \xrightarrow{f} M \xrightarrow{g} M'' \to 0$ of A-module morphisms to be an exact sequence, it is necessary and sufficient that, for any left A-module N, the sequence

$$0 \to {}_A\mathbf{Mod}(M'', N) \xrightarrow{g^*} {}_A\mathbf{Mod}(M, N) \xrightarrow{f^*} {}_A\mathbf{Mod}(M', N)$$

is exact. Furthermore, a sequence $0 \to N' \xrightarrow{f} N \xrightarrow{g} N''$ of left A-module morphisms is exact if and only if, for any left A-module M, the sequence

$$0 \to {}_A\mathbf{Mod}(M, N') \xrightarrow{f_*} {}_A\mathbf{Mod}(M, N) \xrightarrow{g_*} {}_A\mathbf{Mod}(M, N'')$$

is exact (cf. Exercise 1.2).

1.2 Direct products and direct sums

Let $\{M_\lambda\}_{\lambda \in \Lambda}$ be a family of left A-modules. Pick one element x_λ from each M_λ and write the resulting set $x = \{x_\lambda\}_{\lambda \in \Lambda}$, calling x_λ the λ-component of x. Let P be the set of all $x = \{x_\lambda\}_{\lambda \in \Lambda}$ constructed in the above manner. For $x=\{x_\lambda\}_{\lambda\in\Lambda}$, $y=\{y_\lambda\}_{\lambda\in\Lambda}\in P$ and $a\in A$, we define the operations

$$x+y=\{x_\lambda+y_\lambda\}_{\lambda\in\Lambda},\quad ax=\{ax_\lambda\}_{\lambda\in\Lambda},$$

which make P a left A-module. The map $p_\lambda : P\to M_\lambda$ which assigns to $x=\{x_\lambda\}_{\lambda\in\Lambda}\in P$ the λ-component x_λ of x is a left A-module morphism, and we call p_λ the canonical projection from P to M_λ. The pair $(P, \{p_\lambda\}_{\lambda\in\Lambda})$ consisting of P and the family of canonical projections p_λ $(\lambda\in\Lambda)$ is called the **direct product** of the family of left A-modules $\{M_\lambda\}_{\lambda\in\Lambda}$, and is written $P=\prod_{\lambda\in\Lambda} M_\lambda$. For $\Lambda=\{1,2,\ldots,n\}$, this is sometimes written $M_1\times\ldots\times M_n$. The direct product $(P, \{p_\lambda\}_{\lambda\in\Lambda})$ has the following property.

(P) Given a pair $(N, \{q_\lambda\}_{\lambda\in\Lambda})$ consisting of an arbitrary left A-module N and a family of left A-module morphisms $q_\lambda : N\to M_\lambda$ $(\lambda\in\Lambda)$, there exists a unique left A-module morphism $f : N\to P$ which satisfies $p_\lambda\circ f=q_\lambda$ $(\lambda\in\Lambda)$.

Hence the map which assigns to each $f\in{}_A\mathbf{Mod}(N, P)$, the element $\{p_\lambda\circ f\}_{\lambda\in\Lambda}\in\prod_{\lambda\in\Lambda}{}_A\mathbf{Mod}(N, M_\lambda)$ is a bijection. Furthermore, if A is a commutative ring, we have

$${}_A\mathbf{Mod}(N, P)\cong\prod_{\lambda\in\Lambda}{}_A\mathbf{Mod}(N, M_\lambda)$$

as left A-modules. The element of ${}_A\mathbf{Mod}(N, P)$ which corresponds to $\{f_\lambda\}_{\lambda\in\Lambda}\in\prod_{\lambda\in\Lambda}{}_A\mathbf{Mod}(N, M_\lambda)$ is written $\prod_{\lambda\in\Lambda} f_\lambda$ and is said to be the direct product of the A-module morphisms $\{f_\lambda\}_{\lambda\in\Lambda}$. A left A-module P with the above property is unique up to isomorphism, and the direct product of the family of left A-modules $\{M_\lambda\}_{\lambda\in\Lambda}$ is characterized by property (P).

Let S be the subset of P consisting of all those elements whose λ-components are zero except for a finite number of λs. Then S turns out to be a left A-submodule of P. When Λ is a finite set, we have $S=P$. Given $x_\lambda\in M_\lambda$, let $i_\lambda(x_\lambda)$ stand for the element of S whose λ-component is x_λ and all other components zero. Then the map $i_\lambda : M_\lambda\to S$ is a left A-module injection and is called the canonical embedding of M_λ into S. Identifying $i_\lambda(x_\lambda)$ with x_λ and regarding M_λ as a left A-submodule of S, the element $x=\{x_\lambda\}_{\lambda\in\Lambda}$ of S can be written in the form $\sum_{\lambda \in \Lambda} x_\lambda$ which, by definition, is a finite sum. The pair $(S, \{i_\lambda\}_{\lambda \in \Lambda})$ consisting of S and the family of left A-module morphisms i_λ $(\lambda \in \Lambda)$ is called the **direct sum** of the family $\{M_\lambda\}_{\lambda \in \Lambda}$ of A-modules, and is denoted $S = \coprod_{\lambda \in \Lambda} M_\lambda$ or $S = \bigoplus_{\lambda \in \Lambda} M_\lambda$. The direct sum $(S, \{i_\lambda\}_{\lambda \in \Lambda})$ has the following property.

(S) Given a pair $(N, \{j_\lambda\}_{\lambda \in \Lambda})$ consisting of an arbitary left A-module N and a family of left A-module morphisms $j_\lambda : M_\lambda \to N$, there exists a unique left A-module morphism $f : S \to N$ such that $f \circ i_\lambda = j_\lambda (\lambda \in \Lambda)$.

Thus the map that assigns to each $f \in {}_A\mathbf{Mod}(S, N)$ the element $\{f \circ i_\lambda\}_{\lambda \in \Lambda} \in \prod_{\lambda \in \Lambda} {}_A\mathbf{Mod}(M_\lambda, N)$ is a bijection. Moreover, if A is a commutative ring, this is an isomorphism

$$ {}_A\mathbf{Mod}(S, N) \cong \prod_{\lambda \in \Lambda} {}_A\mathbf{Mod}(M_\lambda, N) $$

of left A-modules. The element of ${}_A\mathbf{Mod}(S, N)$ corresponding to $\{f_\lambda\}_{\lambda \in \Lambda} \in \prod_{\lambda \in \Lambda} {}_A\mathbf{Mod}(M_\lambda, N)$ is written $\coprod_{\lambda \in \Lambda} f_\lambda$ and is called the direct sum of A-module morphisms $\{f_\lambda\}_{\lambda \in \Lambda}$. There exists a left A-module S which is unique up to isomorphism with the above property, and the direct sum of a family of left A-modules $\{M_\lambda\}_{\lambda \in \Lambda}$ can be characterized as the left A-module which satisfies property (S).

Free A-modules Let A be a ring. Given a set Λ, we assign to each $\lambda \in \Lambda$ a left A-module A_λ which is isomorphic to A, and denote the direct sum of the family of left A-modules $\{A_\lambda\}_{\lambda \in \Lambda}$ by $F_A(\Lambda) = \coprod_{\lambda \in \Lambda} A_\lambda$. If we identify the image under i_λ of the identity element 1_λ of A_λ with λ, then we have $\Lambda \subset F_A(\Lambda)$, and an element x of $F_A(\Lambda)$ can be written uniquely in the form $x = \sum_{\lambda \in \Lambda} x_\lambda \lambda$ where $x_\lambda \in A$ and $x_\lambda = 0$ except for a finite number of λ. In this situation, $F_A(\Lambda)$ is said to be the **free left A-module** generated by Λ. Letting Map (Λ, M) be the family of all maps from Λ to a left A-module M, we obtain a one-to-one correspondence

$$ {}_A\mathbf{Mod}(F_A(\Lambda), M) \cong \mathrm{Map}(\Lambda, M). $$

Now, Map (Λ, M) has a left A-module structure. Moreover, when A is a commutative ring, the two left A-modules above are isomorphic as left A-modules under the same correspondence.

Bases Let A be a ring and S a subset of a left A-module M. If S satisfies conditions (1) and (2) below, S is said to be a **basis** for the A-module M.

(1) For any finite subset $\{s_1, \ldots, s_n\}$ of S,

$$\sum_{i=1}^{n} a_i s_i = 0, \quad a_i \in A \ (1 \leqq i \leqq n) \Rightarrow a_i = 0 \ (1 \leqq i \leqq n).$$

(2) Given an arbitrary element $x \in M$, a finite subset $\{s_1, \ldots, s_n\}$ of S can be chosen appropriately so that x can be written in the form

$$x = \sum_{i=1}^{n} a_i s_i \quad \text{where} \quad a_i \in A \ (1 \leqq i \leqq n).$$

A set which satisfies condition (1) is called a **linearly independent** set over A, and a set which satisfies condition (2) is called a **system of generators** of M over A. In particular, a left A-module which has a finite system of generators is said to be **finitely generated**. If $F_A(\Lambda)$ is a free A-module over Λ, Λ becomes a basis for $F_A(\Lambda)$ when we regard Λ as a subset of $F_A(\Lambda)$. Conversely, an A-module which has a basis S is isomorphic to the free A-module $F_A(S)$. Consequently, for a left A-module to be a free left A-module, it is necessary and sufficient that it have a basis. Moreover, when A is a commutative ring, the number of elements in a basis for a finitely generated free A-module is constant regardless of the choice of a basis. This number is called the **rank** of the free A-module. Given a field k, a k-vector space has a basis, and is therefore a free k-module. We call the rank of a finitely generated k-vector space V its **dimension**, and write dim V.

EXERCISE 1.4 Let V_1 and V_2 be subspaces of a finite dimensional k-vector space V. Then

(i) $V_1 \subset V_2 \Rightarrow \dim V_1 \leqq \dim V_2$.

(ii) $V_1 \subset V_2$ and $\dim V_1 = \dim V_2 \Rightarrow V_1 = V_2$.

(iii) $\dim V_1 + \dim V_2 = \dim(V_1 + V_2) + \dim(V_1 \cap V_2)$, where $V_1 + V_2$ is the subspace of V generated by $V_1 \cap V_2$.

EXERCISE 1.5 If $f: V \to V'$ is a k-linear map, then $\dim V = \dim(\operatorname{Im} f) + \dim(\operatorname{Ker} f)$.

Completely reducible modules Let M be a left A-module. Given any left A-submodule N of M, if there exists a left A-submodule N' of M such that $M = N \oplus N'$ (the direct sum of N and N'), then M is said to be **completely reducible**. A completely reducible left A-module can be expressed as the direct sum of its irreducible left A-submodules. Further, a left A-submodule of a completely reducible left A-module is also completely reducible. (See for instance Hattori [1], Theorems 15.7, 15.8, or Curtis–Reiner [5], 15.2, 15.3.)

1.3 Tensor products

Suppose we are given a ring A, a right A-module M, and a left A-module N. Let $F(M \times N)$ be the free $\mathbb{Z}$-module generated by the set

$$M \times N = \{(x, y);\quad x \in M,\quad y \in N\}$$

and let $K(M \times N)$ be the $\mathbb{Z}$-submodule of $F(M \times N)$ generated by all elements of the type

$$(x + x', y) - (x, y) - (x', y),$$
$$(x, y + y') - (x, y) - (x, y'),$$
$$(xa, y) - (x, ay),$$

for all $x, x' \in M$, $y, y' \in N$, $a \in A$. In these circumstances, we call the factor group $F(M \times N)/K(M \times N)$ the **tensor product** of M and N over A, and denote it by $M \otimes_A N$ or simply by $M \otimes N$. The residue class which contains (x, y) is written $x \otimes y$. For $x, x' \in M$, $y, y' \in N$, $a \in A$, we have by definition

$$(x + x') \otimes y = x \otimes y + x' \otimes y,$$
$$x \otimes (y + y') = x \otimes y + x \otimes y',$$
$$xa \otimes y = x \otimes ay.$$

An element of $M \otimes_A N$ can be written in the form $\sum_{i=1}^{n} x_i \otimes y_i$ where $x_i \in M$, $y_i \in N$. If M is a two-sided (B, A)-module (resp. N is a two-sided

(A, C)-module), then the definition

$$b(x \otimes y) = bx \otimes y, \quad b \in B, x \in M, y \in N$$
$$(\text{resp.} \quad (x \otimes y)c = x \otimes yc, \quad c \in C, x \in M, y \in N),$$

makes $M \otimes_A N$ a left B-module (resp. right C-module). When A is a commutative ring, we can define

$$a(x \otimes y) = ax \otimes y = x \otimes ay, \quad a \in A, \quad x \in M, \quad y \in N,$$

which makes $M \otimes_A N$ an A-module.

Bilinear maps Let A be a commutative ring and let M, N, T be A-modules. A map $f : M \times N \to T$ such that

$$f(ax + bx', y) = af(x, y) + bf(x', y),$$
$$f(x, ay + by') = af(x, y) + bf(x, y'),$$
$$a, b \in A, \quad x, x' \in M, \quad y, y' \in N,$$

is called a **bilinear map**. The set of all bilinear maps from $M \times N$ to T is denoted $B_A(M \times N, T)$. By the definition of tensor products, the map from $M \times N$ to the A-module $M \otimes_A N$

$$\varphi : M \times N \to M \otimes_A N$$

given by $(x, y) \mapsto x \otimes y$ is bilinear, and is called the **canonical bilinear map**. Suppose we are given an A-module T and an A-module morphism $g : M \otimes_A N \to T$. Setting $f = g \circ \varphi$, we see that $f : M \times N \to T$ is a bilinear map. Moreover, the map

$$\Phi : {}_A\mathbf{Mod}\,(M \otimes_A N, T) \to B_A(M \times N, T)$$

which carries g to f is a bijection. In fact, for $f \in B_A(M \times N, T)$, we define $g(x \otimes y) = f(x, y)$, thereby obtaining an A-module morphism $g : M \otimes_A N \to T$. The map defined by $f \mapsto g$ turns out to be the inverse of Φ. The set $B_A(M \times N, T)$ admits an A-module structure making Φ an A-module isomorphism. Further, for any A-module T, the tensor product $M \otimes_A N$ which makes Φ a bijection is unique up to isomorphism. Moreover, the tensor product $M \otimes_A N$ is characterized by the above mentioned property. When M is a free A-module with a basis $\{e_\lambda\}_{\lambda \in \Lambda}$, an element of $M \otimes_A N$ can be written uniquely in the form $\sum_{\lambda \in \Lambda} e_\lambda \otimes y_\lambda$ where $y_\lambda \in N$, $\lambda \in \Lambda$ and $y_\lambda = 0$ except for a finite number of λs. Furthermore, if M and N are free A-modules, so is $M \otimes_A N$.

EXERCISE 1.6 Given a commutative ring A and A-modules M, N, P, prove that

(i) $A \otimes_A M \cong M$,

(ii) $M \otimes_A N \cong N \otimes_A M$,

(iii) $(M \otimes_A N) \otimes_A P \cong M \otimes_A (N \otimes_A P)$.

EXERCISE 1.7 Given a commutative ring A and A-modules M, N, let M^*, N^* be the dual A-modules of M, N respectively. Show that

$$ {}_A\mathbf{Mod}\,(M, N^*) \cong {}_A\mathbf{Mod}\,(N, M^*) \cong B_A(M \times N, A). $$

Remark The map $F : M \mapsto M^*$ from the category ${}_A\mathbf{Mod}$ to itself is a contravariant functor, and F is adjoint to itself.

EXERCISE 1.8 Given a field k, let M, N be k-vector spaces with dual k-vector spaces M^*, N^* respectively.

(i) Define a map $\varphi : M^* \otimes N \to \mathbf{Mod}_k(M, N)$ for $f \in M^*$, $y \in N$, $x \in M$ by $\varphi(f \otimes y)(x) = f(x)y$. Then φ is a k-linear injection. Moreover, φ is a k-linear isomorphism if M or N is finite dimensional.

(ii) Define a map $\rho : M^* \otimes N^* \to (M \otimes N)^*$ for $f \in M^*$, $g \in N^*$, $x \in M$, $y \in N$ by $\rho(f \otimes g)\ (x \otimes y) = f(x)g(y)$. Then ρ is a k-linear injection, and is furthermore a k-linear isomorphism when both M and N are finite dimensional.

EXERCISE 1.9 Let V, V' be k-vector spaces, and let W, W' be subspaces of V, V' respectively. Then

(i) The canonical embedding $W \otimes W' \to V \otimes V'$ is a k-linear injection.

(ii) $(V \otimes W') \cap (W \otimes V') = W \otimes W'$.

(iii) The kernel of the canonical projection $f : V \otimes V' \to V/W \otimes V'/W'$ is given by

$$ \operatorname{Ker} f = V \otimes W' + W \otimes V'. $$

Change of rings of scalars Let A, B be rings and F a two-sided

(A, B)-module. If N is a right A-module, $N \otimes_A F$ is a right B-module. If M is a right B-module, $\mathbf{Mod}_B(F, M)$ becomes a right A-module when we define

$$(u+v)(x) = u(x) + v(x), \quad (ua)(x) = u(ax), \quad u, v \in \mathbf{Mod}_B(F, M), \quad a \in A.$$

In this manner, we obtain maps from $\mathbf{Mod}_A$ to $\mathbf{Mod}_B$ and from $\mathbf{Mod}_B$ to $\mathbf{Mod}_A$ which we respectively denote by

$$\Phi : N \mapsto N \otimes_A F, \quad \Psi : M \mapsto \mathbf{Mod}_B(F, M). \tag{1.2}$$

In this situation, we obtain the bijective correspondence

$$\mathbf{Mod}_A(N, \mathbf{Mod}_B(F, M)) \cong \mathbf{Mod}_B(N \otimes_A F, M). \tag{1.3}$$

In fact, to a right A-module morphism $g : N \to \mathbf{Mod}_B(F, M)$, there corresponds a map $f : N \times F \to M$ given by

$$f(y, x) = g(y)(x), \quad y \in N, \quad x \in F,$$

which has the following properties.

(1) $f(y+y', x) = f(y, x) + f(y', x)$,

(2) $f(y, x+x') = f(y, x) + f(y, x')$,

(3) $f(ya, x) = f(y, ax)$,

(4) $f(y, xb) = f(y, x)b$,

where $x, x' \in F$, $y, y' \in N$, $a \in A$, $b \in B$.
Thus the right B-module morphism $\varphi(g) : N \otimes_A F \to M$ is determined uniquely, and we obtain the map

$$\varphi : \mathbf{Mod}_A(N, \mathbf{Mod}_B(F, M)) \to \mathbf{Mod}_B(N \otimes_A F, M).$$

Since the inverse of φ can be constructed similarly, we conclude that φ is a bijection.

Remark Φ and Ψ are covariant functors from $\mathbf{Mod}_A$ to $\mathbf{Mod}_B$ and from $\mathbf{Mod}_B$ to $\mathbf{Mod}_A$ respectively, and are adjoint.

EXERCISE 1.10 Given a commutative ring A, let $M' \to M \to M'' \to 0$ be an exact sequence of A-modules. For any A-module N,

$$M' \otimes_A N \to M \otimes_A N \to M'' \otimes_A N \to 0$$

is an exact sequence of A-modules.

EXERCISE 1.11 For a commutative ring A, if M_λ ($\lambda \in \Lambda$) and N are A-modules, we have

$$\left(\coprod_{\lambda \in \Lambda} M_\lambda\right) \otimes N \cong \coprod_{\lambda \in \Lambda} (M_\lambda \otimes N).$$

Let A and B be rings and let $u: A \to B$ be a ring morphism. A left (resp. right) B-module M becomes a left (resp. right) A-module when we take the map $\varphi: A \times M \to M$ (resp. $\varphi: M \times A \to M$) defined by $\varphi(a, x) = u(a)x$ (resp. $\varphi(x, a) = xu(a)$) as the structure map. We denote this A-module by u_*M. On the other hand, B is a two-sided B-module when we regard B as acting on itself via ring multiplication. However, when B is regarded merely as a left B-module, we obtain the left A-module u_*B. Thus $F = u_*B$ can be regarded as a two-sided (A, B)-module. Likewise, from a right A-module N, we construct the right B-module $u^*N = N \otimes_A F$. In this situation, for a right B-module M and a right A-module N, we obtain the bijective correspondence

$$\mathbf{Mod}_A(N, u_*M) \cong \mathbf{Mod}_B(u^*N, M). \tag{1.4}$$

Evidently, for a right B-module M, if we define a map

$$f: \mathbf{Mod}_B(F, M) \to u_*M$$

by $f(v) = v(1)$ for $v \in \mathbf{Mod}_B$ (F, M), we see that f is a B-module isomorphism, so that $\mathbf{Mod}_B$ (F, M), $\cong u_*M$. Hence, from (1.3), we get the bijective correspondence (1.4). Furthermore, $u_*: \mathbf{Mod}_B \to \mathbf{Mod}_A$ and $u^*: \mathbf{Mod}_A \to \mathbf{Mod}_B$ are covariant functors which are adjoint.

Let A, B be commutative rings. Given a ring morphism $u: A \to B$, an A-module N, and a B-module M, an A-module morphism $f: N \to u_*M$ is sometimes called an A-u-semi-linear map.

EXERCISE 1.12 Suppose $\mathfrak{a}$ is an ideal of a commutative ring A and $u: A \to A/\mathfrak{a}$ the canonical ring morphism. If M is an A-module, show that

$$\mathfrak{a}M = \left\{\sum_i a_i x_i \quad \text{(finite sum)}; \quad a_i \in \mathfrak{a}, \quad x_i \in M\right\}$$

is an A-submodule of M and $u^*M \cong M/\mathfrak{a}M$ (cf. Exercise 1.10).

The tensor product of A-module morphisms Suppose we are given right A-modules M, M', left A-modules N, N', and A-module morphisms $f : M \to M'$, $g : N \to N'$. The map

$$\varphi : M \times N \to M' \otimes_A N'$$

defined by $\varphi(x, y) = f(x) \otimes g(y)$ for $x \in M$, $y \in N$ satisfies

$$\varphi(x + x', y) = \varphi(x, y) + \varphi(x', y),$$
$$\varphi(x, y + y') = \varphi(x, y) + \varphi(x, y'),$$
$$\varphi(xa, y) = \varphi(x, ay),$$

where $\qquad x, x' \in M, \quad y, y' \in N, \quad a \in A.$

Thus the $\mathbb{Z}$-module morphism

$$h : M \otimes_A N \to M' \otimes_A N'$$

defined by $h(x \otimes y) = f(x) \otimes g(x)$ for $x \in M$, $y \in N$ is determined uniquely. The morphism h is said to be the tensor product of f and g, and is written $f \otimes g$. By definition, we have

$$(f \otimes g)(x \otimes y) = f(x) \otimes g(y), \quad x \in M, \quad y \in N.$$

In particular, if A is a commutative ring, $f \otimes g$ is an A-module morphism.

EXERCISE 1.13 Let M, M', M'' be right A-modules and N, N', N'' left A-modules. If $u : M \to M'$, $u' : M' \to M''$, $v : N \to N'$, $v' : N' \to N''$ are A-module morphisms, then $(u' \circ u) \otimes (v' \circ v) = (u' \otimes v') \circ (u \otimes v)$.

EXERCISE 1.14 For a field k, let M, M', N, N' be k-linear spaces. If M and N, or M' and N' are finite dimensional, then we have

$$\mathbf{Mod}_k(M, M') \otimes \mathbf{Mod}_k(N, N') \cong \mathbf{Mod}_k(M \otimes N, M' \otimes N').$$

1.4 Projective limits and inductive limits

Let A be a commutative ring and let Λ be an ordered set. Given any λ, $\mu \in \Lambda$, if there exists $\nu \in \Lambda$ such that $\lambda \leqq \nu$ and $\mu \leqq \nu$, then we call Λ a directed set. Suppose now that Λ is a directed set. If a family of A-modules $\{M_\lambda\}_{\lambda \in \Lambda}$ and a family of A-module morphisms

$\{u_{\lambda\mu}: M_\mu \to M_\lambda\}_{(\lambda,\mu)\in\Lambda\times\Lambda,\lambda\leqq\mu}$ satisfy the following two conditions

(1) $\lambda\leqq\mu\leqq\nu(\lambda,\mu,\nu\in\Lambda)$ implies $u_{\lambda\nu}=u_{\lambda\mu}\circ u_{\mu\nu}$,

(2) $u_{\lambda\lambda}=1_{M_\lambda}$ (the identity map of M_λ),

then $\{M_\lambda, u_{\lambda\mu}\}_{\lambda,\mu\in\Lambda,\lambda\leqq\mu}$ is called a **projective system** of A-modules.

Given a projective system as defined above, if we write

$$M=\left\{x=\{x_\lambda\}_{\lambda\in\Lambda}\in\prod_{\lambda\in\Lambda}M_\lambda\,;u_{\lambda\mu}(x_\mu)=x_\lambda(\lambda\leqq\mu)\right\},$$

M becomes an A-submodule of the A-module $P=\prod_{\lambda\in\Lambda}M_\lambda$. When we let $u_\lambda: M\to M_\lambda$ denote the restriction of the canonical projection $p_\lambda: P\to M_\lambda$ to M, then the A-module M and the family of A-module morphisms $\{u_\lambda: M\to M_\lambda\}_{\lambda\in\Lambda}$ have the following properties.

(1) $u_\lambda=u_{\lambda\mu}\circ u_\mu \quad (\lambda\leqq\mu)$.

(2) If an A-module N and a family of A-module morphisms $v_\lambda: N\to M_\lambda$ $(\lambda\in\Lambda)$ satisfy $v_\lambda=u_{\lambda\mu}\circ v_\mu$ $(\lambda\leqq\mu)$, then there exists a unique A-module morphism $v: N\to M$ such that for each $\lambda\in\Lambda$, $v_\lambda=u_\lambda\circ v$.

The pair consisting of the A-module M and the family of A-module morphisms $\{u_\lambda\}_{\lambda\in\Lambda}$ is called the **projective limit** of the projective system $\{M_\lambda, u_{\lambda\mu}\}_{\lambda,\mu\in\Lambda,\lambda\leqq\mu}$ and is written $M=\varprojlim_\lambda M_\lambda$.

When $\{M_\lambda, \mu_{\lambda\mu}\}_{\lambda,\mu\in\Lambda,\lambda\leqq\mu}$ is a projective system of A-modules, then for any A-module N,

$$\{{}_A\mathbf{Mod}(N, M_\lambda), (u_{\lambda\mu})_*\}_{\lambda,\mu\in\Lambda,\lambda\leqq\mu}$$

is also a projective system of A-modules, and we have

$${}_A\mathbf{Mod}\left(N, \varprojlim_\lambda M_\lambda\right)\cong\varprojlim_\lambda {}_A\mathbf{Mod}(N, M_\lambda).$$

Dually, if a family $\{M_\lambda\}_{\lambda\in\Lambda}$ of A-modules and a family $\{u_{\mu\lambda}: M_\lambda\to M_\mu\}_{\lambda,\mu\in\Lambda,\lambda\leqq\mu}$ satisfy the conditions

(1) $\lambda\leqq\mu\leqq\nu(\lambda,\mu,\nu\in\Lambda)$ implies $u_{\nu\lambda}=u_{\nu\mu}\circ u_{\mu\lambda}$,

(2) $u_{\lambda\lambda}=1_{M_\lambda}$ (the identity map of M_λ),

then $\{M_\lambda, u_{\mu\lambda}\}_{\lambda,\mu\in\Lambda,\lambda\leqq\mu}$ is called an **inductive system** of A-modules.

Let us define a relation on $S=\coprod_{\lambda\in\Lambda}M_\lambda$ as follows. For $x=\sum_\lambda x_\lambda$, $y=\sum_\lambda y_\lambda\in S$, define a relation $\sim$ by writing $x\sim y$ when there exists

$\mu \in \Lambda$ such that $\lambda \leqq \mu$ for any $\lambda \in \Lambda$ such that $x_\lambda \neq 0$ or $y_\lambda \neq 0$, and the equality $\sum_{\lambda \in \Lambda} u_{\mu\lambda}(x_\lambda) = \sum_{\lambda \in \Lambda} u_{\mu\lambda}(y_\lambda)$ holds in M_μ. The relation $\sim$ is an equivalence relation. Further, $x \sim y$ and $x' \sim y'$ imply $x + y \sim x' + y'$ and $ax \sim ay$ for $a \in A$. Thus, by letting $M = S/\sim$ be the set of equivalence classes, M becomes an A-module. Let the composite of the canonical embedding $i_\lambda : M_\lambda \to S$ and the canonical projection $\pi : S \to M$ be $u_\lambda : M_\lambda \to M$. Then the A-module M together with the family $\{u_\lambda : M_\lambda \to M\}_{\lambda \in \Lambda}$ of A-module morphisms have the following properties.

(1) $u_\lambda = u_\mu \circ u_{\mu\lambda} \quad (\lambda \leqq \mu)$.

(2) If an arbitrary A-module N and a family of A-module morphisms $\{v_\lambda : M_\lambda \to N\}_{\lambda \in \Lambda}$ satisfy $v_\lambda = v_\mu \circ u_{\mu\lambda}$, then there exists a unique A-module morphism $v : M \to N$ such that, for each $\lambda \in \Lambda$, $v_\lambda = v \circ u_\lambda$.

The A-module M together with the family $\{u_\lambda\}_{\lambda \in \Lambda}$ of A-module morphisms is called the **inductive limit** of the inductive system $\{M_\lambda, u_{\mu\lambda}\}_{\lambda, \mu \in \Lambda, \lambda \leqq \mu}$, and is denoted $M = \varinjlim_\lambda M_\lambda$.

If $\{M_\lambda, u_{\mu\lambda}\}_{\lambda, \mu \in \Lambda, \lambda \leqq \mu}$ is an inductive system, then, for any A-module N, $\{{}_A\mathbf{Mod}\,(M_\lambda, N), (u_{\mu\lambda})^*\}_{\lambda, \mu \in \Lambda, \lambda \leqq \mu}$ is a projective system and we have

$$ {}_A\mathbf{Mod}\left(\varinjlim_\lambda M_\lambda, N\right) \cong \varprojlim_\lambda {}_A\mathbf{Mod}\,(M_\lambda, N). $$

Remark The notions of projective and inductive systems and their limits can be defined likewise for other algebraic systems such as commutative rings, groups, etc. (cf. Appendix A.3).

EXAMPLE 1.1 Let p be a prime number and let $\mathbb{N}$ be the set of all natural numbers viewed as an ordered set with respect to size. For $n \in \mathbb{N}$, set $A_n = \mathbb{Z}/p^n\mathbb{Z}$ and for $n, m \in \mathbb{N}$ with $m \leqq n$, let $u_{mn} : A_n \to A_m$ be the canonical ring morphism. Then $\{A_n, u_{mn}\}_{m, n \in \mathbb{N}, m \leqq n}$ is a projective system of commutative rings from which we can define the commutative ring $\varprojlim_n A_n$. The elements of this ring are called p-adic integers.

EXAMPLE 1.2 Given a group G, let $\{N_\lambda\}_{\lambda \in \Lambda}$ be the set of all normal subgroups of G with finite indices. Defining $\lambda \leqq \mu$ when

$N_\mu \subseteq N_\lambda$, Λ becomes an ordered set. Set $G_\lambda = G/N_\lambda$. For $\lambda \leqq \mu$, let $u_{\lambda\mu}: G_\mu \to G_\lambda$ be the canonical group morphism. Then $\{G_\lambda, u_{\lambda\mu}\}_{\lambda,\mu\in\Lambda,\lambda\leqq\mu}$ becomes a projective system of groups and we call $\hat{G} = \varprojlim_\lambda G_\lambda$ the **pro-finite completion** of G. Moreover, $\hat{G}$ is a topological group when viewed as a subgroup of the topological group $\prod_\lambda G_\lambda$, where each G_λ is regarded as a discrete group. If $\bigcap_{\lambda\in\Lambda} N_\lambda = \{e\}$, then $\hat{G}$ satisfies the Hausdorff separation axiom, and hence G can be identified with a dense subgroup of $\hat{G}$. In general, a group which happens to be the projective limit of finite groups, viewed as a topological group, is called a **pro-finite group**. For instance, the Galois group of a Galois extension of infinite degree is a pro-finite group. A pro-finite group is a totally disconnected compact topological group. Conversely, such a topological group is a pro-finite group.

2 Algebras over a commutative ring

In this section, the notion of algebras over a commutative ring A will be introduced; and we present some examples of such algebras. In passing, we take up some properties of tensor algebras, symmetric algebras and exterior algebras. An algebra over A is defined by giving an A-module morphism which is called its structure map. Coalgebras, which will be defined in the next chapter, are naturally in a dual relationship with algebras. We also touch upon some properties of algebras over fields in §4 and §5.

2.1 Algebras over a commutative ring

Let A be a commutative ring. Given a ring B and a ring morphism $\eta_B: A \to B$, B can be viewed as a left B-module via ring multiplication, from which we construct the left A-module $(\eta_B)_* B$. Its action will be written ax for $a \in A$ and $x \in B$. If we have

$$(ax)y = x(ay) = a(xy), \qquad a \in A, \qquad x, y \in B,$$

then B is said to be an **A-algebra**. Now the map $f: B \times B \to B$ defined by $f(x, y) = xy$ turns out to be bilinear when B is regarded as an A-module. Thus we obtain an A-module morphism

$$\mu_B: B \otimes_A B \to B.$$

From this fact, we see that an A-algebra B can be defined in the following manner. For an A-module B and A-module morphisms $\eta_B : A \to B$, $\mu_B : B \otimes_A B \to B$, the following diagrams

(1) (the associative law)

$$\begin{array}{ccc} B \otimes_A B \otimes_A B & \xrightarrow{1 \otimes \mu_B} & B \otimes_A B \\ {\scriptstyle \mu_B \otimes 1} \downarrow & & \downarrow {\scriptstyle \mu_B} \\ B \otimes_A B & \xrightarrow{\mu_B} & B \end{array}$$

(2) (the unitary property)

$$\begin{array}{ccccc} B \otimes_A B & \xrightarrow{\eta_B \otimes 1} & B \otimes_A B & \xleftarrow{1 \otimes \eta_B} & B \otimes_A A \\ & \searrow{\scriptstyle \sim} & \downarrow {\scriptstyle \mu_B} & {\scriptstyle \sim}\swarrow & \\ & & B & & \end{array}$$

commute. Namely, when we identify $B \otimes_A A$ and $A \otimes_A B$ with B, if the equations

$$\mu_B(1 \otimes \mu_B) = \mu_B(\mu_B \otimes 1), \quad \mu_B(1 \otimes \eta_B) = \mu_B(\eta_B \otimes 1) = 1_B$$

hold, then B is called an A-algebra. Here, μ_B is said to be the **multiplication map** of B, η_B the **unit map** of B, and together, we call μ_B, η_B the **structure maps** of the A-algebra B.

Given A-algebras B, C, if a map $\mu : B \to C$ is a ring morphism as well as an A-module morphism, then we call μ an **A-algebra morphism.** For A-algebras B, C, if we let $\mu_B, \eta_B, \mu_C, \eta_C$ be their respective structure maps, then the A-module morphism $u : B \to C$ is an A-algebra morphism if and only if the diagrams

$$\begin{array}{ccc} B \otimes_A B & \xrightarrow{\mu_B} & B \\ {\scriptstyle u \otimes u} \downarrow & & \downarrow {\scriptstyle u} \\ C \otimes_A C & \xrightarrow{\mu_C} & C \end{array}$$

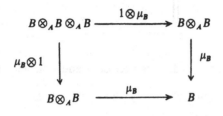

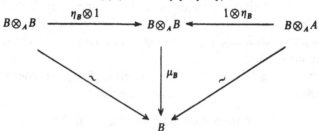

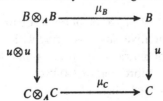

$$\begin{array}{ccc} A & \xrightarrow{\eta_B} & B \\ \| & & \downarrow u \\ A & \xrightarrow{\eta_C} & C \end{array}$$

commute; in other words, it is necessary and sufficient that

$$\mu_C(u \otimes u) = u \circ \mu_B \quad \text{and} \quad \eta_C = u \circ \eta_B.$$

We let $\mathbf{Alg}_A$ and $\mathbf{M}_A$ denote categories that have as objects all A-algebras and all commutative A-algebras respectively. If B, C are A-algebras (resp. commutative A-algebras), the set of all A-algebra morphisms from B to C is written $\mathbf{Alg}_A(B, C)$ (resp. $\mathbf{M}_A(B, C)$ for the commutative case).

Given A-algebras B, C, the A-module $B \otimes_A C$ becomes an A-algebra where the structure maps are given by

$$\mu_{B\otimes C} = (\mu_B \otimes \mu_C)(1 \otimes \tau \otimes 1), \quad \eta_{B\otimes C} = \eta_B \otimes \eta_C.$$

The map τ above denotes the A-module isomorphism $C \otimes B \to B \otimes C$ defined by $x \otimes y \mapsto y \otimes x$. The A-algebra $B \otimes_A C$ is called the **tensor product** of B and C. The multiplication in $B \otimes_A C$ is given by

$$(b \otimes c)(b' \otimes c') = bb' \otimes cc', \qquad b, b' \in B, \qquad c, c' \in C,$$

and the unit element is given by $1 \otimes 1$.

EXAMPLE 1.3 Given a commutative ring A and indeterminates $\{X_\lambda\}_{\lambda \in \Lambda}$, the set of all polynomials in X_λ ($\lambda \in \Lambda$) with coefficients in A is written $A[X_\lambda]_{\lambda \in \Lambda}$ and is a commutative ring via the addition and multiplication defined naturally. Moreover, the canonical embedding $A \to A[X_\lambda]_{\lambda \in \Lambda}$ makes $A[X_\lambda]_{\lambda \in \Lambda}$ an A-algebra.

EXAMPLE 1.4 The set $M_n(A)$ of all $n \times n$ square matrices with coefficients in a commutative ring A is a ring with respect to addition and multiplication of matrices and is an A-module with respect to scalar multiplication. Moreover, $M_n(A)$ becomes an A-algebra when we define $\eta_{M_n(A)} : a \mapsto ae^{(n)}$ where $e^{(n)}$ is the $n \times n$ identity matrix.

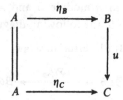

EXAMPLE 1.5 Let $B = \mathrm{Map}(S, A)$ be the set of all maps from a set S to a commutative ring A. For $u, v \in B$, we define the operations

$$(u+v)(x) = u(x) + v(x), \qquad (uv)(x) = u(x)v(x), \qquad x \in S,$$

which make B a ring. Defining the action

$$(au)(x) = au(x), \qquad a \in A, \quad x \in S,$$

B becomes an A-module. Notice that B becomes an A-algebra when we define an A-module morphism $\eta_B : A \to B$ by

$$\eta_B(a)(x) = a, \qquad x \in S.$$

EXAMPLE 1.6 Let A be a commutative ring and G a group (resp. semigroup). Denote the free A-module generated by G by AG. Defining the multiplication on AG by

$$\left(\sum_{x \in G} a_x x\right)\left(\sum_{y \in G} b_y y\right) = \sum_{z \in G}\left(\sum_{xy = z} a_x b_y\right) z,$$

AG becomes a ring. The map $\eta_{AG} : A \to AG$ given by $\eta_{AG}(a) = a1$ where 1 is the identity element of G makes AG an A-algebra. The A-algebra AG is said to be the **group** (resp. **semigroup**) **A-algebra** of G.

Given a commutative ring A, let B, C be A-algebras. The maps $i_1 : B \to B \otimes C$, $i_2 : C \to B \otimes C$ respectively defined by $i_1(b) = b \otimes 1$ and $i_2(c) = 1 \otimes c$ for $b \in B$, $c \in C$ turn out to be A-algebra morphisms, and are said to be the canonical embeddings. Now suppose we are given a commutative A-algebra T and A-algebra morphisms $f : B \to T$ and $g : C \to T$. Then there exists a unique A-algebra morphism $h : B \otimes_A C \to T$ such that $h \circ i_1 = f$ and $h \circ i_2 = g$. In fact, since the map $\varphi : B \times C \to T$ defined by $\varphi(b, c) = f(b)g(c)$ for $b \in B$ and $c \in C$ is bilinear, we obtain the A-module morphism $h : B \otimes_A C \to T$ by defining $h(b \otimes c) = f(b)g(c)$. Now we have

$$h(i_1(b)) = h(b \otimes 1) = f(b)g(1) = f(b), \quad b \in B,$$

$$h(i_2(c)) = h(1 \otimes c) = f(1)g(c) = g(c), \quad c \in C,$$

so that $h \circ i_1 = f$, $h \circ i_2 = g$. Moreover, since

$$\begin{aligned} h((b \otimes c)(b' \otimes c')) &= h(bb' \otimes cc') = f(bb')g(cc') \\ &= f(b)g(c)f(b')g(c') \\ &= h(b \otimes c)h(b' \otimes c'), \end{aligned}$$

it follows that h is an A-algebra morphism. Clearly, h is determined uniquely by f and g. Thus, for any commutative A-algebra T, we have the bijective correspondence

$$\mathbf{Alg}_A(B \otimes_A C, T) \cong \mathbf{Alg}_A(B, T) \times \mathbf{Alg}_A(C, T).$$

Remark If B and C are commutative A-algebras, then the A-algebra $B \otimes_A C$ is the direct sum of B and C in the category $\mathbf{M}_A$ of commutative A-algebras.

Let $\{B_\lambda\}_{\lambda \in \Lambda}$ be A-algebras. We define multiplication on the (A-module) direct product $\prod_{\lambda \in \Lambda} B_\lambda$ by

$$\{b_\lambda\}_{\lambda \in \Lambda} \{b_\lambda'\}_{\lambda \in \Lambda} = \{b_\lambda b_\lambda'\}_{\lambda \in \Lambda}.$$

This makes $\prod_{\lambda \in \Lambda} B_\lambda$ an A-algebra which we call the **direct product** of A-algebras $\{B_\lambda\}_{\lambda \in \Lambda}$. If $\Lambda = \{1, 2, \ldots, n\}$, $\prod_{i=1}^{n} B_i$ is sometimes written $B_1 \times \cdots \times B_n$.

2.2 Filtered algebras and graded algebras

Set $I = \{0\} \bigcup \mathbb{N}$. I is a semigroup with respect to addition. Let A be a commutative ring and B an A-algebra. If a family $\{B_i\}_{i \in I}$ of A-submodules of B satisfies the conditions

$$B_0 \subset B_1 \subset B_2 \subset \cdots, \qquad B_i B_j \subset B_{i+j}, \qquad \bigcup_{i \in I} B_i = B, \qquad 1 \in B_0,$$

then B is said to be a **filtered A-algebra**, and $\{B_i\}_{i \in I}$ is called a **filtration** on B. If a family of A-submodules $\{B_{(i)}\}_{i \in I}$ of the A-module B satisfies the conditions

$$B = \coprod_{i \in I} B_{(i)}, \qquad B_{(i)} B_{(j)} \subset B_{(i+j)},$$

then B is called a **graded A-algebra**.

Now let B, C be filtered A-algebras, and let $\{B_i\}_{i\in I}$, $\{C_i\}_{i\in I}$ be their respective filtrations. When an A-algebra morphism $u: B \to C$ is such that $u(B_i) \subseteq C_i (i \in I)$, then u is said to be a **filtered A-algebra morphism**. If $B = \coprod_{i\in I} B_{(i)}$ and $C = \coprod_{i\in I} C_{(i)}$ are graded A-algebras and if, furthermore, an A-algebra morphism $u: B \to C$ satisfies $u(B_{(i)}) \subseteq C_{(i)} (i \in I)$, then u is said to be a **graded A-algebra morphism**. From the definition, we see that B_0 and $B_{(0)}$ are A-subalgebras of B.

Let $\{B_i\}_{i\in I}$ be a filtration on a filtered A-algebra B. Set $B_{(0)} = B_0$, $B_{(i)} = B_i/B_{i-1}$ $(i \in I,\ i > 0)$ and define a multiplication on the direct sum of A-modules $\operatorname{gr} B = \coprod_{i\in I} B_{(i)}$ in the following manner. We let $g_1 : B_i \to B_{(i)} = B_i/B_{i-1}$ be the canonical A-module projections, and we define maps $f_{ij} : B_{(i)} \times B_{(j)} \to B_{(i+j)}$ by

$$f_{ij}(g_i(x), g_j(y)) = g_{i+j}(xy), \qquad x \in B_i, \quad y \in B_j.$$

We see readily that the f_{ij} are bilinear. Thus we obtain A-module morphisms

$$\mu_{ij} : B_{(i)} \otimes_A B_{(j)} \to B_{(i+j)}.$$

Therefore, for $x = \sum_{i\in I} g_i(x_i)$, $y = \sum_{j\in I} g_j(y_j) \in \operatorname{gr} B$, we can define

$$\begin{aligned}\mu_{\operatorname{gr} B}(x \otimes y) &= \mu_{\operatorname{gr} B}\left(\sum_{i,j} g_i(x_i) \otimes g_j(y_j)\right)\\ &= \sum \mu_{ij}(g_i(x) \otimes g_j(y_j)),\end{aligned}$$

which yields the A-module morphism

$$\mu_{\operatorname{gr} B} : \operatorname{gr} B \otimes \operatorname{gr} B \to \operatorname{gr} B.$$

Furthermore, the image of the unit map η_B of the A-algebra B is $\eta_B(A) \subset B_0 = B_{(0)} \subset \operatorname{gr} B$, which gives rise to the A-module morphism $\eta_{\operatorname{gr} B} : A \to \operatorname{gr} B$. With $\mu_{\operatorname{gr} B}$ and $\eta_{\operatorname{gr} B}$ as the structure maps, $\operatorname{gr} B$ becomes a graded A-algebra. We call $\operatorname{gr} B$ the **associated graded A-algebra of the filtered A-algebra B**.

Tensor algebras Let A be a commutative ring and M an A-module. For each $n \in I$, define an A-module $T_n(M)$ inductively by

$$T_{(0)}(M) = A, \qquad T_{(n+1)} = T_{(n)}(M) \otimes M, \qquad n \in I.$$

Set $T(M) = \coprod_{n\in I} T_{(n)}(M)$ where $T_{(n)}(M)$ is called the ***n*th homogeneous component** of $T(M)$. For $p, q \in \mathbb{N}$, set $x = x_1 \otimes \cdots \otimes x_p \in T_{(p)}(M)$ and $y = y_1 \otimes \cdots \otimes y_q \in T_{(q)}(M)$ for $x_i \in M$, $y_j \in M$ $(1 \leqq i \leqq p,\ 1 \leqq j \leqq q)$, and write $x \otimes y = x_1 \otimes \cdots \otimes x_p \otimes y_1 \otimes \cdots \otimes y_q \in T_{(p+q)}(M)$.

In case either p or $q = 0$, set $a \otimes x = ax$ for $a \in T_{(0)}(M)$, $x \in T_{(p)}(M)$. If $p = q = 0$, set $a \otimes b = ab$ for $a, b \in T_{(0)}(M)$. In this manner, we get an A-module morphism

$$\mu_{pq} : T_{(p)}(M) \otimes T_{(q)}(M) \to T_{(p+q)}(M)$$

for each $(p, q) \in I \times I$. The above morphism naturally gives rise to an A-module morphism

$$\mu : T(M) \otimes T(M) \to T(M).$$

Moreover, if we let η be the canonical embedding $i_0 : T_{(0)}(M) = A \to T(M)$, $T(M)$ becomes a graded A-algebra with μ and η as the structure maps. We call $T(M)$ the **tensor A-algebra** over M.

Now let B be an arbitrary A-algebra and let $f : T(M) \to B$ be any A-algebra morphism. Then the restriction of f to $T_{(1)}(M) = M$ is an A-module morphism from M to B. Conversely, given any A-module morphism $g : M \to B$, we can define maps $g_n : T_{(n)}(M) \to B$ by

$$g_n(x_1 \otimes \cdots \otimes x_n) = f(x_1) f(x_2) \cdots f(x_n), \quad x_i \in M, \quad (1 \leqq i \leqq n), \quad n \in \mathbb{N},$$
$$g_0(a) = a1,$$

and since the g_n $(n \in I)$ are A-module morphisms, we obtain the A-module morphism $f = \coprod_{n\in I} g_n : T(M) \to B$, which turns out to be an A-algebra morphism. Therefore, for an A-algebra B, we have a bijective correspondence

$$\mathbf{Alg}_A(T(M), B) \cong \mathbf{Mod}_A(M, B). \tag{1.5}$$

EXERCISE 1.15 Given an A-module morphism $f : M \to N$, if we define maps $T_{(n)}(f) : T_{(n)}(M) \to T_{(n)}(N)$ by $T_{(n)}(f)\ (x_1 \otimes \cdots \otimes x_n) = f(x_1) \otimes \cdots \otimes f(x_n)$ for $x_i \in M$ $(1 \leqq i \leqq n)$, then $T_{(n)}(f)$ is an A-module morphism which gives rise to an A-module morphism $T(f) : T(M) \to T(N)$. Moreover, $T(f)$ is a graded A-algebra morphism.

Remark The correspondence $M \mapsto T(M)$ is a covariant functor from the category $\mathbf{Mod}_A$ to $\mathbf{Alg}_A$, and is adjoint to the covariant functor from $\mathbf{Alg}_A$ to $\mathbf{Mod}_A$ which forgets the A-algebra structure and merely regards the object as an A-module.

Symmetric algebras Let S_n be the group of all permutations on n letters $\{1, 2, \ldots, n\}$, i.e. the symmetric group of degree n. If we let $\mathfrak{a}_{(n)}$ be the A-submodule of $T_n(M)$ generated by all elements of the form

$$x_1 \otimes x_2 \otimes \cdots \otimes x_n - x_{\sigma(1)} \otimes x_{\sigma(2)} \otimes \cdots \otimes x_{\sigma(n)},$$
$$x_i \in M \ (1 \leqq i \leqq n), \quad \sigma \in S_n,$$

and if we set $\mathfrak{a} = \coprod_{n \in \mathbb{N}} \mathfrak{a}_{(n)}$, we see that $\mathfrak{a}$ is an ideal of $T(M)$ and that $S(M) = T(M)/\mathfrak{a}$ is a commutative A-algebra, which we call the **symmetric A-algebra** over M. Set $S_{(n)}(M) = T_{(n)}(M)/\mathfrak{a}_{(n)}$ $(n \in \mathbb{N})$ and $S_{(0)}(M) = A$. Then it follows that $S(M) = \coprod_{n \in I} S_{(n)}(M)$, so that $S(M)$ becomes a graded A-algebra. Both $S_{(0)}(M) = A$ and $S_{(1)}(M) = M$ can be identified with A-submodules of $S(M)$.

Given an arbitrary commutative A-algebra C and an A-algebra morphism $f : S(M) \to C$, the restriction $g : M \to C$ of f to M is an A-module morphism. Conversely, given an A-module morphism $g : M \to C$, g can be extended uniquely by (1.5) to an A-algebra morphism $f : T(M) \to C$. Since C is a commutative A-algebra, it follows that $\mathfrak{a} \subset \operatorname{Ker} f$. Therefore f induces an A-algebra morphism from $S(M) = T(M)/\mathfrak{a}$ to C. Consequently, we obtain a bijective correspondence

$$\mathbf{M}_A(S(M), C) \cong \mathbf{Mod}_A(M, C). \tag{1.6}$$

EXERCISE 1.16 Given A-modules M, N, let $M \oplus N$ be the direct sum of M and N. Show that

$$S(M \oplus N) \cong S(M) \otimes_A S(N).$$

Exterior algebras For $n \geqq 2$, let $\mathfrak{b}_{(n)}$ be the A-submodule of $T_{(n)}(M)$ generated by all elements of the form $x_1 \otimes \cdots \otimes x_n$ for $x_i \in M$ $(1 \leqq i \leqq n)$ such that $x_i = x_j$ for some $i \neq j$ $(1 \leqq i, j \leqq n)$. Set $\mathfrak{b}_{(0)} = \mathfrak{b}_{(1)} = \{0\}$.

Then $\mathfrak{b} = \coprod_{n \in I} \mathfrak{b}_{(n)}$ is an ideal of $T(M)$ and the A-algebra $\wedge(M) = T(M)/\mathfrak{b}$ is said to be the **exterior algebra** over M. Writing $\wedge_{(n)}(M) = T_{(n)}(M)/\mathfrak{b}_{(n)}$, it follows that $\wedge(M) = \coprod_{n \in I} \wedge_{(n)}(M)$ so that $\wedge(M)$ is a graded A-algebra. We identify $\wedge_{(0)}(M) = A$ and $\wedge_{(1)}(M) = M$ with A-submodules of $\wedge(M)$, and let the image of $x_1 \otimes \ldots \otimes x_n$ under the canonical projection from $T_{(n)}(M)$ to $\wedge_{(n)}(M)$ be denoted $x_1 \wedge \cdots \wedge x_n$. Then for $x, y, z \in M$, it follows that

$$x \wedge x = 0 \text{ (therefore } x \wedge y = -y \wedge x),$$
$$(ax + by) \wedge z = a(x \wedge z) + b(y \wedge z), \quad a, b \in A.$$

If M is a free A-module of rank n, then $\wedge_{(i)}(M)$ is a free A-module of rank $\binom{n}{i}$ $\left(\text{where in case } i > n, \text{ we set } \binom{n}{i} = 0\right)$. Therefore $\wedge(M)$ is a free A-module of rank 2^n. Moreover, if $\{x_1, \ldots, x_n\}$ is a basis for M and $y_i = \sum_{j=1}^{n} a_{ij} x_j$ for $a_{ij} \in A$, then we have

$$y_1 \wedge \cdots \wedge y_n = \det(a_{ij}) x_1 \wedge \cdots \wedge x_n.$$

Here, $\det(a_{ij})$ stands for the determinant of the $n \times n$ square matrix (a_{ij}).

3 Lie algebras

Group rings and the universal enveloping algebras of Lie algebras happen to be key examples of Hopf algebras. In this section, our concern is mainly with universal enveloping algebras. For basic properties of Lie algebras, the reader is referred to books such as [6] and [7]. Theorem 1.3.5 (due to Hochschild) dealing with the complete reducibility of representations of p-Lie algebras is applied in the proofs or related theorems concerning representations of algebraic groups in Chapter 5.

3.1 Lie algebras

Given a $\mathbb{Z}$-module L, if there exists a map $\varphi : L \times L \to L$ satisfying the following conditions, then L is said to be a **Lie ring**. For $x, y \in L$, we let

$[x, y]$ denote $\varphi(x, y)$. Then for $x, y, z \in L$, we have

(1) $[x+y, z] = [x, z] + [y, z]$, $[x, y+z] = [x, y] + [x, z]$,

(2) $[x, x] = 0$,

(3) $[[x, y], z] + [[z, x], y] + [[y, z], x] = 0$ (the Jacobi law).

Let A be a commutative ring. If L is a Lie ring as well as an A-module, L is said to be an A-**Lie algebra** when the following condition is satisfied.

(4) $[ax, y] = [x, ay] = a[x, y]$, $\quad x, y \in L, \quad a \in A$.

By the above definition, φ is a bilinear map, so it induces an A-module morphism $\mu_L : L \otimes L \to L$. The morphism μ_L is called the **structure map** of the A-Lie algebra L. Moreover, $\mu_L(x \otimes y)$ is sometimes called the **product** of x and y.

EXAMPLE 1.7 Let B be a ring. If for $x, y \in B$, we define $[x, y] = xy - yx$, then B becomes a Lie ring. We denote such a Lie ring by B_L. Given a commutative ring A, if B is an A-algebra, then B_L is an A-Lie algebra. In fact, we have

$$\begin{aligned}[ax, y] &= (ax)y - y(ax) = x(ay) - (ay)x \\ &= [x, ay] = a[x, y].\end{aligned}$$

A Lie ring L for which $[x, y] = 0$ for any $x, y \in L$ is said to be **commutative**. If B is a commutative ring, then B_L is a commutative Lie ring.

EXAMPLE 1.8 Let B be a ring and M a two-sided B-module. A map $D : B \to M$ satisfying the conditions

(1) $D(x+y) = D(x) + D(y)$,

(2) $D(xy) = xD(y) + D(x)y$, $\quad x, y \in B$,

is said to be a **derivation** from B to M. Substituting $x = y = 1$ in (2), we obtain $D(1) = 2D(1)$. Therefore $D(1) = 0$. Given a commutative ring A and an A-algebra B, if a derivation D is also an A-module morphism, then D is said to be an A-**derivation** from B to M. In such a situation, for $a \in A$ and $x \in B$, we have

$$D(ax) = aD(x) + D(a1)x = aD(x).$$

Therefore $D(a1) = 0$ for $a \in A$. Now define $D_u : B \to M$ for $u \in M$ by $D_u(x) = xu - ux$. Then D_u turns out to be an A-derivation from B to M which is said to be an **inner derivation**. Let $\mathrm{Der}_A(B, M)$ denote the set of all A-derivations from B to M. When in particular $M = B$, we simply write $\mathrm{Der}_A(B)$. Given $D_1, D_2 \in \mathrm{Der}_A(B, M)$ and $a \in A$, if we define

$$(D_1 + D_2)(x) = D_1(x) + D_2(x), \quad (aD)(x) = aD(x), \quad x \in B,$$

$\mathrm{Der}_A(B, M)$ becomes an A-module. If $D_1, D_2 \in \mathrm{Der}_A(B)$, it follows that $[D_1, D_2] = D_1D_2 - D_2D_1 \in \mathrm{Der}_A(B)$. Hence, if we let $[D_1, D_2]$ be the product of D_1 and D_2, then $\mathrm{Der}_A(B)$ becomes an A-Lie algebra.

Even when B is an A-algebra on which the associative law does not generally hold, $\mathrm{Der}_A(B)$ can be defined similarly making $\mathrm{Der}_A(B)$ an A-Lie algebra. For instance, when L is an A-Lie algebra, an element D of $\mathrm{Der}_A(L)$ is an A-module morphism from L to L which satisfies the condition

$$D([x, y]) = [D(x), y] + [x, D(y)], \qquad x, y \in L.$$

In particular for $x \in L$, we set $D_x(y) = [x, y]$ for $y \in L$. Then, by Jacobi's law, $D_x \in \mathrm{Der}_A(L)$, and such an A-derivation is said to be 'inner'.

EXERCISE 1.17 Let k be a field and let A be a k-algebra.

(i) If $D \in \mathrm{Der}_k(A)$, then

$$D^n(xy) = \sum_{i=0}^{n} \binom{n}{i} D^i(x) D^{n-1}(y), \qquad x, y \in A;$$

if in particular, the characteristic of k is $p > 0$, then $D^p \in \mathrm{Der}_k(A)$.

(ii) Suppose k is of characteristic 0 and that $D \in \mathrm{Der}_k(A)$ is nilpotent, namely, there exists a natural number n such that $D^n = 0$. Then

$$\exp D = 1 + D + \frac{1}{2}D^2 + \frac{1}{3!}D^3 + \cdots$$

is an automorphism of the k-algebra A (i.e. $\exp D$ is a k-algebra isomorphism from A onto A).

Now let L, L' be A-Lie algebras. If an A-module morphism

$f : L \to L'$ satisfies

$$f([x, y]) = [f(x), f(y)], \qquad x, y \in L,$$

then f is said to be an A-**Lie algebra morphism**. If f is also a bijection, f is called an A-**Lie algebra isomorphism**. In such a case, L and L' are said to be isomorphic, and we write $L \cong L'$. The set of all A-Lie algebra morphisms from L to L' is written $\mathbf{Lie}_A(L, L')$ and we denote the category of A-Lie algebras by $\mathbf{Lie}_A$.

When an A-submodule M of an A-Lie algebra L satisfies

$$x, y \in M \Rightarrow [x, y] \in M,$$

then M becomes an A-Lie algebra under the same operation as in L. Such an M is called an A-**sub-Lie algebra** of L. When an A-submodule N of L is such that

$$x \in N \quad \text{and} \quad y \in L \Rightarrow [x, y] \in N,$$

then N is said to be an **ideal** of L. Now the factor module L/N inherits an A-Lie algebra structure, and the canonical projection $\pi : L \to L/N$ is an A-Lie algebra morphism. We call L/N a **factor** A-**Lie algebra**. Given an A-Lie algebra morphism $f : L \to L'$,

$$\operatorname{Ker} f = \{x \in L;\ f(x) = 0\}, \qquad \operatorname{Im} f = \{f(x) \in L';\ x \in L\}$$

are respectively an ideal of L and an A-sub-Lie algebra of L', and, further, we have $L/\operatorname{Ker} f \cong \operatorname{Im} f$.

Let L be a Lie ring and M a $\mathbb{Z}$-module. If a map $f : L \times M \to M$ satisfies the following conditions, M is said to be an L-**module**. Set $f(x, s) = xs$ for $x \in L$ and $s \in M$. Then

(1) $x(s + t) = xs + xt, \qquad x, y \in L, \qquad s, t \in M,$

(2) $(x + y)s = xs + ys,$

(3) $[x, y]s = x(ys) - y(xs).$

Now let A be a commutative ring. If L is an A-Lie algebra, M is an A-module, and the map $f : L \times M \to M$ satisfies (1), (2), (3) above as well as

(4) $(ax)s = a(xs) = x(as), \qquad a \in A,\ x \in L,\ s \in M,$

then M is said to be an L-**module**. In these circumstances, since f is a bilinear map, we obtain an A-module morphism

$$\varphi : L \otimes_A M \to M$$

which is called the **structure map** of the L-module M. For an L-module M, we may define $\rho(x)s = xs$ and get the A-module morphism

$$\rho : L \to \operatorname{End}_A(M)$$

The map ρ is an A-Lie algebra morphism from L into $\operatorname{End}_A(M)_L$, and we call ρ a **representation** of L. The structure map μ_L of the A-Lie algebra L determines an L-module structure on L, and the representation of L obtained in this manner is called the **adjoint representation** of L.

3.2 The universal enveloping algebra of a Lie algebra

Given a commutative ring A, let L be an A-Lie algebra and let $T(L)$ be the tensor A-algebra over the A-module L.

Let $\mathfrak{c}$ denote the ideal of $T(L)$ generated by elements of the form $x \otimes y - y \otimes x - [x, y]$ for $x, y \in L$. The A-algebra $U(L) = T(L)/\mathfrak{c}$ is called the **universal enveloping A-algebra** of L. The restriction $i : L \to U(L)$ of the canonical projection $T(L) \to U(L)$ to L is said to be the canonical embedding. By definition, we have $i([x, y]) = i(x)i(y) - i(y)i(x)$ for $x, y \in L$, so that i is an A-Lie algebra morphism from L into $U(L)_L$. The intersection $\mathfrak{c}_n = \mathfrak{c} \cap T_n(L)$ $(n \in I)$ is an A-submodule of $T_n(L)$, and we have $\mathfrak{c} = \bigcup_{n \in I} \mathfrak{c}_n$. Thus, setting $U_n(L) = T_n(L)/\mathfrak{c}_n$, we can identify $U_n(L)$ with an A-submodule of $U(L)$, so that $U(L) = \bigcup_{n \in I} U_n(L)$. Moreover, if we let μ_{ij} be the restriction of the multiplication map $\mu : T(L) \otimes_A T(L) \to T(L)$ to $T_i(L) \otimes_A T_j(L)$, then we can conclude that $\mu_{ij}(\mathfrak{c}_i \otimes \mathfrak{c}_j) \subseteq \mathfrak{c}_{i+j}$. This show that μ_{ij} induces the map

$$\bar{\mu}_{ij} : U_i(L) \otimes_A (L) \to U_{i+j}(L).$$

It turns out that $\bar{\mu}_{ij}$ is precisely the restriction of the multiplication map of $U(L)$ to $U_i(L) \otimes U_j(L)$. Therefore $\{U_n(L)\}_{n \in I}$ is a filtration on the A-algebra $U(L)$, and $U(L)$ is a filtered A-algebra. The universal enveloping A-algebra of L together with its canonical embedding $i : L \to U(L)$ have the following property.

(U) Given an arbitrary A-algebra B and an A-algebra map $\varphi : L \to B_L$, there exists a unique A-algebra morphism $\psi : U(L) \to B$ such that $\psi \circ i = \varphi$.

By the above property, the map which associates ψ with $\psi \circ i$ gives rise to a bijective correspondence

$$\mathbf{Alg}_A(U(L), B) \cong \mathbf{Lie}_A(L, B_L). \tag{1.7}$$

There exists a pair $(U(L), i)$ unique up to isomorphism that possesses the above property, and the universal enveloping A-algebra of L may be characterized by property (U).

Let L, L' be A-Lie algebras, and let $f : L \to L'$ be an A-Lie algebra morphism. We let $(U(L), i), (U(L'), i')$ denote the universal enveloping A-algebras of L and L' respectively. Since the A-module morphism $i' \circ f$ is an A-Lie algebra morphism from L to $U(L')_L$, we conclude from property (U) that there exists a unique A-algebra morphism $\varphi : U(L) \to U(L')$ such that $\varphi \circ i = i'$.

The correspondence $L \mapsto U(L)$ is a covariant functor from $\mathbf{Lie}_A$ to $\mathbf{Alg}_A$, and is adjoint to the covariant functor $B \mapsto B_L$ from $\mathbf{Alg}_A$ to $\mathbf{Lie}_A$.

Given A-algebras L_1 and L_2, we may define a multiplication of A-Lie algebras component-wise on the (A-module) direct sum $L_1 \oplus L_2$ by

$$[(x_1\ x_2), (y_1, y_2)] = ([x_1, y_1], [x_2, y_2]), \quad x_1\ y_1 \in L_1, \quad x_2, y_2 \in L_2,$$

whereby $L = L_1 \oplus L_2$ becomes an A-Lie algebra. We call this the direct sum of the A-Lie algebras L_1 and L_2. Let $(U, i), (U_1, i_1), (U_2, i_2)$ respectively be the universal enveloping A-algebras of L, L_1, L_2. To each of the canonical embeddings $f_1 : L_1 \to L$ and $f_2 : L_2 \to L$, there exist unique A-algebra morphisms $\varphi_1 : U_1 \to U$ and $\varphi_2 : U_2 \to U$ such that $\varphi_1 \circ i_1 = i$ and $\varphi_2 \circ i_2 = i$. We employ these morphisms in order to define an A-module morphism $\varphi : U_1 \otimes_A U_2 \to U$ by

$$\varphi(x \otimes y) = \varphi_1(x)\varphi_2(y), \qquad x \in U_1, \qquad y \in U_2,$$

so that φ becomes an A-algebra morphism. In turn, defining an A-module morphism $f : L \to U_1 \otimes_A U_2$ by

$$f(x_1, x_2) = f_1(x_1) \otimes 1 + 1 \otimes f_2(x_2), \qquad x_1 \in L_1, \qquad x_2 \in L_2,$$

f becomes an A-Lie algebra morphism from L to $(U_1 \otimes_A U_2)_L$. Thus there exists a unique A-algebra morphism $\psi : U \to U_1 \otimes_A U_2$ which satisfies the condition $\psi \circ i = f$. Now, since $\varphi \circ f = i$, it follows that $\varphi \circ \psi \circ i = \varphi \circ f = i$. Moreover, $\psi \circ \varphi \circ f = \psi \circ i = f$. Thus $\varphi \circ \psi$ and $\psi \circ \varphi$ are identity maps and φ is an A-algebra isomorphism. Therefore we conclude that

$$U(L_1 \oplus L_2) \cong U(L_1) \otimes_A U(L_2)$$

Henceforth, we shall take advantage of the above isomorphism to identify $U(L_1 \oplus L_2)$ with $U(L_1) \otimes_A U(L_2)$.

EXERCISE 1.8 Let M, N be k-Lie algebras and $\rho : M \to \mathrm{Der}_k(N)$ a k-Lie algebra morphism. Prove that when a multiplication is defined on the (k-linear space) direct sum $N \oplus M$ by

$$[(a, x), (b, y)] = ([a, b] + \rho(x)b - \rho(y)a, [x, y]),$$
$$a, b \in N, \quad x, y \in M,$$

then $N \oplus M$ becomes a k-Lie algebra. What we defined is called the **semi-direct sum** of M and N and is written $N \oplus_\rho M$.

Given an A-Lie algebra L and an A-sub-Lie algebra M, let their universal enveloping A-algebras be $(U(L), i)$ and $(U(M), i')$ respectively. The canonical embedding $f : M \to L$ gives rise to a corresponding A-algebra morphism $\varphi : U(M) \to U(L)$ which is injective by Theorem 1.3.1 to be introduced in the next subsection. Thus $U(M)$ can be identified with an A-subalgebra of $U(L)$. In this situation, $U(M)$ is the A-subalgebra of $U(L)$ generated by 1 and $i(M)$.

Let M be an ideal of the A-Lie algebra L. The left ideal of $U(L)$ generated by $i(M)$ coincides with the right ideal generated by $i(M)$, and hence is a two-sided ideal of $U(L)$. We denote this ideal by N. Let $f : L \to L/M$ be the canonical A-Lie algebra morphism. We shall prove that the A-algebra morphism $\varphi : U(L) \to U(L/M)$ corresponding to f is surjective and that $\mathrm{Ker}\, \varphi = N$. Since $i(M) \subset \mathrm{Ker}\, \varphi$, it follows that $N \subset \mathrm{Ker}\, \varphi$. Therefore φ can be written as a composite of the canonical projection $\pi : U(L) \to U(L)/N$ and the A-algebra morphism $\psi : U(L)/N \to U(L/M)$, so that we can write $\varphi = \psi \circ \pi$. Furthermore, since $M \subset \mathrm{Ker}\,(\pi \circ i)$, there exists an A-Lie algebra morphism $\sigma : L/M \to (U(L)/N)_L$ such that $\pi \circ i = \sigma \circ f$. Meanwhile, by property (U) of $(U(L/M), i')$, there exists a unique A-algebra morphism $\psi' : U(L/M) \to U(L)/N$ which satisfies the condition that $\sigma = \psi' \circ i'$.

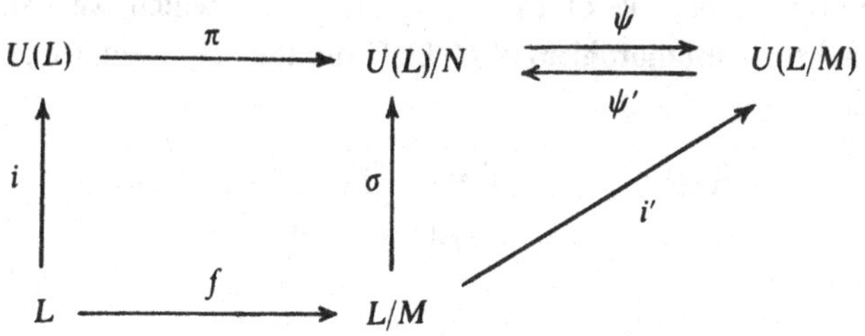

In these circumstances, we have $\psi \circ \sigma \circ f = \psi \circ \pi \circ i = \varphi \circ i = i' \circ f$, so that $\psi \circ \sigma = i'$. Thus $\psi' \circ \psi \circ \sigma = \psi' \circ i' = \sigma$ and $\psi \circ \psi' \circ i' = \psi \circ \sigma = i'$, and hence $\psi' \circ \psi$ and $\psi \circ \psi'$ are identity maps. This fact allows us to conclude that

$$U(L)/N \cong U(L/M),$$

as was to be proved.

Regarding an A-algebra B as an A-module, we consider the A-algebra obtained by defining the product of $x, y \in B$ by $(xy)^{\text{op}} = yx$ which we denote B^{op} and call the **opposite A-algebra** of B. Similarly, if L is an A-Lie algebra, we regard L as an A-module, and denote by L^{op} the A-Lie algebra obtained by defining the product of $x, y \in L$ by $[x, y]^{\text{op}} = [y, x]$, and call it the **opposite A-Lie algebra** of L. We shall next show that in the case of the universal enveloping A-algebra of an A-Lie algebra L, we have the isomorphism $U(L^{\text{op}}) \cong U(L)^{\text{op}}$. The canonical embedding $i : L \to U(L)$ induces the A-Lie algebra morphism $i^{\text{op}} : L^{\text{op}} \to U(L)_L{}^{\text{op}}$. Therefore, due to property (U) there exists a unique A-algebra morphism $\varphi' : U(L^{\text{op}}) \to U(L)^{\text{op}}$ such that $\varphi' \circ i = i^{\text{op}}$. The morphism φ' may be viewed as an A-algebra morphism from $U(L^{\text{op}})^{\text{op}}$ to $U(L)$. By substituting L^{op} for L, we obtain another A-algebra morphism $\psi' : U(L)^{\text{op}} \to U(L^{\text{op}})$. Hence we have the equalities: $i^{\text{op}} = \varphi' \circ i = \varphi' \circ \varphi \circ i^{\text{op}}$ and $i = \varphi \circ \varphi' \circ i = \varphi \circ i^{\text{op}} = i$. Therefore $\varphi' \circ \varphi$ and $\varphi \circ \varphi'$ are identity maps, proving that indeed

$$U(L^{\text{op}}) \cong U(L)^{\text{op}}.$$

Now the A-module morphism $f : L \to L^{\text{op}}$ defined by $f(x) = -x$ for

$x \in L$ is an isomorphism of A-Lie algebras. For this reason the corresponding A-algebra morphism $\varphi : U(L) \to U(L^{\mathrm{op}})$ is an isomorphism. Since we have shown that $U(L^{\mathrm{op}}) \cong U(L)^{\mathrm{op}}$, φ induces an anti-automorphism ψ of the A-algebra $U(L)$, which we call the **principal anti-automorphism** of $U(L)$. From the definition, we get the equality

$$\psi(i(x_1)i(x_2)\cdots i(x_n)) = (-1)^n i(x_1)i(x_2)\cdots i(x_n),$$
$$x_j \in L (1 \leqq j \leqq n).$$

3.3 The Poincaré–Birkhoff–Witt theorem

THEOREM 1.3.1 Given an A-Lie algebra L, regard its universal enveloping A-algebra $U(L)$ as a filtered A-algebra. Let gr $U(L)$ be the associated graded A-algebra of $U(L)$. If L is a free A-module, then

$$\mathrm{gr}\ U(L) \cong S(L).$$

Moreover, the canonical embedding $i : L \to U(L)$ is an injection.

We first provide a lemma in order to prove the theorem. Let $\{x_\lambda\}_{\lambda \in \Lambda}$ be a basis for a free A-module L. We introduce an ordering in Λ, and regard Λ as a totally ordered set. Let the polynomial ring over A with indeterminates $\{z_\lambda\}_{\lambda \in \Lambda}$ be denoted $P = A[z_\lambda]_{\lambda \in \Lambda}$. Also, given a finite sequence $M = \{\lambda_1, \ldots, \lambda_n\}$ consisting of elements of Λ, set

$$x_M = x_{\lambda_1} \otimes \cdots \otimes x_{\lambda_n} \in T_{(n)}(M), \quad |x_M| = \sum_{i=1}^{n} \lambda_i,$$
$$z_M = z_{\lambda_1} z_{\lambda_2} \cdots z_{\lambda_n} \in P, \text{ and } z_\phi = 1.$$

The set $\{z_M;\ M = \{\lambda_1, \ldots, \lambda_n\}, \lambda_1 \leqq \lambda_2 \leqq \ldots \leqq \lambda_n\}$ is a basis for the A-module P, and, by setting

$$P_n = \{f \in A[z_\lambda]_{\lambda \in \Lambda};\ f = \sum_M a_M z_M,\ \ a_M \in A,\ \ \max |z_M| \leqq n\},$$

we see that $\{P_n\}_{n \in I}$ is a filtration on P, thus making P a filtered A-algebra such that $P \cong S(L)$.

LEMMA 1.3.2 For each integer $n \geqq 0$, there exists a unique

A-module morphism $f_n : L \otimes P_n \to P$ satisfying the conditions below. As a matter of notation, when $M = \{\lambda_1, \ldots, \lambda_n\}$ and $\lambda \leqq \lambda_i$ $(1 \leqq i \leqq n)$, we write $\lambda \leqq M$.

$$(\alpha_n)\quad f_n(x_\lambda \otimes z_M) = z_\lambda z_M, \qquad \lambda \leqq M, \quad z_M \in P_n,$$

$$(\beta_n)\quad f_n(x_\lambda \otimes z_M) - z_\lambda z_M \in P_m, \qquad z_M \in P_m, \quad m \leqq n,$$

$$(\gamma_n)\quad f_n(x_\lambda \otimes f_n(x_\mu \otimes z_N)) = f_n(x_\mu \otimes f_n(x_\lambda \otimes z_N)) + f_n([x_\lambda, x_\mu] \otimes z_n), \quad z_N \in P_{n-1}.$$

Moreover, the restriction of f_n to $L \otimes P_{n-1}$ coincides with f_{n-1}.

Proof The restriction of f_n to $L \otimes P_{n-1}$ satisfies conditions (α_{n-1}), (β_{n-1}) and (γ_{n-1}). We will use induction on n to prove the existence of f_n.

If $n = 0$, by setting $f_0(x_\lambda \otimes 1) = z_\lambda$, we see that (α_0) holds, whereas (β_0) and (γ_0) are self-evident. Therefore, assuming the existence and uniqueness of f_{n-1}, we need only show that f_{n-1} can be extended uniquely to f_n.

In case $\lambda \leqq M$, the verification of (α_n) is trivial if we set $f_n(x_\lambda \otimes z_M) = z_\lambda z_M$. For $\lambda \nleqq M$, we take the minimum element μ of M and set $N = M - \{\mu\}$. Then $M = \{\mu, N\}$, where $\mu < \lambda$ and $\mu \leqq N$. In such a case, we get $z_M = z_\mu z_N = f_{n-1}(x_\mu \otimes z_N)$ from (α_{n-1}), and $f_n(x_\lambda \otimes z_N) = f_{n-1}(x_\lambda \otimes z_N) = z_\lambda z_N + w$ for $w \in P_{n-1}$ from (β_{n-1}). Therefore,

$$f_n(x_\mu \otimes f_n(x_\lambda \otimes z_N)) + f_n([x_\lambda, x_\mu] \otimes z_N) = z_\mu z_\lambda z_N + f_{n-1}(x_\mu \otimes w) + f_{n-1}([x_\lambda, x_\mu] \otimes z_N),$$

$$f_n(x_\lambda \otimes z_M) = f_n(x_\lambda \otimes f_{n-1}(x_\mu \otimes z_N)).$$

Therefore, defining

$$f_n(x_\lambda \otimes z_M) = z_\mu z_\lambda z_N + f_{n-1}(x_\mu \otimes w) + f_{n-1}([x_\lambda, x_\mu] \otimes z_N),$$

conditions (α_n), (β_n) hold, and, when $\mu < \lambda$ and $\mu \leqq N$, (γ_n) holds. It thus suffices to show that f_n satisfies (γ_n) in all possible cases. This is trivial for $\lambda = \mu$. When $\lambda < \mu$ and $\lambda \leqq N$, noting that $[x_\lambda, x_\mu] = -[x_\mu, x_\lambda]$, it follows readily that (γ_n) holds. So let us exclude the cases $\lambda \leqq N$ and $\mu \leqq N$. If we let ν be the minimum element of N and set $Q = N - \{\nu\}$, then we obtain $N = \{\nu, Q\}$ where $\nu \leqq Q$, $\nu < \lambda$ and $\nu < \mu$. For the sake of simplicity, we set $f_n(x \otimes z) = xz$ for $x \in L$ and $z \in P$. Then, by the inductive hypothesis, we get

$$\begin{aligned} x_\mu z_N = x_\mu(x_\nu z_Q) &= x_\nu(x_\mu z_Q) + [x_\mu, x_\nu]z_Q \\ &= x_\nu(z_\mu z_Q + w) + [x_\mu, x_\nu]z_Q, \qquad w \in P_{n-2}. \end{aligned}$$

Since $\nu \leqq Q$ and $\nu < \mu$, we can apply (γ_n) to $x_\nu(z_\mu z_Q)$. By the inductive hypothesis, we can also apply (γ_n) to $x_\lambda(x_\nu w)$. Therefore (γ_n) can be applied to $x_\lambda(x_\nu(x_\mu z_Q))$ thus yielding

$$\begin{aligned} x_\lambda(x_\mu z_N) &= x_\lambda(x_\nu(x_\mu z_Q)) + x_\lambda([x_\mu, x_\nu]z_Q) \\ &= x_\nu(x_\lambda(x_\mu z_Q)) + [x_\lambda, x_\nu](x_\mu z_Q) + [x_\mu, x_\nu](x_\lambda z_Q) \\ &\quad + [x_\lambda,[x_\mu, x_\nu]]z_Q. \end{aligned}$$

Since this equation is valid even when λ and μ are interchanged, it follows that

$$\begin{aligned} &x_\lambda(x_\mu z_N) - x_\mu(x_\lambda z_N) \\ &\quad = x_\nu(x_\lambda(x_\mu z_Q) - x_\mu(x_\lambda z_Q)) + [x_\lambda, [x_\mu, x_\nu]]z_Q - [x_\mu, [x_\lambda, x_\nu]]z_Q \\ &\quad = x_\nu([x_\lambda, x_\mu]z_Q) + [x_\lambda, [x_\mu, x_\nu]]z_Q + [x_\mu,[x_\nu, x_\lambda]]z_Q \\ &\quad = [x_\lambda, x_\mu]x_\nu z_Q + ([x_\nu, [x_\lambda, x_\mu]] + [x_\lambda, [x_\mu, x_\nu]] + [x_\mu, [x_\nu, x_\lambda]])z_Q \\ &\quad = [x_\lambda, x_\mu]z_N \end{aligned}$$

which proves (γ_n).

COROLLARY 1.3.3 *P has an L-module structure with structure map $f : L \otimes P \to P$ satisfying the properties*

(1) $f(x_\lambda \otimes z_M) = z_\lambda z_M \qquad (\lambda \leqq M)$,

(2) $f(x_\lambda \otimes z_M) \equiv z_\lambda z_M \qquad (\text{mod } P_n), \qquad |M| = n.$

Proof Since $P = \bigcup_{n \in I} P_n$, if we define $f(x \otimes z) = f_n(x \otimes z)$ for $x \in L$, $z \in P_n$, we get an A-module morphism $f : L \otimes P \to P$. Regardless of the way the n for which $x \in P_n$ is chosen, f will be uniquely determined. By (γ_n) it follows that

$$f([x, y] \otimes z) = f(x \otimes f(y \otimes z)) - f(y \otimes f(x \otimes z)).$$

Therefore P becomes an L-module with f as the structure map. Moreover, it is clear from (α_n) and (β_n) that conditions (1) and (2) are satisfied.

COROLLARY 1.3.4 Given an A-Lie algebra L, suppose that L is a free A-module when regarded as an A-module. If $t \in T_n(L) \cap \mathfrak{c}$ and t_n is the nth homogeneous component of t, namely if $t = t_n + t'$ for $t' \in T_{n-1}(L)$, then $t_n \in \mathfrak{a}$.

Proof By Corollary 1.3.3, P is an L-module, and we obtain a representation $\rho : L \to (\mathrm{End}_A(P))_L$ of L. Property (U) implies that there exists a unique A-algebra morphism $\sigma : U(L) \to \mathrm{End}_A(P)$ such that $\sigma \circ i = \rho$. Let the composite of the canonical A-algebra morphism $\pi : T(L) \to U(L)$ and σ be denoted $\varphi : T(U) \to \mathrm{End}_A(P)$. If $t \in T_n(L) \cap \mathfrak{c}$, then $\varphi(t) = 0$. If we set $t_n = \sum_{i=1}^{r} z_{M_i}$, $|M_i| = n (1 \leqq i \leqq r)$, then, by Corollary 1.3.3, we get $\varphi(t) \equiv \sum_{i=1}^{r} z_{M_i} (\mathrm{mod}\ P_{n-1})$. Thus $\sum_{i=1}^{r} z_{M_i}$ is 0 in P. Meanwhile, since $P \cong S(L)$, it follows that $t_n \in \mathfrak{a}$.

Proof of Theorem **1.3.1** In order to prove that the canonical map $\varphi : S(L) \to \mathrm{gr}\ U(L)$ is an isomorphism, it suffices to show that φ is injective. Let the composite of the canonical A-algebra morphism $\pi : T(L) \to S(L)$ and the map φ be denoted $\psi : T(L) \to \mathrm{gr}\ U(L)$. Then φ is an injection $\Leftrightarrow t \in T_n(L)$ and $\psi(t) \in U_{n-1}(L)$ imply $t \in \mathfrak{a}$. In turn, if $\psi(t) \in U_{n-1}(L)$, then there exists $t' \in T_{n-1}(L)$ such that $t - t' \in \mathfrak{c}$. Since t is the nth homogeneous component of $t - t'$, it follows from Corollary 1.3.4 that $t \in \mathfrak{a}$. We claim that $i : L \to U(L)$ is an injection. Since the restriction of φ to L is injective, we have $L \cong U_1/U_0$. Therefore $U_1 = L \oplus A$, which proves our claim.

3.4 p-k-Lie algebras

Let k be a field of characteristic $p > 0$ and let A be a k-Lie algebra. Denote the k-Lie algebra $\mathrm{Der}_k(A)$ by L. If $D \in L$, then $D^p \in L$. Therefore we get a map $D \mapsto D^p$ from L to L. In these circumstances, we have

(1) $(cD)^p = c^p D^p, \qquad c \in k, \quad D \in L,$

(2) $(D_1 + D_2)^p = D_1{}^p + D_2{}^p + \sum_{i=1}^{p-1} s_i(D_1, D_2),$

(3) $(\mathrm{ad}\ D)^p = \mathrm{ad}\ D^p,$

where ad D is a k-linear map from L to L defined by $X \mapsto [D, X]$ for

$X \in L$, and $s_i(D_1, D_2)$ is the homogeneous polynomial of degree p in D_1 and D_2 obtained in the following manner.

For a polynomial ring $A[\xi]$ in one variable over a ring A,

$$(\xi a + b)^p = \xi^p a^p + b^p + \sum_{i=1}^{p-1} s_i(a, b)\xi^i, \qquad a, b \in A, \tag{1.8}$$

where $s_i(a, b)$ above is expressed in terms of a and b via the k-algebra operation on A. However, we will show that it can also be expressed via the operation on the k-Lie algebra A_L. For $a \in A$, consider the maps from A into A given by

$$a_L : x \mapsto ax,$$
$$a_R : x \mapsto xa,$$
$$\operatorname{ad} a : x \mapsto ax - xa = [a, x] = (a_L - a_R)(x).$$

Since $a_L \circ a_R = a_R \circ a_L$, it follows that $(\operatorname{ad} a)^p = (a_L)^p - (a_R)^p$. Therefore

$$(\operatorname{ad} a)^{p-1} = (a_L - a_R)^{p-1} = \sum_{i=0}^{p-1} (a_L)^i \circ (a_R)^{p-1-i},$$

which gives

$$(\operatorname{ad} a)^{p-1}(x) = \sum_{i=0}^{p-1} a^i x a^{p-1-i}.$$

Now, by differentiating both sides of (1.8) with respect to ξ, we get

$$\sum_{i=1}^{p-1} i s_i(a, b)\xi^{i-1} = \sum_{i=0}^{p-1} (\xi a + b)^i a (\xi a + b)^{p-1-i}$$
$$= (\operatorname{ad}(\xi a + b))^{p-1}(a) \tag{1.9}$$

where $s_i(a, b)$ is the element of A_L determined by equation (1.9).

We employ the polynomials $s_i(a, b)$ in order to define an extension of the map $D \mapsto D^p$ of the k-Lie algebra $L = \operatorname{Der}_k(A)$ to an arbitrary k-Lie algebra in the following manner. Let k be a field with characteristic $p > 0$. Given a k-Lie algebra L, if there exists a map $x \mapsto x^{[p]}$ from L to itself which satisfies the following properties, then we call L a **p-k-Lie algebra**.

(1) $(cx)^{[p]} = c^p x^{[p]}$, $\qquad c \in k, \quad x \in L,$

(2) $(x + y)^{[p]} = x^{[p]} + y^{[p]} + \sum_{i=1}^{p-1} s_i(x, y)$, $\quad x, y \in L,$

(3) $(\operatorname{ad} x)^p(y) = (\operatorname{ad} x^{[p]})(y)$.

By definition, both $\mathrm{Der}_k(A)$ and $\mathrm{End}_k(V)_L$ (where V is a finite dimensional k-linear space) are p-k-Lie algebras with respect to the pth power map. Suppose that we are now given two p-k-Lie algebras L and L'. We call a k-Lie algebra morphism $p : L \to L'$ satisfying $\rho(x^{[p]}) = \rho(x)^{[p]}$, a p-k-**Lie algebra morphism**. Moreover, a p-k-Lie algebra morphism

$$\rho : L \to \mathrm{End}_k(V)_L$$

is said to be a **representation** of the p-k-Lie algebra L and the L-module V is called a p-L-**module**.

Let $(U(L), i)$ be the universal enveloping k-algebra of the p-k-Lie algebra L. Let $\mathfrak{p}$ denote the two-sided ideal of $U(L)$ generated by all those elements of the form $x^p - x^{[p]}$ for $x \in L$, and set $\bar{U}(L) = U(L)/\mathfrak{p}$. Let $\pi : U(L) \to \bar{U}(L)$ be the canonical k-algebra morphism and write $\pi(a) = \bar{a}$. The composite $\sigma = \pi \circ i : L \to \bar{U}(L)_L$ becomes a p-k-Lie algebra morphism. We call the pair consisting of $\bar{U}(L)$ and σ the p-**universal enveloping** k-**algebra** of L. If L is a finite dimensional k-linear space with a basis $\{x_1, \ldots, x_n\}$, then $\bar{U}(L)$ is a p^n-dimensional k-linear space spanned by $\{x_1^{e_1} x_2^{e_2} \cdots x_n^{e_n};\ 0 \leqq e_i < p\}$. The pair $(\bar{U}(L), \sigma)$ has the following properties.

(1) $\bar{U}(L)$ is a k-algebra; $\bar{U}(L)_L$ is a p-k-Lie algebra with respect to the pth power map. Moreover, $\sigma : L \to \bar{U}(L)_L$ is a p-k-Lie algebra morphism.

(2) If a k-algebra A has the same properties as $\bar{U}(L)$ in (1), then there exists a unique k-algebra morphism $\nu : \bar{U}(L) \to A$ such that $\rho = \nu \circ \sigma$.

A k-algebra $\bar{U}(L)$ satisfying conditions (1) and (2) above is unique up to isomorphism. Moreover, the p-universal enveloping k-algebra of L is characterized by these properties.

The representations of a p-k-Lie algebra L and those of $\bar{U}(L)$ are in one-to-one correspondence, and we have the following theorem apropos representations of a p-k-Lie algebra.

THEOREM 1.3.5 (Hochschild) The following conditions are equivalent for a finite dimensional p-k-Lie algebra L.

(i) Any representation of the p-k-Lie algebra L is completely reducible.

(ii) L is a commutative p-k-Lie algebra and $kL^{[p]} = L$.

Proof (ii)$\Rightarrow$(i) Let $\{x_1, \ldots, x_n\}$ be a basis for L and set $x_i^{[p]} = y_i$

$(1 \leqq i \leqq n)$. Since $kL^{[p]} = L$, $\{y_1, \ldots, y_n\}$ is also a basis for L. Since L is commutative, $\bar{U}(L)$ is a commutative k-algebra where $x_i^p = y_i$ $(1 \leqq i \leqq n)$. Given $u \in \bar{U}(L)$ such that $u \neq 0$, we have

$$u = \sum a_{e_1 \ldots e_n} x_1^{e_1} \cdots x_n^{e_n}$$

$$\Rightarrow u^p = \sum (a_{e_1 \ldots e_n})^p y_1^{e_1} \cdots y_n^{e_n} \neq 0, \qquad a_{e_1 \ldots e_n} \in k.$$

Therefore $\bar{U}(L)$ has no nilpotent elements other than 0. Since $\bar{U}(L)$ is a finite dimensional commutative k-algebra, it follows that $\bar{U}(L)$ is a semi-simple k-algebra and that an arbitrary $\bar{U}(L)$-module is completely reducible (cf. Theorems 1.4.4, 1.5.5). Therefore any representation of a p-k-Lie algebra L is completely reducible.

(i)$\Rightarrow$(ii) Let $\{x_1, \ldots, x_n\}$ be a basis for L, denote by M the n-dimensional k-linear space $\coprod_{i=1}^{n} kt_i$, and let $f : L \to M$ be a p-semi-linear map from L to M defined by

$$f\left(\sum_{i=1}^{n} a_i x_i\right) = \sum_{i=1}^{n} a_i^p t_i, \qquad a_i \in k \; (1 \leqq i \leqq n).$$

We employ f in order to define a p-k-Lie algebra structure on the k-linear space $E = L \oplus M$ in the following manner.

$$[(x, m), (x', m')] = ([x, x'], 0), \qquad x, x' \in L, \quad m, m' \in M,$$

$$(x, m)^{[p]} = (x^{[p]}, f(x)), \qquad x \in L, \qquad m \in M.$$

Now the canonical projection $p : (x, m) \mapsto x$ and the canonical embedding $i : x \mapsto (x, 0)$ are p-k-Lie algebra morphisms respectively from E to L and L to E. These morphisms induce k-algebra morphisms of p-universal enveloping k-algebras

$$\varphi : \bar{U}(E) \to \bar{U}(L), \qquad \psi : \bar{U}(L) \to \bar{U}(E).$$

The ideal $\operatorname{Ker} \varphi = M'$ of $\bar{U}(E)$ is generated by M, and we obtain the (k-linear space) direct sum $M' = \bar{U}(E)^+ M \oplus M$. Here, $\bar{U}(E)^+$ denotes the kernel of the canonical projection $\bar{U}(E) \to k$. Let $\gamma : M' \to M$ be the canonical projection from M' to M. For $u, v \in \bar{U}(L)$, we have $\psi(u)\psi(v) - \psi(uv) \in \operatorname{Ker} \varphi = M'$. Thus the map given by

$$g : (u, v) \mapsto \gamma(\psi(u)\psi(v) - \psi(uv))$$

is a k-linear map from $\bar{U}(L) \times \bar{U}(L)$ into M. For the above map, we have

$$g(uv, w) = g(u, vw) \qquad u, v, w \in \bar{U}(L).$$

Now we define a $\bar{U}(L)$-module structure on the k-linear space $S = \bar{U}(L) \oplus M$. Denote by $u \rightharpoonup s$ the action of $u \in \bar{U}(L)$ on $s \in S$ and define

$$u \rightharpoonup (v + m) = uv + g(u, v) \qquad u, v \in \bar{U}(L), \quad m \in M.$$

Then S becomes a $\bar{U}(L)$-module. In fact

$$\begin{aligned}(au + a'u') \rightharpoonup (v + m) &= (au + a'u')v + g(au + a'u', v)\\ &= a(u \rightharpoonup (v + m)) + a'(u' \rightharpoonup (v + m)),\\ u' \rightharpoonup (u \rightharpoonup (v + m)) &= u' \rightharpoonup (uv + g(u, v))\\ &= u'(uv) + g(u', uv)\\ &= (u'u)v + g(u'u, v)\\ &= (uu') \rightharpoonup (v + m).\end{aligned}$$

We note that S is a completely reducible $\bar{U}(L)$-module by hypothesis and M is a $\bar{U}(L)$-submodule of S. Therefore there exists a $\bar{U}(L)$-submodule Q of S which satisfies the condition $S = M \oplus Q$. Thus an arbitrary $v \in \bar{U}(L) \subseteq S$ may be written in the form

$$v = q(v) + h(v), \qquad q(v) \in Q, \quad h(v) \in M.$$

We will employ the property of the k-linear map $h : \bar{U}(L) \to M$ in order to deduce the assertion of the theorem. Since

$$u \rightharpoonup v = uv + g(u, v) = q(uv) + h(uv) + g(u, v) \in \bar{U}(L),$$

it follows that $h(uv) = -g(u, v)$. On the other hand, for $x, y \in L$, we have

$$\begin{aligned}g(x, y) &= \gamma(\psi(x)\psi(y) - \psi(xy))\\ &= \gamma(\psi(y)\psi(x) - \psi([x, y]) - \psi(xy))\\ &= \gamma(\psi(y)\psi(x) - \psi(yx)) = g(y, x).\end{aligned}$$

Therefore $h([x, y]) = h(xy) - h(yx) = g(y, x) - g(x, y) = 0$, and hence $h([L, L]) = 0$. Moreover,

$$\begin{aligned}h(x^{[p]}) = h(x^p) = -g(x^{p-1}, x) &= \gamma(\psi(x^p) - \psi(x^{p-1})\psi(x))\\ &= \gamma((x^{[p]}, 0) - (x, 0)^{[p]}) = -f(x).\end{aligned}$$

So for arbitrary elements $a_i \in k$ $(1 \leqq i \leqq n)$, we have

$$h\left(\sum_{i=1}^{n} a_i x_i^{[p]}\right) = -f\left(\sum_{i=1}^{n} a_i^{1/p} x_i\right) = -\sum_{i=1}^{n} a_i t_i.$$

Thus, if $h\left(\sum_{i=1}^{n} a_i x_i^{[p]}\right) = 0$, then $a_i = 0$ $(1 \leqq i \leqq n)$ and $\{x_i^{[p]}, 1 \leqq i \leqq n\}$ is linearly independent modulo $[L, L]$. Therefore $[L, L] = 0$ and $\{x_i^{[p]}, 1 \leqq i \leqq n\}$ is linearly independent over k. We thus conclude that L is commutative and $kL^{[p]} = L$.

4 Semi-simple algebras

In this section, we prove Wedderburn's structure theorem and Wedderburn–Malcev's decomposition theorem for semi-simple k-algebras. These theorems will be applied in the next chapter which will deal with the dual case, namely the structure of co-semi-simple k-coalgebras (not necessarily finite dimensional). Concerning semi-simple k-algebras, the reader is asked to refer to books such as [1] or [5].

4.1 The radical

The intersection of all maximal left ideals of a ring A is called the **radical** of A and is written rad A. Let M be a left A-module and set ann $M = \{a \in A : aM = 0\}$. Then ann M is a two-sided ideal of A and is called the **annihilator** of M. Similarly, for $x \in M$, we let ann $(x) = \{a \in A; ax = 0\}$. The set ann (x) is a left ideal of A.

THEOREM 1.4.1 The radical of A is a two-sided ideal of A. Moreover, rad $A = \cap$ ann M, where M ranges over the elements of the set $\mathfrak{M}$ of all irreducible left A-modules.

Proof Let M be an irreducible left A-module. For $x \in M$, $x \neq 0$, we have $M = Ax$. Therefore the assignment $a \mapsto ax$ defines an A-module surjection $\varphi : A \to M$. Thus Ker $\varphi = \mathfrak{a}$ is a left ideal of A. Since $A/\mathfrak{a}$ is an irreducible left A-module, $\mathfrak{a}$ is a maximal left ideal. Therefore, rad $A \subset \mathfrak{a} = \text{ann}\,(x)$. Since $\text{ann}\, M = \bigcap_{\substack{x \in M \\ x \neq 0}} \text{ann}\,(x)$, it follows that rad $A \subset$ ann M. Conversely, let $\mathfrak{a}$ be a maximal left ideal of A. Since

$\operatorname{ann}(A/\mathfrak{a}) \subset \mathfrak{a}$, we have

$$\operatorname{rad} A = \bigcap_{\text{maximal left ideals } \mathfrak{a}} \mathfrak{a} \supseteqq \bigcap_{M \in \mathfrak{M}} \operatorname{ann} M.$$

Therefore $\operatorname{rad} A = \bigcap_{M \in \mathfrak{M}} \operatorname{ann} M$. Since $\bigcap_{M \in \mathfrak{M}} \operatorname{ann} M$ is a two-sided ideal, so is $\operatorname{rad} A$.

THEOREM 1.4.2 Let $a \in A$. Then $a \in \operatorname{rad} A \Leftrightarrow$ for all $x \in A$, $1 - xa$ has a left inverse.

Proof Since $1 = xa + (1 - xa)$, if $a \in \operatorname{rad} A$, $1 - xa$ is not contained in any maximal left ideal. Therefore $A(1 - xa) = A$, so that $1 - xa$ has a left inverse. Conversely, if $a \notin \operatorname{rad} A$, then for some maximal left ideal $\mathfrak{a}$, we have $a \notin \mathfrak{a}$. Thus $Aa + \mathfrak{a} = A$. Therefore there exists $x \in A$ and $b \in \mathfrak{a}$ such that $xa + b = 1$. In other words, there exists $x \in A$ for which $1 - xa$ does not have a left inverse.

COROLLARY 1.4.3 A left ideal $\mathfrak{n}$ consisting of nilpotent elements of a ring A is contained in $\operatorname{rad} A$.

Proof Let $a \in \mathfrak{n}$. Given any $x \in A$, there is a natural number n for which $(xa)^n = 0$. The element $1 - xa$ has a left inverse $1 + xa + (xa)^2 + \cdots + (xa)^{n-1}$. Therefore $a \in \operatorname{rad} A$.

Given an ideal $\mathfrak{a}$ of a ring A, if there exists a natural number n such that $\mathfrak{a}^n = \{0\}$, then $\mathfrak{a}$ is said to be a **nilpotent ideal**.

THEOREM 1.4.4 Let k be a field. Then the radical of a finite dimensional k-algebra A is the maximal nilpotent ideal of A.

Proof Write $\operatorname{rad} A = R$. Since A is finite dimensional, there is a natural number n for which $R^n = R^{n+1}$. We assert that $R^n = \{0\}$. Suppose not, i.e. that $R^n \neq \{0\}$. Now, from among all those left ideals $\mathfrak{a}$ such that $R^n(\mathfrak{a}) \neq \{0\}$, pick one that is minimal. Since $R\mathfrak{a} \subset \mathfrak{a}$ and $R^n(R\mathfrak{a}) = R^n\mathfrak{a} \neq \{0\}$, the minimality of $\mathfrak{a}$ forces $R\mathfrak{a} = \mathfrak{a}$. Thus we have $\mathfrak{a} = \{0\}$ by the next lemma. This contradicts the choice of $\mathfrak{a}$. Therefore $R^n = \{0\}$. Conversely, since a nilpotent ideal of A consists of nilpotent elements, Corollary 1.4.3 implies that R is the maximal nilpotent ideal of A.

LEMMA 1.4.5 (Nakayama) Given a ring A, let M be a finitely generated A-module and $\mathfrak{a}$ a left ideal of A contained in rad A. Then, $\mathfrak{a}M = M$ implies $M = \{0\}$.

Proof Suppose $M \neq \{0\}$, and pick a set of elements $\{u_1, \ldots, u_n\}$ of M so that n is the least possible number for which it is a system of generators for M. Since $u_n \in M = \mathfrak{a}M$, we can write $u_n = a_1u_1 + \cdots + a_nu_n$ where $a_i \in \mathfrak{a}$ $(1 \leqq i \leqq n)$. Therefore

$$(1 - a_n)u_n = a_1u_1 + a_2u_2 + \cdots + a_{n-1}u_{n-1}.$$

Since $a_n \in \mathfrak{a} \subseteq \text{rad}\, A$, Theorem 1.4.2 implies that $1 - a_n$ has a left inverse. Thus u_n is contained in the A-submodule of M generated by $\{u_1, \ldots, u_{n-1}\}$. This contradicts the choice of $\{u_1, \ldots, u_n\}$. Therefore we conclude that $M = \{0\}$.

4.2 Semi-simple k-algebras

Let k be a field. A finite dimensional k-algebra A is called a **semi-simple k-algebra** when rad $A = \{0\}$. We provide some lemmas which deal with the structures of such k-algebras. We note that a non-commutative ring all of whose non-zero elements are units is called a **skew field**, and a ring with no two-sided ideals other than $\{0\}$ and itself is called a **simple ring**. In particular, a skew field is a simple ring.

LEMMA 1.4.6 Let A be a commutative k-algebra and let M, N be irreducible A-modules. If $f: M \to N$ is an A-module morphism and f is not the zero morphism, then it is an isomorphism. In particular, $\text{End}_A(M)$ is a skew field. If k is an algebraically closed field and M is finite dimensional over k, then $\text{End}_A(M) \cong k$.

Proof Let $f \in \mathbf{Mod}_A(M, N)$. Since $\text{Ker} f$ and $\text{Im} f$ are respectively A-subcomodules of M and N, the fact that f is not the zero morphism implies that $\text{Ker} f \neq M$ and $\text{Im} f \neq \{0\}$. Thus $\text{Ker} f = \{0\}$ and $\text{Im} f = N$. Therefore f is an isomorphism. Since an arbitrary non-zero element of $\text{End}_A(M)$ is an isomorphism (and hence, a unit), $\text{End}_A(M)$

is a skew field. Now suppose that k is an algebraically closed field and M is a finite dimensional module over k. For $f \in \mathrm{End}_A(M)$ and $f \neq 0$, since $\mathrm{End}_A(M)$ is finite dimensional over k, it follows that $1, f, f^2, \ldots$ are not linearly independent over k. Let $F(X) \in k[X]$ be the polynomial of the least degree satisfying $F(f) = 0$. Since k is algebraically closed, we have $F(X) = \prod_{i=1}^{n} (X - \alpha_i) = 0$ where $\alpha_i \in k$ $(1 \leqq i \leqq n)$. Thus $F(f) = \prod_{i=1}^{n} (f - \alpha_i 1) = 0$. Hence there exists $\alpha \in k$ such that $f = \alpha 1_M$. The assignment $f \mapsto \alpha$ gives the isomorphism $\mathrm{End}_A(M) \cong k$.

LEMMA 1.4.7 If D is a skew field, then $M_n(D)$ is a simple ring.

Proof Let $e_{ij}^{(n)} \in M_n(D)$ be an $n \times n$ square matrix whose (i, j) entry is 1 and all of whose other entries are 0, and let $e^{(n)}$ be the $n \times n$ identity matrix. Since $e^{(n)} = e_{11}^{(n)} + \cdots + e_{nn}^{(n)}$, it follows that $M_n(D) = M_n(D)e_{11}^{(n)} + \cdots + M_n(D)e_{nn}^{(n)}$ and $e_{ii}^{(n)} M_n(D) e_{ii}^{(n)} \cong D$. Let $\mathfrak{a} \neq \{0\}$ be a two-sided ideal of $M_n(D)$. Pick an element of $\mathfrak{a}$ such that $a = \sum_{i,j} a_{ij} e_{ij}^{(n)} \neq 0$. If $a_{ij} \neq 0$, then we have $e_{ii}^{(n)} a e_{jj}^{(n)} = a_{ij} e_{ij}^{(n)} \in \mathfrak{a}$. Thus $e_{ij}^{(n)} \in \mathfrak{a}$. Therefore $e_{ki}^{(n)} e_{ij}^{(n)} e_{jl}^{(n)} = e_{kl}^{(n)} \in \mathfrak{a}$ $(1 \leqq k, l \leqq n)$, which allows us to conclude that $\mathfrak{a} = M_n(D)$.

THEOREM 1.4.8 (Wedderburn's structure theorem) Given a finite dimensional k-algebra A, the following conditions are equivalent.

(i) A is semi-simple.
(ii) All left A-modules are completely reducible.
(iii) $A \cong M_{n_1}(D_1) \times \cdots \times M_{n_r}(D_r)$.

The $D_1, \ldots, D_r$ above are finite dimensional k-algebras and skew fields whereas $M_n(D)$ is the k-algebra of all $n \times n$ square matrices with entries in D. In particular, if k is an algebraically closed field, then $D_i \cong k$ $(1 \leqq i \leqq r)$.

Proof (i)$\Rightarrow$(ii) Since A is semi-simple, $\mathrm{rad}\, A = \bigcap_{\text{maximal left ideals } \mathfrak{a}} \mathfrak{a} = \{0\}$. In turn, A is finite dimensional over k, so we can choose a finite number of maximal left ideals $\mathfrak{a}_1, \ldots, \mathfrak{a}_n$ such that $\bigcap_{i=1}^{n} \mathfrak{a}_i = \{0\}$. Pick n

to be the smallest natural number for which the above condition holds. In this situation, the canonical left A-module morphism

$$A \to \coprod_{i=1}^{n} A/\mathfrak{a}_i$$

is injective, and $A/\mathfrak{a}_i$ is irreducible, so that A can be identified with a left A-submodule of the completely reducible left A-module $\coprod_{i=1}^{n} A/\mathfrak{a}_i$. Therefore A is a completely reducible A-module. Now let M be an arbitrary left A-module. We can pick a free left A-module P and a left A-module surjection $f: P \to M$. The left A-module P is a direct sum of left A-modules A, so it is completely reducible. Since $\operatorname{Ker} f$ is a left A-submodule of P, there exists a left A-submodule Q of P which satisfies the condition $P = (\operatorname{Ker} f) \oplus Q$. Thus $M \cong P/\operatorname{Ker} f \cong Q$ is completely reducible.

(ii)$\Rightarrow$(iii) Since A is a completely reducible left A-module, there exist (non-isomorphic) simple left ideals $\mathfrak{a}_1, \ldots, \mathfrak{a}_r$ so that A can be written

$$A \cong n_1\mathfrak{a}_1 \oplus \cdots \oplus n_r\mathfrak{a}_r,$$

as a (left A-module) direct sum. By Lemma 1.4.6, $\mathfrak{a}_i \neq \mathfrak{a}_j \Rightarrow \operatorname{Mod}_A(\mathfrak{a}_i, \mathfrak{a}_j) = \{0\}$. Thus

$$\operatorname{End}_A(A) \cong \operatorname{End}_A(n_1\mathfrak{a}_1) \times \cdots \times \operatorname{End}_A(n_r\mathfrak{a}_r).$$

Letting $\mathfrak{a}^{(i)}$ $(1 \leqq i \leqq n)$ be left A-modules which are isomorphic to $\mathfrak{a}$, we set $n\mathfrak{a} = \mathfrak{a}^{(1)} + \cdots + \mathfrak{a}^{(n)}$ and let $\theta_i : \mathfrak{a} \to \mathfrak{a}^{(i)}$ be left A-module isomorphisms $(1 \leqq i \leqq n)$. Now, $D = \operatorname{End}_A(\mathfrak{a})$ is a skew field, and for $f \in \operatorname{End}_A(n\mathfrak{a})$, we write $\varphi_{ij} = \theta_j' f \theta_i$ where θ_j' is the composite of the canonical projection from $n\mathfrak{a}$ to $\mathfrak{a}^{(j)}$ and θ_j^{-1}. The map which carries f to $\varphi = (\varphi_{ij})_{1 \leqq i, j \leqq n}$,

$$\operatorname{End}_A(n\mathfrak{a}) \to M_n(D)$$

is a k-algebra isomorphism. Therefore $\operatorname{End}_A(A) \cong A^{\mathrm{op}}$ is isomorphic to the direct product of k-algebras $M_{n_i}(D_i)$ $(1 \leqq i \leqq r)$. Hence the same thing can be said of A. If k is an algebraically closed field, Lemma 1.4.6 implies that $D_i \cong k$ $(1 \leqq i \leqq r)$.

(iii)$\Rightarrow$(i) If $A \cong M_{n_1}(D_1) \times \cdots \times M_{n_r}(D_r)$, then $\operatorname{rad} A$ is isomorphic to the direct product of $\operatorname{rad} M_{n_i}(D_i)$ $(1 \leqq i \leqq r)$. Meanwhile, since each $M_{n_i}(D_i)$ is simple, we have $\operatorname{rad} A = \{0\}$. Therefore A is semi-simple.

4.3 **Wedderburn–Malcev's decomposition theorem**

Given a finite dimensional k-algebra A, if $A \otimes_k K$ is a semi-simple K-algebra for any extension field K of k, we call A a **separable k-algebra**. A separable k-algebra is a semi-simple k-algebra, and conversely, for an algebraically closed field k, a semi-simple k-algebra is separable.

THEOREM 1.4.9 Let B be a finite dimensional k-algebra and suppose that $A = B/\text{rad}\, B$ is separable. Then there exists a semi-simple k-subalgebra S of B such that B may be written as the direct sum of k-modules, $B = S \oplus \text{rad}\, B$. Given two such k-subalgebras S_1 and S_2 of B, there exists an element $n \in \text{rad}\, B$ such that

$$(1-n)^{-1} S_1 (1-n) = S_2.$$

We proceed to prove this theorem by utilizing cohomology theory of k-algebras. We need some preparation for this.

Cohomology groups of k-algebras Let A be a k-algebra and let M be a two-sided A-module. Set $\otimes^n A = A \otimes \cdots \otimes A$, the n-fold tensor product of A over k, and set $\otimes^0 A, = k$. For $f \in \mathbf{Mod}_k(\otimes^n A, M)$ $(n \geqq 0)$, define $\delta f \in \mathbf{Mod}_k(\otimes^{n+1} A, M)$ by

$$\begin{aligned}\delta f(a_1 \otimes \cdots \otimes a_{n+1}) = {} & a_1 f(a_2 \otimes \cdots \otimes a_{n+1}) \\ & + \sum_{i=1}^{n} (-1)^i f(a_1 \otimes \cdots \otimes a_i a_{i+1} \otimes \cdots \otimes a_{n+1}) \\ & + (-1)^{n+1} f(a_1 \otimes \cdots \otimes a_n) a_{n+1}.\end{aligned}$$

A routine check shows that $\delta\delta f = 0$. In particular, if $n = 0$, then $f \in M$ and $\delta f(a) = af - fa$. For $n = 1$, we get $f \in \text{Mod}_k(A, M)$ and $\delta f(a, b) = af(b) - f(ab) + f(b)a$. Thus, $\delta f = 0 \Leftrightarrow f \in \text{Der}_k(A, M)$. If $f \in M$, then $\delta\delta f = 0$ so that $\delta f \in \text{Der}_k(A, M)$. Moreover, δf is an inner derivation. Now set.

$$\begin{aligned} Z^n(A, M) &= \{ f \in \mathbf{Mod}_k(\otimes^n A, M); \quad \delta f = 0 \}, \\ B^n(A, M) &= \{ \delta f \in \mathbf{Mod}_k(\otimes^n A, M); \quad f \in \mathbf{Mod}_k(\otimes^{n-1} A, M) \}. \end{aligned}$$

Then $B^n(A, M) \subset Z^n(A, M)$, and the elements of $Z^n(A, M)$ and $B^n(A, M)$ are respectively called n-cocycles and n-coboundaries. The

k-module

$$H^n(A, M) = Z^n(A, M)/B^n(A, M)$$

is called the **nth cohomology group** of A with coefficients in M.

Now $H^1(A, M)$ is a factor k-module of $\mathrm{Der}_k(A, M)$ by the k-submodule consisting of inner k-derivations. Thus, to say $H^1(A, M) = \{0\}$ is equivalent to saying that an arbitrary k-derivation is 'inner'.

Next we proceed to show that $H^2(A, M)$ corresponds to extensions of the k-algebra A. Let A and B be k-algebras. Given a k-algebra surjection $\rho : B \to A$, if $\operatorname{Ker} \rho = M$ satisfies $M^2 = \{0\}$, (B, ρ) is called an **extension** of A by M. Two such extensions $(B, \rho), (B', \rho')$ of A by M are said to be equivalent when there exists a k-algebra morphism $\varphi : B \to B'$ such that $\rho' \circ \varphi = \rho$, in which case we write $(B, \rho) \sim (B', \rho')$.

Given an extension (B, ρ) of A by M, we see that M is a two-sided ideal of B as well as a two-sided B-module. For $a \in A$, if we pick $b \in B$ so that $\rho(b) = a$ and define $am = bm$, then am is determined uniquely by a regardless of the way in which b was chosen. In fact, $\rho(b) = \rho(b') = a$ implies $b - b' \in M$, and since $M^2 = \{0\}$, it follows that $(b - b')m = bm - b'm = 0$, so that $bm = b'm$. Hence M may be regarded as a two-sided A-module. Now picking a k-module morphism $\sigma : A \to B$ which satisfies the condition $\rho \circ \sigma = 1_A$ and defining a k-module morphism $f_\sigma : A \otimes A \to M$ by

$$f_\sigma(a \otimes b) = \sigma(ab) - \sigma(a)\sigma(b), \qquad a, b \in A,$$

it is readily verified that $f_\sigma \in Z^2(A, M)$. Pick another k-module morphism $\sigma' : A \to B$ such that $\rho \circ \sigma' = 1_A$ and construct $f_{\sigma'}$. We calculate

$$\begin{aligned}(f_\sigma - f_{\sigma'})(a \otimes b) &= \sigma(ab) - \sigma(a)\sigma(b) - \sigma'(ab) - \sigma'(a)\sigma'(b) \\ &= \sigma(ab) - \sigma'(ab) - \sigma(a)(\sigma(b) - \sigma'(b)) \\ &\quad + (\sigma(a) - \sigma'(a))\sigma'(b) \\ &= \delta(\sigma - \sigma')(a \otimes b).\end{aligned}$$

This shows that an element of $H^2(A, M)$ is determined by an extension of A by M. Conversely, given $f \in Z^2(A, M)$, if we define a multiplication on the direct sum $A \oplus M$ of k-modules by

$$(a, u)(b, v) = (ab, av + ub + f(a \otimes b)), \quad a, b \in A, u, v \in M,$$

$A \otimes M$ becomes a k-algebra which we denote by B_f. The canonical projection $\rho : (a, u) \mapsto a$ is a k-algebra morphism from B_f to A such that $\operatorname{Ker} \rho = M$ and $M^2 = \{0\}$. Therefore (B_f, ρ) is an extension of A by M. Moreover, if there exists $g \in \operatorname{Mod}_k(A, M)$ such that for $f, f' \in Z^2(A, M)$ the equality $f - f' = \delta g$ holds, then the map from B_f to $B_{f'}$

$$\varphi : (a, u) \mapsto (a, u + g(a))$$

is a k-algebra morphism which gives the equivalence $(B_f, \rho) \sim (B_{f'}, \rho')$. Thus each element of $H^2(A, M)$ determines one equivalence class of extensions of A by M. It can be shown that this correspondence is a surjection.

Given an extension (B, ρ) of A by M, we say that the extension (B, ρ) 'splits' when there exists a k-algebra morphism $\sigma : A \to B$ such that $\rho \circ \sigma = 1_A$. If this is the case, B has a k-subalgebra $\sigma(A)$ isomorphic to A and admits a decomposition into the (k-module) direct sum : $B = \sigma(A) \oplus M$. If $f \in B^2(A, M)$ is such that $f = \delta g$ for $g \in \operatorname{Mod}_k(A, M)$, then $\sigma : a \mapsto (a, -g(a))$ is a k-algebra morphism from A to B_f satisfying $\rho \circ \sigma = 1_A$. Therefore the extension (B_f, ρ) splits.

Let A be a k-algebra and M a two-sided A-module. Then $N = \mathbf{Mod}_k(A, M)$ becomes a two-sided A-module in the following manner

$$(af)(a') = af(a'), \qquad a, a' \in A, \quad f \in N,$$

$$(fa)(a') = f(aa') - f(a)a'.$$

For $f \in \mathbf{Mod}_k(\otimes^n A, M)$, define $\bar{f} \in \mathbf{Mod}_k(\otimes^{n-1} A, N)$ by

$$\bar{f}(a_1 \otimes \cdots \otimes a_{n-1})(a_n) = f(a_1 \otimes \cdots \otimes a_{n-1} \otimes a_n).$$

Then, for $n \geqq 1$,

$$\begin{aligned} \overline{\delta f}(a_1 \otimes \cdots \otimes a_n)(a_{n+1}) &= \delta f(a_1 \otimes \cdots \otimes a_{n+1}) \\ &= a_1 \bar{f}(a_2 \otimes \cdots \otimes a_n)(a_{n+1}) \\ &\quad + \sum_{i=1}^{n-1} (-1)^i \bar{f}(a_1 \otimes \cdots \otimes a_i a_{i+1} \otimes \cdots \otimes a_n)(a_{n+1}) \\ &\quad + (-1)^n (\bar{f}(a_1 \otimes \cdots \otimes a_{n-1}) a_n)(a_{n+1}). \end{aligned}$$

Thus $\overline{\delta f} = \delta \bar{f}$. This allows us to conclude that

$$H^n(A, M) \cong H^{n-1}(A, \mathbf{Mod}_k(A, M)) \qquad (1.10)$$

for $n \geqq 2$.

LEMMA 1.4.10 Let A be a separable k-algebra and let M be a two-sided A-module. Then

$$H^n(A, M) = \{0\} \qquad (n \geqq 1).$$

Proof Given any two-sided A-module M, as a result of (1.10), we need only prove that $H^1(A, M) = \{0\}$. Since A is separable, there exists a finite dimensional extension field K of k which satisfies the condition $A_K = A \otimes_k K \cong M_{n_1}(K) \times \cdots \times M_{n_r}(K)$. Given a k-module morphism $f: A \to M$, we define a K-module morphism f_K: $A_K \to M_K = M \otimes_k K$ by $f_K(a \otimes c) = f(a) \otimes c$ for $a \in A$ and $c \in K$. Then $\delta f = 0$ implies $\delta f_K = 0$. Let $e_{ij}{}^{(p)}$ $(i, j = 1, 2, \ldots, p)$ be matrix units of $M_{n_p}(K)$. Then

$$e_{ij}{}^{(p)} e_{lm}{}^{(q)} = \delta_{pq} \delta_{jl} e_{im}{}^{(q)}, \qquad \sum_{p=1}^{r} \sum_{i=1}^{n_p} e_{ii}{}^{(p)} = 1.$$

Thus, for $f \in Z^1(A, M)$, we set $u = \sum_{p=1}^{r} \sum_{i=1}^{n_p} e_{i1}{}^{(p)} f_K(e_{1i}{}^{(p)}) \in M_K$ and get

$$\begin{aligned} & e_{lm}{}^{(q)} u - u e_{lm}{}^{(q)} \\ &= \sum_{p=1}^{r} \sum_{i=1}^{n_p} e_{lm}{}^{(q)} e_{i1}{}^{(p)} f_K(e_{1i}{}^{(p)}) \\ &\quad - \sum_{p=1}^{r} \sum_{i=1}^{n_p} e_{i1}{}^{(p)} \{ f_K(e_{1i}{}^{(p)} e_{lm}{}^{(q)}) - e_{1i}{}^{(p)} f_K(e_{lm}{}^{(q)}) \} \\ &= f_K(e_{lm}{}^{(q)}). \end{aligned}$$

Thus, for any $m \in M_K$, we have $f_K(m) = mu - um$. In particular, when $m \in M$, we obtain $f(m) = mu - um$. Moreover, such a u may be chosen from M. Thus $f = \delta u$.

***Proof of Theorem* 1.4.9** If $(\text{rad } B)^2 = \{0\}$, then we use Lemma 1.4.10 and set $M = \text{rad } B$ to obtain $H^2(A, M) = \{0\}$. It follows that an extension of A by M splits, and hence we get the existence of a

k-subalgebra S. We prove this for the general case by using induction on the dimension of the k-algebra B. Suppose that $(\mathrm{rad}\, B)^2 \neq \{0\}$ and that for a k-algebra with dimension less than $\dim B$, a k-subalgebra S exists. Set $\mathrm{rad}\, B = N$. Since $B/N^2/N/N^2 \cong A$, we use the inductive hypothesis on B/N^2 to obtain the existence of a k-subalgebra S_1 of B which satisfies

$$B = S_1 + N, \qquad S_1 \cap N = N^2.$$

Since $N \neq N^2$, it follows that $S_1 \subsetneq B$. Moreover, since $S_1/N^2 \cong A$, we can apply the inductive hypothesis on S_1 to obtain the existence of a k-subalgebra S of S_1 such that

$$S_1 = S + N^2, \qquad S \cap N^2 = \{0\}.$$

For this S, we also have

$$B = S + N, \qquad S \cap N = \{0\},$$

which shows that S is a k-subalgebra of B satisfying the desired conditions of the theorem.

Now let S_1 and S_2 be k-subalgebras of B satisfying the conditions of the theorem. Pick k-algebra morphisms $\sigma_1, \sigma_2 : A \to B$ such that $\sigma_i(A) = S_i$ and $\rho \circ \sigma_i = 1_A$ $(i = 1, 2)$. Defining

$$na = n\sigma_2(a), \qquad an = \sigma_1(a)n, \qquad a \in A,\ n \in N,$$

N becomes a two-sided A-module. The image of the k-module morphism $f = \sigma_1 - \sigma_2$ is contained in N. Moreover, for $a, b \in A$,

$$\begin{aligned} f(ab) &= \sigma_1(ab) - \sigma_2(ab) \\ &= \sigma_1(a)(\sigma_1(b) - \sigma_2(b)) + (\sigma_1(a) - \sigma_2(a))\sigma_2(b) \\ &= af(b) + f(a)b. \end{aligned}$$

Therefore, $f \in Z^1(A, N)$. Lemma 1.4.10 implies $Z^1(A, N) = B^1(A, N)$ so there exists $n \in N$ satisfying $f(a) = an - na$ for $a \in N$. We thus have $\sigma_1(a) - \sigma_2(a) = an - na = \sigma_1(a)n - n\sigma_2(a)$. Therefore $\sigma_1(a)(1 - n) = (1 - n)\sigma_2(a)$ for $a \in A$. Since n is nilpotent, $1 - n$ is a unit and $\sigma_2(a) = (1 - n)^{-1}\sigma_1(a)\,(1 - n)$. Thus, $(1 - n)^{-1}\, S_1(1 - n) = S_2$.

5 Finitely generated commutative algebras

Finitely generated commutative k-algebras play a key role in algebraic geometry. The results in this section will become necessary in Chapter 5 where it will be applied to algebraic groups. Due to limitations of space, some of the proofs have been omitted. They may be found, for instance, in Nagata [2] or Bourbaki [4].

Let A be a commutative ring. Denote by nil A the set of all nilpotent elements of A; this turns out to be an ideal of A. This ideal is called the **nilradical** of A. Given an ideal $\mathfrak{a}$ of A,

$$\{f \in A;\ \text{there exists a natural number } n \text{ for which } f^n \in \mathfrak{a}\}$$

is an ideal containing $\mathfrak{a}$ called the **radical** of $\mathfrak{a}$, which we denote by $\sqrt{\mathfrak{a}}$. An ideal such that $\mathfrak{a} = \sqrt{\mathfrak{a}}$ is said to be a **radical ideal**. Accordingly, $\sqrt{\{0\}}$ is the ideal consisting of all nilpotent elements of A and is simply nil A. When $\{0\}$ is a radical ideal, or equivalently, when nil $A = \{0\}$, A is said to be **reduced**. If $\mathfrak{a}$ is a radical ideal, $A/\mathfrak{a}$ is reduced, and, in particular, $A/\text{nil } A$ is reduced.

THEOREM 1.5.1 The radical of an ideal $\mathfrak{a}$ of a commutative ring A is the intersection of all the prime ideals containing $\mathfrak{a}$.

Proof Denote by $\mathfrak{a}'$ the intersection of all prime ideals containing $\mathfrak{a}$. For $f \in \sqrt{\mathfrak{a}}$, there exists a natural number n for which $f^n \in \mathfrak{a}$. If we let $\mathfrak{p}$ be any prime ideal containing $\mathfrak{a}$, then $f^n \in \mathfrak{a} \subset \mathfrak{p}$. Thus $f \in \mathfrak{p}$, so that $\sqrt{\mathfrak{a}} \subset \mathfrak{a}'$. In the reverse direction, letting $f \notin \sqrt{\mathfrak{a}}$, we get a family

$$\Sigma = \{\mathfrak{b} : \mathfrak{b} \text{ is an ideal of } A \text{ such that } f^n \notin \mathfrak{b},\ n \in \mathbb{N}\}$$

of ideals of A such that $\mathfrak{a} \in \Sigma$. Let $\mathfrak{p}$ be a maximal element in Σ. (Its existence is guaranteed by Zorn's lemma.) Now we will show that $\mathfrak{p}$ is a prime ideal. For $x, y \in A$, if $x \notin \mathfrak{p}$ and $y \notin \mathfrak{p}$, then $Ax + \mathfrak{p},\ Ay + \mathfrak{p} \notin \Sigma$. Thus there exist $a, b \in A$ and $n, m \in \mathbb{N}$ such that $ax \equiv f^n$, $by \equiv f^m$ (mod $\mathfrak{p}$). If this is the case, then $abxy \equiv f^{n+m}$ (mod $\mathfrak{p}$) and $f^{n+m} \notin \mathfrak{p}$, so that $xy \notin \mathfrak{p}$. Hence $\mathfrak{p}$ is a prime ideal. This allows us to conclude that $f \notin \mathfrak{a}'$ and that $\sqrt{\mathfrak{a}} \supset \mathfrak{a}'$.

COROLLARY 1.5.2 The nilradical nil A of a commutative ring A is the intersection of all prime ideals of A. In particular, nil $A \subset$ rad A.

Let k be a field and let A be a commutative k-algebra. When there exists a finite number of elements $\{a_1, \ldots, a_n\}$ of A such that any element of A can be expressed as a linear combination over k of products of powers of $a_1, \ldots, a_n$, A is called a **finitely generated** k-**algebra**. For instance, the polynomial ring over k in n variables $A = k[X_1, \ldots, X_n]$ is a finitely generated k-algebra.

Let $A \subseteq B$ be commutative rings. If for $b \in B$, there exist a natural number n and elements $a_1, \ldots, a_n$ of A such that $b^n + a_1 b^{n-1} + \cdots + a_{n-1} b + a_n = 0$, then b is said to be integral over A. If all elements of B are integral over A then we say that B is integral over A.

THEOREM 1.5.3 (Hilbert's normalization theorem) Let A be a finitely generated commutative k-algebra. Then there exist elements $z_1, \ldots, z_t$ of A such that

(i) A is integral over $k[z_1, \ldots, z_t]$.

(ii) $z_1, \ldots, z_t$ are algebraically independent over k. (cf. [2] Theorem 4.0.3, or [4] Chapter V, §3, No. 1, Theorem 1.)

As an application of this theorem, we get the following result.

THEOREM 1.5.4 Suppose that A is a finitely generated k-algebra as well as an integral domain. If we let $0 = \mathfrak{p}_0 \subset \mathfrak{p}_1 \subset \cdots \subset \mathfrak{p}_t$ be an ascending chain of prime ideals such that there are no prime ideals strictly between $\mathfrak{p}_i$ and $\mathfrak{p}_{i+1}$ where $\mathfrak{p}_t$ is a maximal ideal, then t is equal to the transcendence degree of the quotient field $Q(A)$ of A over k which is denoted by $\text{trans.deg}_k Q(A)$. For a maximal ideal $\mathfrak{m}$ of A, $A/\mathfrak{m}$ is an algebraic extension field of k. If in particular, k is an algebraically closed field, then $A/\mathfrak{m} \cong k$. (cf. [2] Theorems 4.1.1, 4.1.2, or [4] Chapter V, §3, No. 3, Proposition 2, No. 4, Theorem 3.)

THEOREM 1.5.5 (Hilbert's *Nullstellensatz*) The radical of an ideal $\mathfrak{a}$ of a finitely generated commutative k-algebra A is the intersection of all maximal ideals containing $\mathfrak{a}$. In particular, nil A = rad A.

Given a prime ideal $\mathfrak{p}$ of a commutative ring A, when the descending chain of prime ideals beginning with $\mathfrak{p}$, $\mathfrak{p} = \mathfrak{p}_0 \supset \mathfrak{p}_1 \supset \cdots \supset \mathfrak{p}_r$, is of maximum length, then r is called the **height** of $\mathfrak{p}$ and is written $\text{ht}(\mathfrak{p})$. The upper limit (infinity permitted) of the heights of prime ideals of a commutative ring A is called the **Krull**

dimension of A and is denoted Kdim A. If A is a finitely generated k-algebra and an integral domain, then Kdim $A = $ trans. $\deg_k Q(A)$, thanks to Theorem 1.5.4. Moreover, we have the following theorem.

THEOREM 1.5.6 Let $A \subset B$ be commutative rings, and suppose that B is integral over A. Then, given any prime ideal $\mathfrak{p}$ of A, there exists a prime ideal $\mathfrak{q}$ of B such that $\mathfrak{q} \cap A = \mathfrak{p}$. If $\mathfrak{p}$ is a maximal ideal, then so is $\mathfrak{q}$.

COROLLARY 1.5.7 If a commutative ring B is integral over a subring A, then $\mathrm{Kdim}\, A = \mathrm{Kdim}\, B$.

(cf. [2] Theorems 2.4.4, 2.4.6, 2.4.9, or [4] Chapter V, §2, No. 1, Proposition 1, Theorem 1 and their Corollaries.)

We employ the Artin–Rees lemma below to prove the theorem which will follow.

LEMMA 1.5.8 (Artin–Rees) Let A be a finitely generated commutative k-algebra, M a finitely generated A-module, N an A-submodule of M, and $\mathfrak{a}$ an ideal of A. Then there exists a natural number r such that

$$\mathfrak{a}^n M \cap N = \mathfrak{a}^{n-r}(\mathfrak{a}^r M \cap N) \qquad (\forall n \geqq r).$$

(cf. [2] Theorem 3.0.6, or [4] Chapter III, §3, No. 1, Corollary 1 of Proposition 1.)

THEOREM 1.5.9 (Krull's intersection theorem) Let A be a finitely generated commutative k-algebra.

(i) If A is an integral domain, then for a proper ideal $\mathfrak{a}$ of A, we have $\bigcap_{n=0}^{\infty} \mathfrak{a}^n = \{0\}$.

(ii) If $\mathfrak{a}$ is an ideal contained in rad A, then $\bigcap_{n=0}^{\infty} \mathfrak{a}^n = \{0\}$.

(cf. [4] Chapter III, §3, No. 2, Corollary of Proposition 5.)

2
Hopf algebras

Throughout this chapter, we assume that k is a field.

1 Bialgebras and Hopf algebras

In this section, we give the definition of Hopf algebras and present some examples. We begin by defining k-coalgebras, which are in a dual relationship with k-algebras, then k-bialgebras and k-Hopf algebras as algebraic systems in which the structures of k-algebras and k-coalgebras are interrelated by certain laws.

1.1 **Coalgebras**

We define a k-coalgebra dually to a k-algebra. Given a k-linear space C and k-linear maps $\Delta_C \in \mathbf{Mod}_k(C, C\otimes C)$ and $\varepsilon_C \in \mathbf{Mod}_k(C, k)$, we call the system $(C, \Delta_C, \varepsilon_C)$ or simply C a k-**coalgebra** when the diagrams below commute.

(1) (the coassociative law)

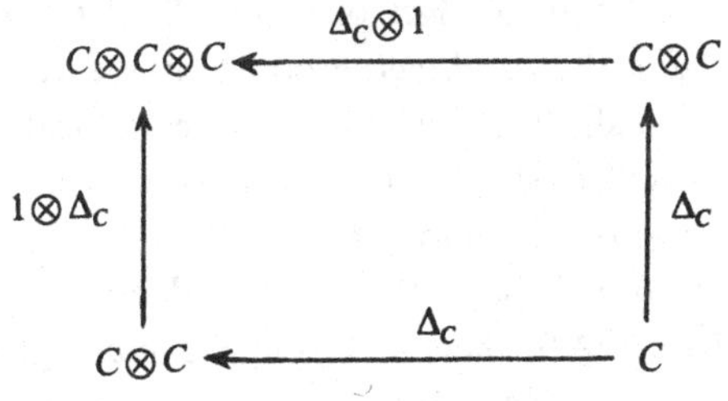

(2) (the counitary property)

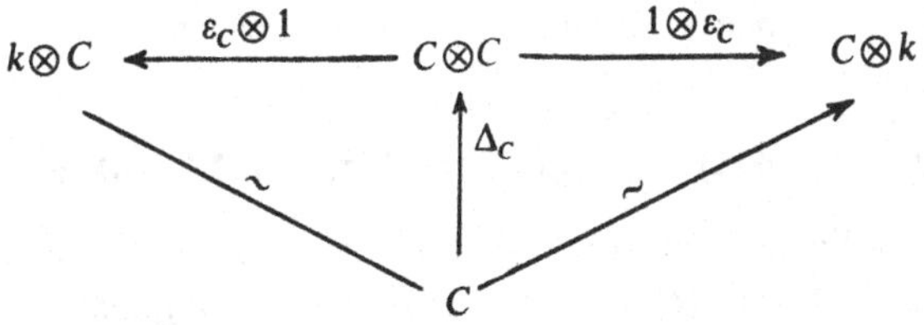

The maps Δ_C, ε_C are respectively called the **comultiplication map** and the **counit map** of C, and together, they are said to be the **structure maps** of the k-coalgebra C. When there is no fear of confusion, we simply write Δ, ε. Now suppose we are given two k-coalgebras $(C, \Delta_C, \varepsilon_C)$ and $(D, \Delta_D, \varepsilon_D)$. A k-linear map $\sigma : C \to D$ satisfying the conditions

$$\Delta_D \circ \sigma = (\sigma \otimes \sigma) \circ \Delta_C, \qquad \varepsilon_D \circ \sigma = \varepsilon_C$$

is called a **k-coalgebra morphism**. The category of k-coalgebras is denoted $\mathbf{Cog}_k$. For k-coalgebras C, D, the set of all k-coalgebra morphisms from C to D is denoted $\mathbf{Cog}_k(C, D)$. For k-linear spaces M, N, we define a k-linear map $\tau : M \otimes N \to N \otimes M$ by $\tau(x \otimes y) = y \otimes x$ for $x \in M$, $y \in N$. (Henceforth, such a map will be denoted by the symbol τ without notice.) If a k-coalgebra $(C, \Delta_C, \varepsilon_C)$ satisfies

$$\tau \circ \Delta_C = \Delta_C,$$

then C is said to be **cocommutative**. For C, $D \in \mathbf{Cog}_k$, the (k-linear space) tensor product $C \otimes D$ becomes a k-coalgebra with structure maps

$$\Delta_{C \otimes D} = (1 \otimes \tau \otimes 1) \circ \Delta_C \otimes \Delta_D,$$
$$\varepsilon_{C \otimes D} = \varepsilon_C \otimes \varepsilon_D,$$

which we call the **tensor product** of C and D. If C, D are cocommutative, then $C \otimes D$ is a direct product in the category of cocommutative k-coalgebras where the canonical projections $\pi_1 : C \otimes D \to C$, $\pi_2 : C \otimes D \to D$ are respectively given by $\pi_1(c \otimes d) = c\varepsilon(d)$ and $\pi_2(c \otimes d) = \varepsilon(c)d$ for $c \in C$, $d \in D$.

When a k-linear subspace D of a k-coalgebra C satisfies the condition $\Delta_C(D) \subset D \otimes D$, then D becomes a k-coalgebra with the restrictions of Δ_C, ε_C to D as the structure maps. Such a D is called a **k-subcoalgebra** of C. The canonical embedding $D \to C$ is a k-coalgebra morphism.

For k-linear spaces M and N, if we define a map $\rho : M^* \otimes N^* \to (M \otimes N)^*$ by

$$\rho(f \otimes g)(x \otimes y) = f(x)g(y), \quad f \in M^*, g \in N^*, x \in M, y \in N,$$

then ρ is a k-linear map which is moreover injective (cf. Exercise 1.8). Given a k-coalgebra (C, Δ, ε), let $C^* = \mathbf{Mod}_k(C, K)$ denote the dual

k-linear space of C. Setting

$$\mu : C^* \otimes C^* \xrightarrow{\rho} (C \otimes C)^* \xrightarrow{\Delta^*} C^*,$$

$$\eta : k \cong k^* \xrightarrow{\varepsilon^*} C^*,$$

(C^*, μ, η) becomes a k-algebra, which we call the **dual k-algebra** of C. In general, ρ is not necessarily an isomorphism. Thus we cannot define a k-coalgebra structure on the dual k-linear space A^* of a k-algebra A in a similar fashion. However, if A is a finite dimensional k-linear space, then $\rho : A^* \otimes A^* \to (A \otimes A)^*$ turns out to be an isomorphism of k-linear spaces; if μ, η are the structure maps of A, then by setting $\Delta = \mu^* \circ \rho^{-1}$, $\varepsilon = \eta^*$ (where we identify k with k^*), we see that $(A^*, \Delta, \varepsilon)$ becomes a k-coalgebra. This is called the **dual k-coalgebra** of A. In the next section, we will define the dual k-coalgebra of an arbitrary k-algebra.

EXAMPLE 2.1 Let S be a set and denote by kS the free k-module generated by S. If we define k-linear maps

$$\Delta : kS \to kS \otimes kS, \qquad \varepsilon : kS \to k$$

for $s \in S$ by $\Delta(s) = s \otimes s$, $\varepsilon(s) = 1$, then $(kS, \Delta, \varepsilon)$ becomes a k-coalgebra. The dual k-linear space $(kS)^*$ of kS can be identified with the set $\mathrm{Map}(S, k)$ of all maps from S to k where the dual k-algebra structure of kS is given by

$$(f + g)(s) = f(s) + g(s),$$
$$(fg)(s) = f(s)g(s),$$
$$(af)(s) = af(s)$$

for $f, g \in \mathrm{Map}(S, k)$, $s \in S$, $a \in k$.

EXAMPLE 2.2 For $S = \{c_0, c_1, c_2, \ldots\}$ we set $C = kS$ and define

$$\Delta c_n = \sum_{i=0}^{n} c_i \otimes c_{n-i}, \qquad \varepsilon(c_n) = \delta_{0n}$$

(where the symbol δ_{ij} denotes the Kronecker delta which is given by: $\delta_{ii} = 1, \delta_{ij} = 0\,(i \neq j)$) to obtain the k-coalgebra (C, Δ, ε). We will briefly look into the structure of the dual k-algebra of C. If we let $x_i \in C^*$

be defined by

$$x_i(c_j) = \langle x_i, c_j \rangle = \delta_{ij}, \qquad i, j = 0, 1, 2, \ldots,$$

then $a \in C^*$ can be written $a = \sum_{i=0}^{\infty} a_i x_i$ where $a_i \in k$. With regard to multiplication on C^*, we have

$$\langle x_i x_j, c_k \rangle = \langle x_i \otimes x_j, \Delta c_k \rangle = \langle x_i \otimes x_j, \sum_{l=0}^{k} c_l \otimes c_{k-l} \rangle$$

$$= \sum_{l=0}^{k} \langle x_i, c_l \rangle \langle x_j, c_{k-l} \rangle = \delta_{j,k-i} = \delta_{i+j,k}.$$

Thus the relation $x_i x_j = x_{i+j}$ holds, and we get $x_i = x_1{}^i$ $(i = 0, 1, 2, \ldots)$. Therefore we see that C^* is isomorphic to the power series ring $k[[x_1]]$ in one variable over k.

EXAMPLE 2.3 Let n be a natural number and set $S = \{s_{ij};\ 1 \leqq i, j \leqq n\}$ and $V = kS$. Defining

$$\Delta(s_{ij}) = \sum_{k=1}^{n} s_{ik} \otimes s_{kj}, \qquad \varepsilon(s_{ij}) = \delta_{ij},$$

we obtain a k-coalgebra (V, Δ, ε). The dual k-linear space V^* of V is a k-linear space of dimension n^2 and if we define $e_{ij} \in V^*$ by

$$\langle e_{ij}, s_{kl} \rangle = \delta_{ik} \delta_{jl},$$

then $\{e_{ij};\ 1 \leqq i, j \leqq n\}$ is a basis for V^* over k. By identifying $x = \Sigma\, x_{ij} e_{ij}$ with the $n \times n$ square matrix (x_{ij}), V^* becomes the k-algebra $M_n(k)$ of all $n \times n$ square matrices with coefficients in k.

Notation for k-coalgebra operations In general, the notation used for operations of k-coalgebras is not as concise as that for operations of k-algebras. The following notation is effective in simplifying various types of operations. Given a k-coalgebra (C, Δ, ε) and $c \in C$, we can write

$$\Delta(c) = \sum_{i=1}^{n} c_{1i} \otimes c_{2i}, \qquad c_{1i}, c_{2i} \in C.$$

We rewrite this formally as

$$\Delta(c) = \sum_{(c)} c_{(1)} \otimes c_{(2)},$$

and for k-linear maps f, g from C to C or k, we write

$$(f\otimes g)\Delta(c)=\sum_{(c)} f(c_{(1)})\otimes g(c_{(2)}).$$

Moreover, since the associative law holds, we have

$$(\Delta\otimes 1)\Delta(c)=(1\otimes\Delta)\Delta(c)=\sum_{(c)} c_{(1)}\otimes c_{(2)}\otimes c_{(3)},$$

and, in general, we define

$$\Delta_1=\Delta,\quad \Delta_n=(\underbrace{1\otimes\cdots\otimes 1}_{n-1 \text{ times}}\otimes\Delta)\Delta_{n-1}\quad (n>1),$$

and write

$$\Delta_n(c)=\sum_{(c)} c_{(1)}\otimes c_{(2)}\otimes\cdots\otimes c_{(n+1)}.$$

Using this method of notation, the counitary property may be expressed by

$$c=\sum_{(c)} c_{(1)}\varepsilon(c_{(2)})=\sum_{(c)}\varepsilon(c_{(1)})c_{(2)}.$$

EXERCISE 2.1 Prove the following equalities.

$$\Delta(c)=\sum_{(c)}\varepsilon(c_{(2)})\otimes\Delta(c_{(1)})=\sum_{(c)}\Delta(c_{(2)})\otimes\varepsilon(c_{(1)}),$$

$$\Delta(c)=\sum_{(c)} c_{(1)}\otimes\varepsilon(c_{(2)})c_{(3)}=\sum_{(c)} c_{(1)}\otimes\varepsilon(c_{(3)})c_{(2)},$$

$$c=\sum_{(c)}\varepsilon(c_{(1)})\varepsilon(c_{(3)})c_{(2)}.$$

THEOREM 2.1.1 Given a k-linear space H, suppose that there are k-linear maps

$$\mu: H\otimes H\to H,\quad \eta: k\to H,\quad \Delta: H\to H\otimes H,\quad \varepsilon: H\to k$$

such that (H,μ,η) is a k-algebra and (H,Δ,ε) is a k-coalgebra. Then the following conditions are equivalent.

(i) μ, η are k-coalgebra morphisms.

(ii) Δ, ε are k-algebra morphisms.

(iii) $\Delta(gh) = \Sigma\, g_{(1)}h_{(1)} \otimes g_{(2)}h_{(2)}$, $\quad \Delta(1) = (1)$,

$\varepsilon(gh) = \varepsilon(g)\varepsilon(h)$, $\quad \varepsilon(1) = 1$.

Proof The conditions under which Δ is a k-algebra morphism are

(1) $\Delta \circ \mu = (\mu \otimes \mu) \circ (1 \otimes \tau \otimes 1) \circ \Delta \otimes \Delta$,

(2) $\Delta \circ \eta = \eta \otimes \eta$ (where k is identified with $k \otimes k$), and the conditions under which ε is a k-algebra morphism are

(3) $\varepsilon \circ \mu = \varepsilon \otimes \varepsilon$ (where k is identified with $k \otimes k$),

(4) $\varepsilon \circ \eta = 1_k$.

On the other hand, μ is a k-coalgebra morphism if it satisfies conditions (1), (3); and η is a k-coalgebra morphism if it satisfies conditions (2) and (4). This fact allows us to conclude that (i)$\Leftrightarrow$(ii). That (ii)$\Leftrightarrow$(iii) is clear from the definition.

When a k-linear space H together with k-linear maps μ, η, Δ, ε satisfy one of the equivalent conditions of Theorem 2.1.1, then $(H, \mu, \eta, \Delta, \varepsilon)$ or simply H is called a k-**bialgebra**. Given two k-bialgebras H and K, when a k-linear map $\sigma : H \to K$ is a k-algebra morphism and is also a k-coalgebra morphism, then σ is called a k-**bialgebra morphism**. The category of k-bialgebras is denoted $\mathbf{Big}_k$. When H, K are k-bialgebras, then the set of all k-bialgebra morphisms from H to K is written $\mathbf{Big}_k(H, K)$.

If a k-linear subspace K of a k-bialgebra H is a k-subalgebra as well as a k-subcoalgebra, then K becomes a k-bialgebra, and is called a k-**sub-bialgebra** of H. Moreover, if a k-bialgebra H is finite dimensional as a k-linear space, then a k-bialgebra structure may be defined on its dual k-module H^*, which we call the **dual** k-**bialgebra** of H.

EXAMPLE 2.4 Let S be a semigroup with an identity element, and let kS be the k-algebra in Example 1.6. As shown in Example 2.1, kS has a k-coalgebra structure. With regard to these two structures, kS admits a k-bialgebra structure. Such a k-bialgebra is called a **semigroup** k-**bialgebra**. In particular, when S is a group, it is called a **group** k-**bialgebra**. The dual k-linear space $(kS)^*$ of kS can be identified with Map (S, k), and, in particular, when S is a finite set, then $(kS)^*$ becomes a dual k-bialgebra of kS. The k-algebra structure of $(kS)^*$ is the one described in Example 1.5, and its k-coalgebra structure is given for $f \in \mathrm{Map}(S, k)$, $x, y \in S$, and the identity element $e \in S$ by

$$\langle \Delta f, x \otimes y \rangle = \langle f, xy \rangle, \qquad \langle \varepsilon f, x \rangle = \langle f, e \rangle.$$

EXAMPLE 2.5 Let L be a k-Lie algebra and let $U(L)$ be its universal enveloping k-algebra. The k-algebra tensor product $U(L) \otimes U(L)$ is isomorphic to the universal enveloping algebra $U(L \oplus L)$ of the (k-Lie algebra) direct sum $L \oplus L$, and the k-Lie algebra diagonal map $x \mapsto x + x$ ($L \to L \oplus L$) induces the k-algebra morphism

$$\Delta : U(L) \to U(L \oplus L) \cong U(L) \otimes U(L),$$

and the k-Lie algebra morphism $L \to \{0\}$ induces the k-algebra morphism

$$\varepsilon : U(L) \to k.$$

In this situation, $(U(L), \Delta, \varepsilon)$ becomes a k-coalgebra, and $U(L)$ has a k-bialgebra structure. We call this the **universal enveloping k-bialgebra** of the k-Lie algebra L.

When an element c of a k-coalgebra C is such that $\varepsilon(c) = 1$ and $\Delta c = c \otimes c$, then c is said to be a **group-like** element. We denote the set of all group-like elements of C by $G(C)$.

THEOREM 2.1.2 Let C be a k-coalgebra.

(i) The elements of $G(C)$ are linearly independent over k. Thus $kG(C)$ may be regarded as a k-subcoalgebra of C.

(ii) When H is a k-bialgebra, $G(H)$ is a semigroup with respect to multiplication, and the k-linear subspace of H generated by $G(H)$ is a k-sub-bialgebra of H isomorphic to the semigroup k-bialgebra $kG(H)$ of $G(H)$.

Proof (i) Suppose that the elements of $G(C)$ are linearly dependent over k, and let $n + 1$ be the least value for a set of elements of $G(C)$ to be linearly dependent. Then there exist $g, g_1, \ldots, g_n \in G(C)$ such that $g_1, \ldots, g_n$ are linearly independent and g can be written

$$g = \lambda_1 g_1 + \cdots + \lambda_n g_n, \qquad \lambda_i \in k, \quad \lambda_i \neq 0 \quad (1 \leqq i \leqq n).$$

Since

$$\Delta g = g \otimes g = \sum_{i,j=1}^{n} \lambda_i \lambda_j g_i \otimes g_j,$$

$$\Delta g = \sum_{i=1}^{n} \lambda_i \Delta g_i = \sum_{i=1}^{n} \lambda_i g_i \otimes g_i,$$

and since $\{g_i \otimes g_j\}_{1 \leqq i, j \leqq n}$ is a set of linearly independent elements of $C \otimes C$, we conclude that $n = 1$ and $g = \lambda_1 g_1$. On the other hand, since $\varepsilon(g) = \lambda_1 \varepsilon(g_1)$ and $\varepsilon(g) = \varepsilon(g_1) = 1$, it follows that $\lambda_1 = 1$. Thus we have $g = g_1$ which contradicts the choice of n. It is apparent that $kG(C)$ is a k-subcoalgebra of C. The claim in (ii) follows readily from (i).

An element c of a k-bialgebra C (resp. a k-coalgebra C which has only one group-like element, i.e. $G(C) = \{g\}$) which satisfies $\Delta c = c \otimes 1 + 1 \otimes c$ (resp. $\Delta c = c \otimes g + g \otimes c$) is called a **primitive element**. The set of all primitive elements of C is denoted $P(C)$. Now we have the following.

THEOREM 2.1.3 If H is a k-bialgebra, then $P(H)$ is a k-linear subspace of H, and for $x, y \in P(H)$, we have $[x, y] = xy - yx \in P(H)$. Thus $P(H)$ has a structure of a k-Lie algebra. Moreover, if $x \in P(H)$, $\varepsilon(x) = 0$. In particular, if the characteristic of k is $p > 0$, then for $x \in P(H)$ we have $x^p \in P(H)$, so that $P(H)$ is a p-k-Lie algebra.

Proof If $x, y \in P(H)$, then

$$\begin{aligned}\Delta([x, y]) &= \Delta x \Delta y - \Delta y \Delta x \\ &= (x \otimes 1 + 1 \otimes x)(y \otimes 1 + 1 \otimes y) \\ &\quad - (y \otimes 1 + 1 \otimes y)(x \otimes 1 + 1 \otimes x) \\ &= [x, y] \otimes 1 + 1 \otimes [x, y].\end{aligned}$$

Thus $[x, y] \in P(H)$. For $x \in P(H)$ we have

$$(1 \otimes \varepsilon)\Delta x = x \otimes 1 + 1 \otimes \varepsilon(x) = x \otimes 1. \quad \text{Hence} \quad \varepsilon(x) = 0.$$

If the characteristic of k is $p > 0$, then

$$\Delta x^p = (x \otimes 1 + 1 \otimes x)^p = \sum_{i=0}^{p} \binom{p}{i} x^i \otimes x^{p-i} = x^p \otimes 1 + 1 \otimes x^p.$$

Therefore $x^p \in P(H)$.

Remark The k-subalgebra H_1 of H generated by $P(H)$ is a k-sub-bialgebra of H, and when the characteristic of k is 0, H_1 is isomorphic to the universal enveloping k-bialgebra $U(P(H))$ of $P(H)$. (cf. Theorem 2.5.3.)

EXAMPLE 2.6 Let $S = \{c_0, c_1, \ldots, c_n, \ldots\}$, and let $C = kS$ be the k-coalgebra defined in Example 2.2. Defining a multiplication by

$$c_i c_j = \binom{i+j}{i} c_{i+j},$$

C becomes a k-algebra. With respect to these two structures, C becomes a k-bialgebra. If k is a field of characteristic 0, then by setting $d_i = \frac{1}{i!} c_i$, we obtain $d_i d_j = d_{i+j}$, which implies $d_i = d_1{}^i$ $(i = 0, 1, 2, \ldots)$. Thus, as a k-algebra, C is isomorphic to the polynomial ring $k[d_1]$ in one variable over k. In this situation, $P(C) = kd_1$ and $C \cong U(P(C))$.

1.2 Hopf algebras

Given a k-coalgebra C and a k-algebra A, set $R = \mathbf{Mod}_k(C, A)$. If $f, g \in R$,

$$f * g = \mu_A \circ (f \otimes g) \circ \Delta_C$$

is said to be the **convolution** of f and g. If $x \in G(C)$, then $(f * g)(x) = f(x)g(x)$, which is simply the definition of the product of two functions on $G(C)$ via multiplication on A. R becomes a k-algebra with structure maps

$$\mu_R(f \otimes g) = f * g, \qquad \eta_R(\alpha) = \alpha \eta_A \circ \varepsilon_C.$$

Given a k-bialgebra H, let H^A and H^C respectively denote H regarded simply as a k-algebra and as a k-coalgebra, and view $R = \mathbf{Mod}_k(H^C, H^A)$ as a k-algebra via convolution as defined above. When the identity map 1 of H is a regular element of R with respect to multiplication on R, the inverse S of 1 is called the **antipode** of H. The antipode S is the element which satisfies one of the following equivalent equations.

$$S * 1 = 1 * S = \eta \circ \varepsilon,$$

$$\mu \circ (S \otimes 1) \circ \Delta = \mu \circ (1 \otimes S) \circ \Delta = \eta \circ \varepsilon.$$

A k-bialgebra with antipode is called a k-**Hopf algebra**. Let H, K be k-Hopf algebras and let S_H, S_K be the antipodes of H, K, respectively. When a k-bialgebra morphism $\sigma : H \to K$ satisfies the condition

$$S_K \circ \sigma = \sigma \circ S_H,$$

σ is called a k-**Hopf algebra morphism**. The category of k-Hopf algebras is denoted by $\mathbf{Hopf}_K$. Given k-Hopf algebras H, K, the set of all k-Hopf algebra morphisms from H to K is written $\mathbf{Hopf}_k(H, K)$.

EXAMPLE 2.7 Denote by kG the group k-bialgebra of a group G. (cf. Example 2.4.) We define a k-linear map $S : kG \to kG$ by $S(x) = x^{-1}$ for $x \in G$. Then

$$(1 * S)(x) = xS(x) = xx^{-1} = e = \varepsilon(x)e = \eta \circ \varepsilon(x), \quad x \in G.$$

Thus S is the antipode of kG, so that kG becomes a k-Hopf algebra. S is an anti-automorphism of kG as a k-algebra and S^2 is the identity map of kG.

EXAMPLE 2.8 Denote the universal enveloping k-bialgebra of a k-Lie algebra L by $U(L)$. (cf. Example 2.5.) Let S be the principal anti-automorphism of $U(L)$. (cf. Chapter 1,§3.2.) Then

$$\begin{aligned}(1 * S)(x) &= 1S(x) + xS(1)\\ &= -x + x\\ &= 0\\ &= \varepsilon(x)1\\ &= \eta \circ \varepsilon(x), \qquad x \in L.\end{aligned}$$

Thus S is the antipode of $U(L)$, so that $U(L)$ becomes a k-Hopf algebra. This is called the **universal enveloping** k-**Hopf algebra** of the k-Lie algebra L.

THEOREM 2.1.4 The following properties hold for the antipode S of a k-Hopf algebra H.

(i) $S(gh) = S(h)S(g), \qquad g, h \in H.$

(ii) $S(1) = 1$; namely, $S \circ \eta = \eta.$

(iii) $\varepsilon \circ S = \varepsilon.$

(iv) $\tau \circ (S \otimes S) \circ \Delta = \Delta \circ S$; in other words,

$$\Delta S(h) = \sum_{(h)} S(h_{(2)}) \otimes S(h_{(1)}).$$

(v) The following conditions are equivalent.

(1) $h \in H$ implies $\sum_{(h)} S(h_{(2)}) h_{(1)} = \eta \circ \varepsilon(h)$.

(2) $h \in H$ implies $\sum_{(h)} h_{(2)} S(h_{(1)}) = \eta \circ \varepsilon(h)$.

(3) $S \circ S = 1$.

(vi) If H is commutative or cocommutative, then $S^2 = 1$.

Remark (i), (ii) imply that S is an anti-k-algebra morphism; (iii), (iv) imply that S is an anti-k-coalgebra morphism.

Proof (i) Define elements μ, ν, ρ of $R = \mathbf{Mod}_k((H \otimes H)^C, H^A)$ in the following manner. For $g, h \in H$, we write

$$\mu(g \otimes h) = gh, \quad \nu(g \otimes h) = S(h)S(g), \quad \rho(g \otimes h) = S(gh).$$

Now, if we show that $\rho * \mu = \mu * \nu = \eta \circ \varepsilon$, then we get $\rho = \nu$, which would prove (i).

$$\begin{aligned}
(\rho * \mu)(g \otimes h) &= \sum_{(g \otimes h)} \rho((g \otimes h)_{(1)}) \mu((g \otimes h)_{(2)}) \\
&= \sum_{(g)(h)} \rho(g_{(1)} \otimes h_{(1)}) \mu(g_{(2)} \otimes h_{(2)}) \\
&= \sum_{(g)(h)} S(g_{(1)} h_{(1)}) g_{(2)} h_{(2)} \\
&= \sum_{(gh)} S((gh)_{(1)})(gh)_{(2)} \\
&= (S * 1)(gh) = \varepsilon(gh) = \varepsilon(g)\varepsilon(h),
\end{aligned}$$

$$\begin{aligned}
(\mu * \nu)(g \otimes h) &= \sum_{(g)(h)} \mu(g_{(1)} \otimes h_{(1)}) \nu(g_{(2)} \otimes h_{(2)}) \\
&= \sum_{(g)(h)} g_{(1)} h_{(1)} S(h_{(2)}) S(g_{(2)}) \\
&= \sum_{(g)} g_{(1)} \varepsilon(h) S(g_{(2)}) \\
&= \varepsilon(g)\varepsilon(h).
\end{aligned}$$

(ii) Since $\varepsilon(1)=1$, $\Delta(1)=1\otimes 1$, we have

$$\varepsilon(1)=(1*S)(1)=S(1)=1.$$

(iii) From the fact that $\varepsilon\circ\eta\circ\varepsilon(h)=\varepsilon(h)\varepsilon(1)=\varepsilon(h)$ and $\eta\circ\varepsilon(h)=\Sigma\, S(h_{(1)})h_{(2)}$, we obtain $\varepsilon(h)=\varepsilon\circ\eta\circ\varepsilon(h)=\Sigma\,\varepsilon(S(h_{(1)})\varepsilon(h_{(2)}))=\varepsilon\circ S(h)$. Thus $\varepsilon=\varepsilon\circ S$.

(iv) For elements Δ, $\nu=\tau(S\otimes S)\Delta$, $\rho=\Delta\circ S$ of $\mathbf{Mod}_k(H^C, (H\otimes H)^A)$, by showing that $\rho*\Delta=\eta\circ\varepsilon=\Delta*\nu$, we obtain $\rho=\nu$ which would prove (iv).

$$(\rho*\Delta)(h)=\sum_{(h)}\Delta\circ S(h_{(1)})\Delta(h_{(2)})=\Delta\left(\sum_{(h)}S(h_{(1)})h_{(2)}\right)$$

$$=\Delta\circ(\eta\circ\varepsilon(h))=\eta_{H\otimes H}\circ\varepsilon_H(h),$$

$$(\Delta*\nu)(h)=\sum_{(h)}(h_{(1)}\otimes h_{(2)})(S(h_{(4)})\otimes S(h_{(3)}))$$

$$=\sum_{(h)}h_{(1)}S(h_{(4)})\otimes h_{(2)}S(h_{(3)})$$

$$=\sum_{(h)}h_{(1)}S(h_{(3)})\otimes\eta\circ\varepsilon(h_{(2)})$$

$$=\sum_{(h)}h_{(1)}S(h_{(3)})\varepsilon(h_{(2)})\otimes\eta(1)$$

$$=\sum_{(h)}h_{(1)}S(h_{(2)})\otimes\eta(1)$$

$$=\varepsilon(h)\otimes\eta(1)=\eta_{H\otimes H}\circ\varepsilon_H(h).$$

(v) (1) $\Rightarrow$ (3): Since $S*(S\circ S)(h)=\sum_{(h)}S(h_{(1)})\,(S\circ S)\,(h_{(2)})=S\left(\sum_{(h)}S(h_{(2)})h_{(1)}\right)=S\circ\varepsilon(h)=\varepsilon(h)$, we conclude that $S\circ S$ is the inverse of S. Thus $S\circ S=1$.

(3) $\Rightarrow$ (2): We have $\varepsilon(h)=(1*S)(h)=\sum h_{(1)}S(h_{(2)})=S\left(\sum h_{(2)}S(h_{(1)})\right)$, so that $S\circ\eta\circ\varepsilon(h)=(S\circ S)\left(\sum_{(h)}h_{(2)}S(h_{(1)})\right)=\sum_{(h)}h_{(2)}S(h_{(1)})$. On

the other hand, since $S \circ \eta = \eta$, we have $S \circ \eta \circ \varepsilon(h) = \eta \circ \varepsilon(h)$. Therefore

$$\sum_{(h)} h_{(2)} S(h_{(1)}) = \eta \circ \varepsilon(h).$$

We can verify (2) ⇒ (3) ⇒ (1) similarly, so that (1), (2) and (3) are equivalent.

(vi) From the definition of the antipode, (v-1) holds if H is commutative; (v-2) holds if H is cocommutative. Thus we have (v-3). Therefore $S \circ S = 1$.

THEOREM 2.1.5 Let H be a k-Hopf algebra and let R be a commutative k-algebra. Then $G(R) = \mathbf{Alg}_k(H, R)$ becomes a group with respect to convolution.

Proof Given $f, g \in \mathbf{Alg}_k(H, R)$, the fact that μ, f, g, Δ are k-algebra morphisms implies that $f * g = \mu \circ (f \otimes g) \circ \Delta$ is also a k-algebra morphism. Theorem 2.1.4 (i), (ii) imply that since S is a k-algebra anti-automorphism of H, $f \circ S$ becomes a k-algebra morphism from H to R when R is commutative. Further, we have

$$\begin{aligned} f * (f \circ S) &= \mu \circ (f \otimes (f \circ S)) \circ \Delta = \mu \circ (f \otimes f) \circ (1 \otimes S) \circ \Delta \\ &= f \circ \mu \circ (1 \otimes S) \circ \Delta = f \circ \eta \circ \varepsilon = \eta \circ \varepsilon. \end{aligned}$$

Similarly,

$$(f \circ S) * f = \eta \circ \varepsilon.$$

Thus $f \circ S$ is the inverse of f, proving that $G(R)$ is indeed a group.

Remark The map $G : R \mapsto G(R)$ is a covariant functor from the category of commutative k-algebras $\mathbf{M}_k$ to the category of groups $\mathbf{Gr}$. When H is a commutative k-Hopf algebra, G may be regarded as a representable functor.

2 The representative bialgebras of semigroups

In this section, we introduce bialgebras consisting of representative functions of semigroups or groups, which are important examples of bialgebras. An arbitrary k-coalgebra may be embedded into such a k-bialgebra, and, from the property of representative bialgebras, it is possible to derive the general properties of k-coalgebras. By constructing representative k-bialgebras in the category of Lie groups or algebraic groups, we obtain the representative rings of Lie groups or the coordinate rings of algebraic groups. We touch upon such topics in Chapters 3 and 4.

2.1 Pairs of dual k-linear spaces

Given a k-linear space V, let the dual k-linear space $\mathbf{Mod}_k(V, k)$ of V be denoted by V^*.

Given a pair of k-linear spaces (V, X), suppose that there exists a bilinear form

$$B : V \times X \to k, \qquad (x, f) \mapsto \langle x, f \rangle$$

satisfying the properties

(1) $\langle x, f \rangle = 0 \quad \forall x \in V \Rightarrow f = 0$,

(2) $\langle x, f \rangle = 0 \quad \forall f \in X \Rightarrow x = 0$.

Then (V, X) is said to be a **pair of dual k-linear spaces**. In this situation, given $f \in X$, the element of V^*

$$\hat{f} : x \mapsto \langle x, f \rangle$$

is determined uniquely, and the correspondence from X to V^* given by $f \mapsto \hat{f}$ is an injection by (1). Thus X may be regarded as a k-linear subspace of V^*. Therefore we view $X \subset V^*$, and write $\hat{f}$ as f. Similarly, we can consider V as a k-linear subspace of X^*. In particular, for $X = V^*$, (V, V^*) is a pair of dual k-linear spaces when we define $\langle x, f \rangle = f(x)$ for $x \in V$, $f \in V^*$. Thus V can be regarded as a k-linear subspace of V^{**}. For $x \in V$, set

$$x^\perp = \{f \in V^*; \langle x, f \rangle = 0\}, \qquad x^{\perp(X)} = x^\perp \cap X.$$

Then $x^\perp$ and $x^{\perp(X)}$ are respectively k-linear subspaces of V^* and X. In general, for a subset S of V, we set

$$S^\perp = \{f \in V^*; \langle x, f \rangle = 0 \quad \forall x \in S\}, \quad S^{\perp(X)} = S^\perp \cap X.$$

Similarly, for a subset T of V^*, we set

$$T^\perp = \{\xi \in V^{**}; \quad \langle \xi, f \rangle = 0 \quad \forall f \in T\}, \quad T^{\perp(V)} = T^\perp \cap V.$$

Given $f \in V^*$, by letting the family of subsets of V^*

$$\{f + x^{\perp}; x \in V\}$$

be a base for a system of neighborhoods of f, V^* becomes a linear topological space. (A k-linear space whose underlying abelian group is a topological group and such that a family of k-linear subspaces can be chosen for its system of neighborhoods is called a linear topological space.) When viewed as a subspace of the linear topological space V^*, X becomes a linear topological space. This topology is called the V-**topology** of X. The X-topology of V may be defined similarly. In such a case, by choosing a suitable finite dimensional (resp. arbitrary dimensional) k-linear subspace W of V, we can write an open (resp. closed) k-linear subspace of X in the form $W^{\perp(X)}$. Letting $\{X_\lambda\}_{\lambda \in \Lambda}$ be a family of k-linear subspaces of X, we get

$$\left(\sum_{\lambda \in \Lambda} X_\lambda\right)^{\perp(V)} = \bigcap_{\lambda \in \Lambda} X_\lambda^{\perp(V)}.$$

Similarly, if we let $\{V_\lambda\}_{\lambda \in \Lambda}$ be a family of k-linear subspaces of V, then

$$\left(\sum_{\lambda \in \Lambda} V_\lambda\right)^{\perp(X)} = \bigcap_{\lambda \in \Lambda} V_\lambda^{\perp(X)}.$$

LEMMA 2.2.1 Let (V, X) be a pair of dual k-linear spaces. If W, Y are k-linear subspaces of V, X respectively, and if W is finite dimensional, then

$$Y^{\perp(V)} + W = (W^{\perp(X)} \cap Y)^{\perp(V)}.$$

Proof We will proceed by induction on dim $W = n$. For $n = 1$, set $W = ku$. Since we obviously have the inclusion $Y^{\perp(V)} + W \subset (W^{\perp(X)} \cap Y)^{\perp(V)}$, we will verify the reverse inclusion. If $u^{\perp(X)} \cap Y = Y$, then we clearly get equality, so suppose that $u^{\perp(X)} \cap Y \subsetneqq Y$. Since $u^{\perp(X)}$ is a subspace of X and $X/u^{\perp(X)}$ is one dimensional, we can write

$$Y = (u^{\perp(X)} \cap Y) + kg, \quad g \notin u^{\perp(X)}.$$

If $v \in (u^{\perp(X)} \cap Y)^{\perp(V)}$, by setting $a = \langle v, g\rangle / \langle u, g\rangle$, we get

$$\langle v - au, g\rangle = 0, \quad \langle v - au, u^{\perp(X)} \cap Y\rangle = 0.$$

Thus $v - au \in Y^{\perp(V)}$. Therefore $v \in Y^{\perp(V)} + ku$, and we obtain $(u^{\perp(X)} \cap Y)^{\perp(V)} \subset Y^{\perp(V)} + ku$. Supposing now that the theorem holds for $\dim W < n$, we will prove the case $\dim W = n$. Set $W = W_1 + ku$ and $\dim W_1 = n - 1$. Then, by the inductive hypothesis,

$$\begin{aligned} Y^{\perp(V)} + W &= Y^{\perp(V)} + W_1 + ku \\ &= (W_1{}^{\perp(X)} \cap Y)^{\perp(V)} + ku \\ &= (u^{\perp(X)} \cap W_1{}^{\perp(X)} \cap Y)^{\perp(V)} \\ &= (W^{\perp(X)} \cap Y)^{\perp(V)}. \end{aligned}$$

Remark Since V and X are symmetric in Lemma 2.2.1, if W, Y are k-linear subspaces of V, X respectively, and if Y is finite dimensional, we have

$$W^{\perp(X)} + Y = (Y^{\perp(V)} \cap W)^{\perp(X)}.$$

LEMMA 2.2.2 Let (V, X) be a pair of dual k-linear spaces. Let Y be a k-linear subspace of X, and pick a set of elements $\{u_1, \ldots, u_n\}$ of V in such a way that the set $\{u_1 + Y^{\perp(V)}, \ldots, u_n + Y^{\perp(V)}\}$ of $V/Y^{\perp(V)}$ becomes linearly independent over k. Then, given an arbitrary set of elements $\{a_1, \ldots, a_n\}$ of k, there exists $f \in Y$ such that $\langle u_i, f\rangle = a_i$ $(1 \leqq i \leqq n)$.

Proof If we set $W_j = \sum_{i \neq j} ku_i$, then $u_j \notin Y^{\perp(V)} + W_j$. Thus, by Lemma 2.2.1, $u_j \notin (W_j{}^{\perp(X)} \cap Y)^{\perp(V)}$. Hence there must exist $f_j \in Y$ such that $\langle u_i, f_j\rangle = \delta_{ij}$. If we set $f = \sum_{i=1}^{n} a_i f_i$, then $f \in Y$ satisfies the conditions of the lemma.

THEOREM 2.2.3 Let (V, X) be a pair of dual k-linear spaces. The closure $\bar{Y}$ of a k-linear subspace Y of X in the V-topology is $Y^{\perp(V)\perp(X)}$.

Proof Since $Y^{\perp(V)\perp(X)}$ is a closed subspace containing Y, it obviously contains $\bar{Y}$. In order to show that $Y^{\perp(V)\perp(X)} = \bar{Y}$, we verify that an arbitrary neighbourhood $U = f + \bigcap_{i=1}^{n} u_i{}^{\perp(X)}$, where $u_i \in V$ $(1 \leqq i \leqq n)$, of an element f of $Y^{\perp(V)\perp(X)}$ intersects Y non-trivially. Pick a basis $\{v_1, \ldots, v_m\}$ of $W = \sum_{i=1}^{n} ku_i$ so that $v_1, \ldots, v_r \in W \cap Y^{\perp(V)}$, $v_{r+1}, \ldots, v_m \notin Y^{\perp(V)}$. By Lemma 2.2.2, there exists $g \in Y$ such that the conditions

$$\langle g, v_j \rangle = 0 \ (1 \leqq j \leqq r), \quad \langle g, v_1 \rangle = \langle f, v_1 \rangle \ (r+1 \leqq 1 \leqq m)$$

are satisfied. Then $g - f \in W^{\perp(X)}$. Therefore $g \in U \cap Y$.

THEOREM 2.2.4 Let (V, X) be a pair of dual k-linear spaces. If Y, Z are k-linear subspaces of X, where Y is a closed subspace and Z is of finite dimension, then $Y + Z$ is a closed subspace. In particular, a finite dimensional subspace of X is closed.

Proof Since Y is a closed subspace, Theorem 2.2.3 implies that $Y = Y^{\perp(V)\perp(X)}$. On the other hand, by the remark following Lemma 2.2.1, we have

$$Y + Z = Y^{\perp(V)\perp(X)} + Z = (Z^{\perp(V)} \cap Y^{\perp(V)})^{\perp(X)}$$

Thus $Y + Z$ is a closed subspace.

THEOREM 2.2.5 Let (V, X) be a pair of dual k-linear spaces. Each k-linear subspace of V is closed $\Leftrightarrow X = V^*$.

Proof $\Leftarrow$ Suppose $X = V^*$. Then we show that a k-linear subspace W of V is closed. If $W = V$, then W is closed. Thus we can assume that $W \subsetneqq V$. If we pick $x \in V$ such that $x \notin W$, then there exists an element f of $V^* = X$ such that $f \in W^{\perp(X)}$, $\langle f, x \rangle \neq 0$. Thus $x \notin W^{\perp(X)\perp(V)} = \overline{W}$. Hence W is closed.

$\Rightarrow$ We suppose that all k-linear subspaces of V are closed and show that $X = V^*$. Let $f \neq 0$ be an element of V^*. If we set $W = f^{\perp(V)}$, then $W \subsetneqq V$. By hypothesis, since W is a closed k-linear subspace, we have $W^{\perp(X)\perp(V)} = W \subsetneqq V$. Thus $W^{\perp(X)} \neq \{0\}$. If we let $g \neq 0$ be an element of $W^{\perp(X)}$, then by Theorem 2.2.4, we have $g \in W^{\perp(X)} = f^{\perp(V)\perp(X)} = kf$ since kf is a closed subspace. Therefore $f \in kg \subset X$.

Next consider the pair of dual k-linear spaces (V, V^*). Given any

k-linear subspace W of V, the relation $(W^{\perp})^{\perp(V)} = W$ holds by Theorem 2.2.4. Moreover, for a subset T of V^*, we have

$$T \text{ is } \textbf{dense} \text{ in } V^* \Leftrightarrow T^{\perp(V)} = \{0\}.$$

EXERCISE 2.2 (1) Given k-linear spaces V, W, if we define a k-linear map

$$\rho : V^* \otimes W^* \to (V \otimes W)^*$$

by $\langle \rho(f \otimes g), x \otimes y \rangle = \langle f, x \rangle \langle g, y \rangle$ for $f \in V^*, g \in W^*, x \in V, y \in W$, then ρ is injective. (cf. Exercise 1.8.) In such a situation, by regarding $V^* \otimes W^*$ as a k-linear subspace of $(V \otimes W)^*$, $V^* \otimes W^*$ is dense in $(V \otimes W)^*$.

(2) Let X, Y be k-linear subspaces of V^*, W^* respectively. Now if we regard $X \otimes Y \subset V^* \otimes W^*$ as k-linear subspaces of $(V \otimes W)^*$, then

$$(X \otimes Y)^{\perp(V \otimes W)} = (X^{\perp(V)} \otimes W) + (V \otimes Y^{\perp(W)}).$$

(3) Let $u : V \to W$ be a k-linear map and let $u^* : W^* \to V^*$ be its dual k-linear map. Then u^* is a continuous function and the image of a closed subspace of W^* is a closed subspace of V^*. Moreover, if we let S, T be k-linear subspaces of V, W respectively, then

$$u^{*-1}(S^{\perp}) = u(S)^{\perp}, \qquad u^*(T^{\perp}) = u^{-1}(T)^{\perp}.$$

(4) If X is a finite codimensional k-linear subspace of V^*, then X is a closed subspace $\Leftrightarrow X^{\perp} = X^{\perp(V)}$.

2.2 The representative bialgebra of a semigroup

Let G be a semigroup with identity element e, and identify the dual k-algebra $(kG)^*$ of the k-coalgebra kG in Example 2.4 with Map (G, k) and denote it by $M_k(G)$. The following action makes $M_k(G)$ a two-sided kG-module.

$$(xfy)(z) = f(yzx), \qquad f \in M_k(G), \qquad x, y, z \in G.$$

For $f \in M_k(G)$, we denote the left kG-module, the right kG-module, and the two-sided kG-module generated by f respectively by kGf, kfG, and $kGfG$.

THEOREM 2.2.6 The following conditions are equivalent.

(i) $\dim kGf < \infty$.
(ii) $\dim kfG < \infty$.
(iii) $\dim kGfG < \infty$.

Proof Suppose $\dim kGf < \infty$ and let $\{f_1, \ldots, f_n\}$ be a basis for kGf. Since we may write

$$(xf)(y) = \sum_{i=1}^{n} g_i(x) f_i(y), \qquad g_i(x) \in k \, (1 \leqq i \leqq n),$$

we obtain

$$xf = \sum_{i=1}^{n} g_i(x) f_i, \qquad fy = \sum_{i=1}^{n} f_i(y) g_i.$$

Moreover, since kfG is contained in a k-linear subspace of $M_k(G)$ spanned by $\{g_1, \ldots, g_n\}$, we have $\dim kfG < \infty$. Thus (i)$\Rightarrow$(ii). We obtain (ii)$\Rightarrow$(i) similarly. That (iii)$\Rightarrow$(i), (ii) is trivial. Moreover, when $\dim kGf < \infty$, since $f_i \in kGf$ we get $\dim kGf_i < \infty \, (1 \leqq i \leqq n)$. From (ii) we obtain $\dim kf_iG < \infty \, (1 \leqq i \leqq n)$. Therefore $\dim kGfG \leqq \infty$, which proves that (i)$\Rightarrow$(iii).

The k-linear map

$$\pi : M_k(G) \otimes M_k(G) \to M_k(G \times G)$$

defined for $f \otimes g \in M_k(G) \otimes M_k(G)$ by $\pi(f \otimes g)(x, y) = f(x)g(y)$ is injective. In fact, if $\sum_{i=1}^{n} f_i \otimes g_i \in M_k(G) \otimes M_k(G)$ is such that $\pi\left(\sum_{i=1}^{n} f_i \otimes g_i\right) = 0$, then by choosing $\{g_1, \ldots, g_n\}$ to be linearly independent over k, for a fixed $x \in G$ we have $\sum_{i=1}^{n} f_i(x) g_i(y) = 0$ for all $y \in G$, so that $\sum_{i=1}^{n} f_i(x) g_i = 0$. Therefore $f_i(x) = 0 \; (1 \leqq i \leqq n)$. Hence $f_i = 0 \; (1 \leqq i \leqq n)$, from which we obtain $\sum_{i=1}^{n} f_i \otimes g_i = 0$

Furthermore, we define a k-algebra morphism

$$\delta : M_k(G) \to M_k(G \times G)$$

by $\delta f(x, y) = f(xy)$ for $f \in M_k(G)$, $x, y \in G$. Then we have the following.

THEOREM 2.2.7 $\delta f \in \pi(M_k(G) \otimes M_k(G)) \Leftrightarrow \dim kGf < \infty$.

Proof $\Rightarrow$ If $\delta f \in \pi(M_k(G) \otimes M_k(G))$, then we can write $\delta f = \pi\left(\sum_{i=1}^{n} g_i \otimes h_i\right)$. Since $\delta f(x, y) = \sum_{i=1}^{n} g_i(x)h_i(y) = (yf)(x)$, we have $yf = \sum_{i=1}^{n} h_i(y)g_i$, which implies that $\dim kGf < \infty$.

$\Leftarrow$ Let $\dim kGf < \infty$. Letting $\{f_1, \ldots, f_n\}$ be a basis for kGf, we can write $yf = \sum_{i=1}^{n} g_i(y)f_i$. Then $\delta f(x, y) = f(xy) = (yf)(x) = \sum_{i=1}^{n} f_i(x)g_i(y)$. Therefore $\delta f = \pi\left(\sum_{i=1}^{n} f_i \otimes g_i\right)$.

When $f \in M_k(G)$ satisfies the condition that $\dim kGf < \infty$, f is said to be a **representative function** on G. The set of all representative functions on G is denoted by $R_k(G)$. $R_k(G)$ is a k-subalgebra of $M_k(G)$. Now we have

THEOREM 2.2.8 $\delta R_k(G) \subset \pi(R_k(G) \otimes R_k(G))$.

Thus, $\Delta = \pi^{-1} \circ \delta$: $R_k(G) \to R_k(G) \otimes R_k(G)$ is a k-algebra morphism that defines a comultiplication on $R_k(G)$. Moreover, by defining a k-algebra morphism $\varepsilon : R_k(G) \to k$ by $\varepsilon(f) = f(e)$, $(R_k(G), \Delta, \varepsilon)$ becomes a k-coalgebra, and the fact that Δ, ε are k-algebra morphisms implies that $R_k(G)$ is a k-bialgebra. This is called the **representative k-bialgebra** of G. If, in particular, G is a group, then $R_k(G)$ becomes a k-Hopf algebra. Before we go on to prove Theorem 2.2.8, we provide a lemma.

LEMMA 2.2.9 Let S be a set and let V be a finite dimensional k-linear subspace of $\mathrm{Map}(S, k)$. Then it is possible to pick a basis $\{f_1, \ldots, f_n\}$ for V and a subset $\{s_1, \ldots, s_n\}$ of S such that the condition $f_i(s_j) = \delta_{ij}$ is satisfied.

Proof For $s \in S$, $f \in V$, we define a map $\varphi : S \to V^*$ by $\varphi(s)(f) = f(s)$. Then $\varphi(S)$ is dense in V^*. In fact, if $f \in \varphi(S)^{\perp(V)}$, then $\langle f, \varphi(S) \rangle = \langle f, S \rangle = 0$, so that $f = 0$. Letting W denote the k-linear subspace of V^* spanned by $\varphi(S)$, since V^* is of finite dimension, Theorem 2.2.4 implies that W is a closed subspace, and hence we get $W = W^{\perp(V)\perp} = \{0\}^{\perp} = V^*$. Since $\varphi(S)$ spans V^*, we can pick $\{\varphi(S_1), \ldots, \varphi(s_n)\}$ as a basis for V^*. If we let $\{f_1, \ldots, f_n\}$ be its dual basis, we get $f_i(s_j) = \langle f_i, \varphi(s_j) \rangle = \delta_{ij}$, which shows that $\{f_1, \ldots, f_n\}$, $\{s_1, \ldots, s_n\}$ satisfy the requirements of the lemma.

***Proof of Theorem* 2.2.8** Let $f \in R_k(G)$ and set $V_f = kGfG$. Since dim $V_f < \infty$, we can pick a basis $\{f_1, \ldots, f_n\}$ of V_f and a subset $\{x_1, \ldots, x_n\}$ of G such that $f_i(x_j) = \delta_{ij}$ $(1 \leqq i, j \leqq n)$ by Lemma 2.2.9. Then $\delta f(x, y) = \sum_{i=1}^{n} g_i(y) f_i(x)$, and by setting $x = x_j$, we get

$$\delta f(x_j, y) = (yf)(x_j) = g_j(y) = (fx_j)(y),$$

which implies that $g_j \in V_f$. Consequently, we obtain $\delta f \in \pi(R_k(G) \otimes R_k(G))$.

COROLLARY 2.2.10 Let $f \in R_k(G)$ and set $V_f = kGfG$. Then V_f is the smallest k-subcoalgebra of $R_k(G)$ containing f. In particular, each element of $R_k(G)$ generates a finite dimensional k-subcoalgebra.

Proof We have shown in the proof of Theorem 2.2.8 that $f \in R_k(G)$ implies $\Delta f \in V_f \otimes V_f$. Moreover, if $g \in V_f$, then $V_g \subset V_f$ so that $\Delta g \in V_g \otimes V_g \subset V_f \otimes V_f$. Thus V_f is a k-subcoalgebra. Let D denote a k-subcoalgebra of $R_k(G)$ containing f. Writing $\Delta f = \sum_i u_i \otimes v_i \in D \otimes D$, it follows that $xf = \sum_i v_i(x) u_i$ and $fy = \sum_i u_i(y) v_i$ are both elements of D, thereby implying that $kGfG \subset D$. Therefore we conclude that $V_f \subset D$.

COROLLARY 2.2.11 If C, D are k-subcoalgebras of $R_k(G)$, then so is $C \cap D$.

Proof If $f \in C \cap D$, then $V_f \subseteq C \cap D$. Thus $\Delta f \in V_f \otimes V_f \subset (C \cap D) \otimes (C \cap D)$.

EXERCISE 2.3 Let G be a group and let $\rho : G \to GL_n(k)$ be a representation of G. For $\rho(x) = (f_{ij}(x))_{1 \leqq i,\, j \leqq n}$, we have $f_{ij} \in M_k(G)$. Let $V(\rho)$ be the k-linear subspace of $M_k(G)$ spanned by the $f_{ij} (1 \leqq i, j \leqq n)$. Prove that $V(\rho)$ is a finite dimensional two-sided G-submodule and that $R_k(G) = \sum_\rho V(\rho)$ (where ρ ranges over all representations of G).

Remark When G is a finite group, $R_k(G)$ is precisely the dual k-bialgebra of the group k-bialgebra of G. If G is a Lie group (resp. a topological group) and k is either $\mathbb{R}$ or $\mathbb{C}$, let $\mathscr{M}_k(G)$ be the k-algebra consisting of all analytic (resp. continuous) functions on G. Then $\mathscr{R}_k(G) = \mathscr{M}_k(G) \cap R_k(G)$ is a k-sub-bialgebra of $R_k(G)$, called the representative k-bialgebra of G. If we let the representation ρ mentioned in Exercise 2.3 be a Lie group (resp. topological group) representation, then, evidently, $\mathscr{R}_k(G) = \sum_\rho V(\rho)$ holds. The k-bialgebra $\mathscr{R}_k(G)$ plays an important role in the representation theory of Lie groups and topological groups.

2.3 Dual k-bialgebras

We noted in §2.1 that, given an arbitrary k-algebra A, it is not always possible to define a k-coalgebra structure on its dual k-linear space A^*. We make use of the representative bialgebras of semigroups in order to define the dual k-coalgebra of a k-algebra. By this correspondence, we obtain an adjoint functor to the functor which maps any k-coalgebra C to its dual k-algebra C^*. We let A_m denote the underlying semigroup (via multiplication) of a k-algebra A, and we construct the representative k-bialgebra $R_k(A_m)$ of the semigroup A_m. Let A^* be the dual k-linear space of A and set

$$A^\circ = R_k(A_m) \cap A^*.$$

The theorem below shows that A° is a k-subcoalgebra of $R_k(A_m)$, and we call A° the **dual k-coalgebra** of A. If A is finite dimensional, we have $A^\circ = A^*$, which is simply the dual k-coalgebra of A defined in §2.1.

THEOREM 2.2.12 (i) the intersection $R_k(A_m) \cap A^*$ is a k-subcoalgebra of $R_k(A^m)$.

(ii) For $f \in A^*$ to be an element of A°, it is necessary and sufficient that $\operatorname{Ker} f$ contain a finite codimensional ideal of A. Note that we refer to an ideal $\mathfrak{a}$ of A as being of **finite codimension** when $\dim A/\mathfrak{a} < \infty$.

Proof (i) If $f \in R_k(A_m) \cap A^* = A^\circ$, then, since $V_f \subset A^\circ$, A° is a k-subcoalgebra of $R_k(A_m)$.

(ii) Let $f \in A^\circ$ and let $\{f_1, \ldots, f_n\}$ be a basis for V_f. For $\mathfrak{a} = \bigcap_{i=1}^{n} \operatorname{Ker} f_i$, we will show that $\mathfrak{a}$ is a finite codimensional ideal of A and that $\mathfrak{a} \subset \operatorname{Ker} f$. Given $x, y \in A$ and $z \in \mathfrak{a}$, we have $f_i(xzy) = (yf_ix)(z) = 0 \, (1 \leqq i \leqq n)$. Thus $xyz \in \mathfrak{a}$. Consequently, $\mathfrak{a}$ is an ideal of A. Moreover, by definition, $\mathfrak{a} \subseteq \operatorname{Ker} f$. If we define a k-linear map $\varphi : A \to (V_f)^*$ by $\varphi(a)(g) = g(a)$ for $a \in A$, $g \in V_f$, then $\operatorname{Ker} \varphi = \{a \in A; g(a) = 0 \; \forall g \in V_f\} = \mathfrak{a}$. Therefore $\dim A/\mathfrak{a} \leqq \dim (V_f)^* = n$, which shows that $\mathfrak{a}$ is a finite codimensional ideal. Conversely, let $f \in A^*$ be such that $\operatorname{Ker} f$ contains a finite codimensional ideal $\mathfrak{b}$ of A. Let $\pi : A \to A/\mathfrak{b}$ be the canonical k-algebra morphism and pick a subset $\{a_1, \ldots, a_n\}$ of A such that $\{\pi(a_1), \ldots, \pi(a_n)\}$ becomes a basis for $A/\mathfrak{b}$. Then kA_mf is a k-linear subspace of $R_k(A_m)$ spanned by $\{a_1 f, \ldots, a_n f\}$, so that we obtain the result $f \in R_k(A_m) \cap A^*$.

Now let C be a k-coalgebra, and let C^* be its dual k-algebra. We define a k-linear map $\lambda : C \to C^{**}$ by $\langle \lambda(c), f \rangle = \langle f, c \rangle$, $f \in C^*$, $c \in C$.

THEOREM 2.2.13 For λ as defined above, we have $\lambda(C) \subset C^{*\circ}$, and λ induces a k-coalgebra morphism $\lambda_C : C \to C^{*\circ}$.

Proof If we regard C^{**} as a k-linear subspace of $M_k((C^*)_m)$, the elements of C^* have an action on C^{**}. The action of $f \in C^*$ on $\lambda(c) \in C^{**}$ is given by

$$\lambda(c) f = \sum_{(c)} \langle f, c_{(1)} \rangle \lambda(c_{(2)}).$$

In fact, for $g \in C^*$, it turns out that

$$\begin{aligned} \langle \lambda(c) f, g \rangle &= \langle \lambda(c), fg \rangle = \langle fg, c \rangle \\ &= \sum_{(c)} \langle f, c_{(1)} \rangle \langle g, c_{(2)} \rangle \\ &= \sum_{(c)} \langle f, c_{(1)} \rangle \langle \lambda(c_{(2)}), g \rangle. \end{aligned}$$

This implies that $k\lambda(c)C^*$ is finite dimensional and that $\lambda(c) \in R_k((C^*)_m) \cap C^{**} = C^{*\circ}$. On the other hand,

$$\sum_{(c)} \langle \lambda(c_{(1)}), f \rangle \langle \lambda(c_{(2)}), g \rangle = \sum_{(c)} \langle f, c_{(1)} \rangle \langle g, c_{(2)} \rangle.$$

Hence $\Delta \circ \lambda = (\lambda \otimes \lambda) \circ \Delta$. Moreover, since $\varepsilon\lambda(c) = \langle \lambda(c), 1 \rangle = \langle c, 1 \rangle = \varepsilon(c)$, we obtain $\varepsilon \circ \lambda = \varepsilon$. Therefore λ is a k-coalgebra morphism.

By Theorem 2.2.13, an arbitrary k-coalgebra C can be regarded as a k-subcoalgebra of $R_k(G)$ when we write $G = (C^*)_m$. Thanks to Corollaries 2.2.10 and 2.2.11, we get the following result.

COROLLARY 2.2.14 Let C be a k-coalgebra.

(i) A k-subcoalgebra generated by one element $c \in C$ is finite dimensional. Thus C can be expressed as the inductive limit (sum) of finite dimensional k-coalgebras.

(ii) The intersection of two k-subcoalgebras of C is a k-subcoalgebra.

Let C be a k-coalgebra. When $\lambda_C : C \to C^{*\circ}$ is a k-coalgebra isomorphism, we call C a **coreflexive k-coalgebra**.

LEMMA 2.2.15 Let C be a k-coalgebra and let A be its dual k-algebra. Then C is coreflexive $\Leftrightarrow$ any finite codimensional ideal of A is closed.

Proof The map $\lambda_C : C \to A^\circ$ is surjective $\Leftrightarrow$ given any finite codimensional ideal $\mathfrak{a}$ of A, $\mathfrak{a}^\perp = \mathfrak{a}^{\perp(C)}$. Meanwhile, by Exercise 2.2(4), we have $\mathfrak{a}^\perp = \mathfrak{a}^{\perp(C)} \Leftrightarrow \mathfrak{a}$ is a closed subspace.

THEOREM 2.2.16 Given a cocommutative k-coalgebra C, suppose that any finite codimensional ideal of its dual k-algebra $C^* = A$ is finitely generated. Then C is coreflexive.

Proof By Lemma 2.2.15, it suffices to show that a finite codimensional ideal $\mathfrak{a}$ of A is closed. It is enough to prove this for the case

$\mathfrak{a} = Af$ where $f \in A$. Let

$$R_f : A \to A, \qquad g \mapsto gf, \quad g \in A$$

be the right translation of A by f. Then R_f is the dual k-linear map of

$$u : C \to C, \qquad c \mapsto \sum_{(c)} c_{(1)} \langle f, c_{(2)} \rangle, \quad c \in C.$$

Consequently R_f is a continuous function (cf. Exercise 2.2(3)), and we obtain

$$\bar{\mathfrak{a}} = \overline{Af} = \overline{R_f(A)} = R_f(\bar{A}) = \bar{A}f = Af = \mathfrak{a}.$$

Remark In general, when any ideal of a commutative ring R is finitely generated, R is said to be a Noetherian ring. In particular, when C^* is a Noetherian ring, then the hypothesis of Theorem 2.2.16 is satisfied, so that C is coreflexive. For example, the ring of polynomials in n variables over a field k and the power series ring in n variables over a field k are Noetherian rings. (cf. references for Chapter 1, [2] Theorems 3.0.3, 5.1.3, or [4] Chapter III, §2, No. 10, Corollaries of Theorem 2.)

EXAMPLE 2.9 Let C be the k-coalgebra defined in Example 2.2. Then its dual k-algebra C^* is isomorphic to the power series ring $k[[x]]$ in one variable. If we let $f \in C^{*\circ} = k[[x]]^\circ$, then there exists a non-negative integer n such that $\operatorname{Ker} f \supset x^n k[[x]]$. Now, setting

$$c = f(1)c_0 + f(x)c_1 + \cdots + f(x^n)c_{n-1},$$

we have $\lambda(c) = f$, so that λ is surjective. Therefore C is coreflexive. (This is a special case of Theorem 2.2.16.)

Now let A be a k-algebra and let A° be its dual k-coalgebra. When we define a k-linear map $\lambda_A : A \to A^{\circ *}$ by

$$\langle \lambda_A(a), f \rangle = \langle f, a \rangle, \qquad a \in A, \quad f \in A^\circ,$$

then λ_A is a k-algebra morphism. A k-algebra A for which λ_A is injective is said to be **proper**. If λ_A is an isomorphism (resp. a surjection), then A is said to be **reflexive** (resp. **weak-reflexive**).

THEOREM 2.2.17 The following conditions are equivalent for a k-algebra A.

(i) A is a proper k-algebra.

(ii) A° is dense in A^*.

(iii) If we let $\mathscr{F}$ be the set of all finite codimensional ideals of A, then

$$\bigcap_{\mathfrak{a}\in\mathscr{F}} \mathfrak{a} = \{0\}.$$

Proof The fact that (i) and (ii) are equivalent follows readily from the definition. Since $A^\circ = \bigcup_{\mathfrak{a}\in\mathscr{F}} \mathfrak{a}^\perp$, we have $(A^\circ)^{\perp(A)} = \bigcap_{\mathfrak{a}\in\mathscr{F}} \mathfrak{a}^{\perp\perp(A)} = \bigcap_{\mathfrak{a}\in\mathscr{F}} \mathfrak{a}$, from which the equivalence of (ii) and (iii) follows.

3 The duality between algebras and coalgebras

In this section, we investigate the relationship between a coalgebra and its dual algebra as well as between an algebra and its dual coalgebra. In the finite dimensional case, there exists an equivalent dual property between these two algebraic systems. However, this is not true in the general case. A k-coalgebra is the sum (inductive limit) of its finite dimensional k-subcoalgebras, and the properties of a k-coalgebra which can ultimately be reduced to the properties of its finite dimensional k-subcoalgebras can be proven by taking their duals and by applying the theory of finite dimensional k-algebras. Such methods will be used frequently in the sequel.

3.1 Ideals and subcoalgebras

THEOREM 2.3.1 Let C be a k-coalgebra and let C^* be its dual k-algebra. Then we have the following.

(i) If $\mathfrak{a}$ is an ideal of C^*, then $\mathfrak{a}^{\perp(C)}$ is a k-subcoalgebra of C.

(ii) D is a k-subcoalgebra of $C \Leftrightarrow D^\perp$ is an ideal of C^*. In this situation, $C^*/D^\perp \cong D^*$ as k-algebras.

Proof (i) Given $x \in \mathfrak{a}^{\perp(C)}$, it suffices to show that $\Delta x \in \mathfrak{a}^{\perp(C)} \otimes \mathfrak{a}^{\perp(C)}$. Pick $\{z_i\}_{1 \leqq i \leqq n}$ to be linearly independent over k in such a way that $\Delta x = \sum_{i=1}^{n} y_i \otimes z_i$. If $\Delta x \neq \mathfrak{a}^{\perp(C)} \otimes C$, then, for some j, we have

$y_j \notin \mathfrak{a}^{\perp(C)}$. This means that there is an $f \in \mathfrak{a}$ for which $\langle y_j, f \rangle \neq 0$. Now pick $g \in C^*$ which satisfies $\langle g, z_i \rangle = \delta_{ij} (1 \leqq i \leqq n)$. Since $fg \in \mathfrak{a}$, we get

$$0 = \langle x, fg \rangle = \langle \Delta x, f \otimes g \rangle = \langle y_j, f \rangle \langle z_j, g \rangle \neq 0,$$

which is a contradiction. Therefore $\Delta x \in \mathfrak{a}^{\perp(C)} \otimes C$. Similarly, we obtain $\Delta c \in C \otimes \mathfrak{a}^{\perp(C)}$, which implies that $\Delta x \in (\mathfrak{a}^{\perp(C)} \otimes C) \cap (C \otimes \mathfrak{a}^{\perp(C)}) = \mathfrak{a}^{\perp(C)} \otimes \mathfrak{a}^{\perp(C)}$.

(ii) If we let $i : D \to C$ be the canonical embedding, its dual map $i^* : C^* \to D^*$ is a surjection, and Ker $i^* = D^\perp$ is an ideal of C^*. Therefore $C^*/D^\perp \cong D^*$. Conversely, if $\mathfrak{a} = D^\perp$ is an ideal of C^*, then $D = D^{\perp\perp(C)} = \mathfrak{a}^{\perp(C)}$ is a k-subcoalgebra of C by (i).

THEOREM 2.3.2 Let A be a k-algebra and let A° be its dual k-coalgebra.

(i) If $\mathfrak{a}$ is an ideal of A, then $\mathfrak{a}^\perp \cap A^\circ$ is a k-subcoalgebra of A°.

(ii) If D is a k-subcoalgebra of A°, then $D^{\perp(A)}$ is an ideal of A.

Proof (i) It suffices to show that if $f \in \mathfrak{a}^\perp \cap A^\circ$, then $\Delta f \in (\mathfrak{a}^\perp \cap A^\circ) \otimes (\mathfrak{a}^\perp \cap A^\circ)$. Set $\Delta f = \sum_{i=1}^{n} g_i \otimes h_i$, where $\{g_i\}_{1 \leqq i \leqq n}$ is chosen to be linearly independent over k. Pick a subset $\{a_i\}_{1 \leqq i \leqq n}$ of A such that $\langle a_i, g_j \rangle = \delta_{ij}$. For $a \in \mathfrak{a}$, we have $a_i a \in \mathfrak{a}$, so that $0 = \langle a_i a, f \rangle = \sum_{j=1}^{n} \langle a_i, g_j \rangle \langle a, h_j \rangle = \langle a, h_i \rangle$. Thus $h_i \in \mathfrak{a}^\perp$. Hence $\Delta f \in A^\circ \otimes (\mathfrak{a}^\perp \cap A^\circ)$. Similarly, we obtain $\Delta f \in (\mathfrak{a}^\perp \cap A^\circ) \otimes (\mathfrak{a}^\perp \cap A^\circ)$.

(ii) Given $a \in A$, $b \in D^{\perp(A)}$ and $f \in D$, we have $\langle ab, f \rangle = \langle a \otimes b, \Delta f \rangle \subseteq \langle a, D \rangle \langle b, D \rangle = 0$.

Therefore $ab \in D^{\perp(A)}$. Similarly, we have $ba \in D^{\perp(A)}$. Thus $D^{\perp(A)}$ is an ideal of A.

Remark The converse of this theorem does not necessarily hold. For instance, if A is the field of rational functions over k in one variable, namely, $A = k(x)$, then $\mathfrak{a} = kx$ is neither an ideal nor a k-subalgebra. However, $\mathfrak{a}^\perp \cap A^\circ = \{0\}$ is a k-subcoalgebra of A°.

A k-subcoalgebra M of a k-coalgebra C is called a **simple k-subcoalgebra** if it does not have any k-subcoalgebras other than $\{0\}$ and M. Moreover, the sum of all simple k-subcoalgebras of C is called the **coradical** of C and is denoted by corad C. When $C = \text{corad } C$, C is said to be **co-semi-simple**. If C has only one simple k-subcoalgebra, we say that C is **irreducible**; and if all simple k-subcoalgebras of C are one dimensional, C is called a **pointed** ***k*-coalgebra**. A pointed irreducible k-coalgebra is said to be **connected**, and a cocommutative connected k-bialgebra is called a ***k*-hyperalgebra**. For example, the k-coalgebra kS in Example 2.1 is the sum of its simple k-subcoalgebras ks for all $s \in S$ and is a co-semi-simple pointed k-coalgebra. When the universal enveloping k-bialgebra $U(L)$ of the k-Lie algebra L in Example 2.5 is regarded as a k-coalgebra, it has only one simple k-subcoalgebra $k1$. Since $U(L)$ is a pointed irreducible k-coalgebra and is moreover cocommutative, it is a k-hyperalgebra.

THEOREM 2.3.3 If k is an algebraically closed field, then a cocommutative k-coalgebra is a pointed k-coalgebra.

Proof Let D be a simple k-subcoalgebra of a cocommutative k-coalgebra C. Since D is a finite dimensional cocommutative k-coalgebra, D^* is a finite dimensional commutative simple k-algebra, thanks to Theorem 2.3.1. Thus D^* is a finite dimensional extension field of k, and since k is algebraically closed, we have $D \cong k$. Therefore C is a pointed k-coalgebra.

THEOREM 2.3.4 Given a k-coalgebra C, let $\mathscr{M}$ be the set of all simple k-subcoalgebras of C, and let $\mathscr{I}$ be the set of all maximal ideals of C^* which are not dense. If $M \in \mathscr{M}$, then $M^\perp \in \mathscr{I}$, and if $\mathfrak{a} \in \mathscr{I}$, we have $\mathfrak{a}^{\perp(C)} \in \mathscr{M}$. Moreover, these assignments, which turn out to be each other's inverses, set up a one-to-one correspondence between $\mathscr{M}$ and $\mathscr{I}$.

Proof If $M \in \mathscr{M}$, M is finite dimensional and $C^*/M^\perp \cong M^*$. Theorem 2.3.1 implies that M^* is simple, so that $M^\perp$ is a maximal ideal of C^*. Moreover, since $M^{\perp\perp(C)} = M \neq 0$, $M^\perp$ is not dense in C^*. Thus $M^\perp \in \mathscr{I}$. Conversely, if $\mathfrak{a} \in \mathscr{I}$, $\mathfrak{a}^{\perp(C)}$ is a k-subcoalgebra of C which is not equal to $\{0\}$. Consequently there exists a simple k-subcoalgebra

$M \neq \{0\}$ which is contained in $\mathfrak{a}^{\perp(C)}$. Now it follows that $\mathfrak{a} \subseteq \mathfrak{a}^{\perp(C)\perp} \subset M^{\perp} \subsetneqq C^*$, and, since $\mathfrak{a}$ is a maximal ideal, we obtain $\mathfrak{a}^{\perp(C)} = M^{\perp\perp(C)} = M$. Therefore, $M \in \mathcal{M}$. It is apparent that these correspondences are inverses.

3.2 *k*-subalgebras and coideals

When a k-linear subspace V of a k-coalgebra C satisfies the conditions

$$\Delta V \subset V \otimes C + C \otimes V, \qquad \varepsilon(V) = 0,$$

V is said to be a **coideal**. In this situation, a k-coalgebra structure can be defined canonically on the factor space C/V. We call this the **factor** k-**coalgebra** of C by V. Moreover, a k-linear subspace V of C such that

$$\Delta V \subset V \otimes C \qquad (\text{resp. } \Delta V \subset C \otimes V)$$

is called a **right** (resp. **left**) **coideal**. A k-subcoalgebra is a right (or left) coideal.

THEOREM 2.3.5 Let C be a k-coalgebra.

(i) If $\mathfrak{a}$ is a right (resp. left) ideal of C^*, then $\mathfrak{a}^{\perp(C)}$ is a right (resp. left) coideal of C.

(ii) If V is a right (resp. left) coideal of C, then $V^{\perp}$ is a right (resp. left) ideal of C^*, and the converse also holds.

Proof (i) If $\mathfrak{a}$ is a right ideal of C^*, then $\Delta^*(\mathfrak{a} \otimes C^*) \subset \mathfrak{a}$. Thus $\langle \Delta \mathfrak{a}^{\perp(C)}, \mathfrak{a} \otimes C^* \rangle = \langle \mathfrak{a}^{\perp(C)}, \Delta^*(\mathfrak{a} \otimes C^*) \rangle = 0$. Therefore

$$\Delta \mathfrak{a}^{\perp(C)} \subset (\mathfrak{a} \otimes C^*)^{\perp(C)} = \mathfrak{a}^{\perp(C)} \otimes C + C \otimes C^{*\perp(C)} = \mathfrak{a}^{\perp(C)} \otimes C.$$

The proof of (ii) is similar. For the converse, merely apply (i) to $V^{\perp\perp(C)} = V$.

THEOREM 2.3.6 Let C be a k-coalgebra and let C^* be its dual k-algebra.

(i) If I is a k-subalgebra of C^*, then $I^{\perp(C)}$ is a coideal of C.

(ii) If J is a coideal of C, then $J^{\perp}$ is a k-subalgebra of C^*, and the converse also holds.

Proof (i) Let I be a k-subalgebra of C^*. Since I contains the identity element, it follows that $\varepsilon(I^{\perp(C)}) = 0$. Moreover, we have

$\langle\rho(I\otimes I), \Delta I^{\perp(C)}\rangle = \langle\Delta^*\rho(I\otimes I), I^{\perp(C)}\rangle \subset \langle I, I^{\perp(C)}\rangle = 0$. Therefore $\Delta I^{\perp(C)} \subset \rho(I\otimes I)^{\perp(C)} \subset C\otimes I^{\perp(C)} + I^{\perp(C)}\otimes C$. (cf. Exercise 1.8 concerning the map ρ.) Therefore $I^{\perp(C)}$ is an ideal of C.

(ii) Let J be a coideal of C. Then we have $\langle\Delta^*\rho(J^\perp\otimes J^\perp), J\rangle = \langle\rho(J^\perp\otimes J^\perp), \Delta J\rangle \subset \langle\rho(J^\perp\otimes J^\perp), J\otimes C\rangle + \langle\rho(J^\perp\otimes J^\perp), C\otimes J\rangle = 0$. Consequently $\mu_{C*}(J^\perp\otimes J^\perp) \subset J^\perp$. Similarly, we get $\eta_{C*}(1)\in J^\perp$. For the converse, set $J^\perp = I$ and apply (i) to $I^{\perp(C)} = J^{\perp\perp(C)} = J$.

THEOREM 2.3.7 Let A be a k-algebra and let A° be its dual k-coalgebra.

(i) If B is a k-subalgebra of A, then $B^\perp\cap A^\circ$ is a coideal of A°.

(ii) If D is a coideal of A°, then $D^{\perp(A)}$ is a k-subalgebra of A.

Proof (i) Since $1\in B$, it follows that $\varepsilon(B^\perp\cap A^\circ) = \langle B^\perp\cap A^\circ, 1\rangle = 0$. Let $A^\circ = (B^\perp\cap A^\circ)\oplus Y$ be the direct sum decomposition of A° as a k-linear space. Letting $\{f_\lambda\}_{\lambda\in\Lambda}$, $\{g_\mu\}_{\mu\in M}$ respectively be bases for $B^\perp\cap A^\circ$ and Y, we see that $\{g_\mu|_B\}_{\mu\in M}$ is linearly independent over k. In fact, if $\left\langle \sum\alpha_\mu g_\mu, B\right\rangle = 0$, then $\sum\alpha_\mu g_\mu\in B^\perp\cap A^\circ\cap Y = 0$. Thus $\alpha_\mu = 0$ for each μ. In order to show that $B^\perp\cap A^\circ$ is a coideal of A°, it suffices to show that if $f\in B^\perp\cap A^\circ$ and $\Delta f = \sum_\lambda f_\lambda\otimes f_\lambda' + \sum_\mu g_\mu\otimes g_\mu'$ for f_λ', $g_\mu'\in A^\circ$, then $g_\mu'\in B^\perp\cap A^\circ$. Pick a finite subset $\{a_v\}$ of B such that $\langle a_v, g_\mu\rangle = \delta_{v\mu}$ and $\langle a_v, f_\lambda\rangle = 0$. If $a\in B$, then since $a_v a\in B$, it follows that

$$0 = \langle a_v a, f\rangle = \sum_\lambda \langle a_v, f_\lambda\rangle\langle a, f_\lambda'\rangle + \sum_n \langle a_v, g_\mu\rangle\langle a, g_\mu'\rangle$$
$$= \langle a, g_v'\rangle.$$

Therefore we have $g_\mu'\in B^\perp\cap A^\circ$, and hence $B^\perp\cap A^\circ$ is a coideal of A°.

(ii) Since $\varepsilon(D) = 0$, it follows that $\langle D, 1\rangle = 0$. Thus $1\in D^{\perp(A)}$. Now, given a, $b\in D^{\perp(A)}$ and $f\in D$, we have

$$\langle ab, f\rangle = \langle a\otimes b, \Delta f\rangle \subset \langle a\otimes b, D\otimes A^\circ\rangle + \langle a\otimes b, A^\circ\otimes D\rangle = 0.$$

Thus $ab\in D^{\perp(A)}$. Consequently, $D^{\perp(A)}$ is a k-subalgebra.

Remark The converses of Theorem 2.3.6 (i) and Theorem 2.3.7 (i), (ii) do not necessarily hold.

3.3 Radicals and coradicals

We proceed to investigate the relationship between the coradical of a k-coalgebra C and the radical of its dual k-algebra C^*. If C is finite dimensional, it follows readily that $(\text{corad } C)^\perp = \text{rad } C^*$. In order to show that this result holds generally, we define an operation on a k-coalgebra which is dual to the product of ideals of a k-algebra. Let X and Y be k-linear subspaces of a k-coalgebra C, and let σ be the composite of the comultiplication map $\Delta : C \to C \otimes C$ on C and the canonical k-linear map $C \otimes C \to C/X \otimes C/Y$. Set $\text{Ker } \sigma = X \sqcap Y$. (Usually, the symbol $\wedge$ is used in place of $\sqcap$ in this definition. However, in this book, we have employed the latter so as to avoid confusion with the exterior product.) From the definition, we have $X \sqcap Y = \Delta^{-1}(C \otimes Y + X \otimes C)$, so $X \sqcap Y = (X^\perp Y^\perp)^{\perp(C)}$. In particular, we set $\sqcap^0 X = \{0\}$, $\sqcap^1 X = X$, and $\sqcap^n X = (\sqcap^{n-1} X) \sqcap X$. If X is a left coideal and Y is a right coideal, then $X^\perp$ is a left ideal and $Y^\perp$ is a right ideal. In these circumstances, $X^\perp Y^\perp$ becomes a two-sided ideal of C^*, and $X \sqcap Y$ becomes a k-subcoalgebra of C. In particular, if X and Y are k-subcoalgebras of C, then $X \sqcap Y$ as well as $Y \sqcap X$ are k-subcoalgebras of C.

EXERCISE 2.4 (i) If X is a left (resp. right) coideal, show that $X \sqcap Y$ (resp. $Y \sqcap X$) is a right (resp. left) coideal and that $X \subset Y \sqcap X$ (resp. $X \subset X \sqcap Y$).

(ii) If $X \subset \text{Ker } \varepsilon$, show that $\bigcap_{n=0}^{\infty} (\sqcap^n X)$ is a coideal.

(iii) Let C, D be k-coalgebras and let $f : C \to D$ be a k-coalgebra morphism. Then, for k-linear subspaces X and Y of C, show that $f(X \sqcap Y) \subset f(X) \sqcap f(Y)$.

Let D be a k-subcoalgebra of a k-coalgebra C and set $D^\perp = \mathfrak{a}$. For any natural number n, we have $\sqcap^n(D) = (\mathfrak{a}^n)^{\perp(C)}$. Setting $D_n = \sqcap^{n+1} D$, we obtain a chain $D = D_0 \subset D_1 \subset D_2 \subset \cdots$ of k-subcoalgebras of C. Set $D_\infty = \bigcup_{n=0}^{\infty} D_n$. If $D_\infty = C$, D is said to be **conilpotent**.

LEMMA 2.3.8 Let C be a k-coalgebra and let D be a conilpotent k-subcoalgebra of C. Then $D^\perp \subset \text{rad } C^*$.

Proof Given $f \in D^{\perp}$, it suffices to show that $\varepsilon - f$ is an invertible element of C^*. Let $\pi_n : C^* \to D_n^*$ be the dual k-algebra morphism of the canonical embedding $D_n = \sqcap^{n+1} D \to C$ of k-coalgebras. Since D is conilpotent, it follows that $\varprojlim (D_n)^* = C^*$. Meanwhile, for each n, since $m > n$ implies $\pi_n(f^m) = 0$, $\pi_n(\varepsilon + f + \cdots + f^n)$ is the inverse of $\pi_n(\varepsilon - f)$. Therefore $\varepsilon - f$ is invertible in C^*.

THEOREM 2.3.9 Let C be a k-coalgebra. Then

(i) $\operatorname{rad} C^* = (\operatorname{corad} C)^{\perp}$.

(ii) D is a conilpotent k-subcoalgebra of $C \Leftrightarrow D \supset \operatorname{corad} C$.

Proof (i) If $x \in C$, then x is contained in a finite dimensional k-subcoalgebra D of C. Since $R = \operatorname{rad} D^* = (\operatorname{corad} D)^{\perp}$ is the largest nilpotent ideal of D^*, there exists a natural number n such that $R^n = \{0\}$. Since

$$D = (R^n)^{\perp(C)} = \sqcap^n(\operatorname{corad} D) = D_{n-1} \subseteq \sqcap^n(\operatorname{corad} C) = C_{n-1},$$

it follows that $C = \bigcup_{n=0}^{\infty} C_n$. In other words, $\operatorname{corad} C$ is conilpotent. By Lemma 2.3.8, we get $(\operatorname{corad} C)^{\perp} \subset \operatorname{rad} C^*$. Meanwhile, since $(\operatorname{corad} C)^{\perp}$ is the intersection of a family of finite codimensional maximal ideals, we have $(\operatorname{corad} C)^{\perp} \supset \operatorname{rad} C^*$. Therefore $(\operatorname{corad} C)^{\perp} = \operatorname{rad} C^*$.

(ii) Due to (i), $\operatorname{corad} C$ is conilpotent. Therefore, if a k-subcoalgebra D contains $\operatorname{corad} C$, D is also conilpotent. Conversely, if D is conilpotent k-subcoalgebra, then Lemma 2.3.8 implies that $D^{\perp} \subset \operatorname{rad} C^*$. Thus $D = D^{\perp\perp(C)} \supset (\operatorname{rad} C^*)^{\perp(C)} = \operatorname{corad} C$.

COROLLARY 2.3.10 $\sqcap^{n+1}(\operatorname{corad} C) = ((\operatorname{rad} C^*)^{n+1})^{\perp(C)}$.

3.4 The decomposition theorem of k-coalgebras

The decomposition theorem for finite dimensional k-algebras (Theorem 1.4.9) can be extended dually to the decomposition theorem of k-coalgebras (which are not necessarily finite dimensional). Given a simple k-coalgebra N, its dual k-algebra N^* is a simple k-algebra. When N^* is separable, N is said to be **separable**. Given a k-coalgebra C, let $\operatorname{corad} C$ be the coradical of C. When any simple k-subcoalgebra

N of corad C is separable, corad C is said to be a **separable coradical**. If k is an algebraically closed field, corad C is always separable. Given a k-coalgebra C, let D be a k-subcoalgebra. A k-coalgebra morphism $\pi: C \to D$ such that the restriction of π to D is the identity map is called a **projection** from C to D.

THEOREM 2.3.11 Let C be a k-coalgebra and suppose that corad C is separable. Then there exists a coideal I of C such that $C = I \oplus (\text{corad } C)$ (the direct sum as k-linear spaces). In other words, there exists a projection from C to corad C.

Before we prove this theorem, we provide some lemmas.

LEMMA 2.3.12 If a k-coalgebra morphism $\sigma: C \to D$ is surjective, then $\sigma(\text{corad } C) \supset \text{corad } D$. Therefore, any simple k-subcoalgebra of D is contained in $\sigma(\text{corad } C)$.

Proof Let E be a simple k-subcoalgebra of D. Since σ is surjective, for $y \in E$, $y \neq 0$, there exists $x \in C$ such that $\sigma(x) = y$. Let X be the k-subcoalgebra of C generated by x. Then, since $\dim X < \infty$, $X \cap (\text{corad } C) = \text{corad } X$, and $E \cap \sigma(X) \neq \{0\}$, we may assume that C and D are finite dimensional. Now the dual k-algebra morphism $\sigma^*: D^* \to C^*$ of σ is injective, and the radical $(\text{corad } C)^\perp$ of C^* is a nilpotent ideal, so $(\text{corad } C)^\perp \cap D^*$ is likewise a nilpotent ideal of D^*. Thus $(\text{corad } C)^\perp \cap D^* \subset \text{rad } D^*$. Meanwhile, since $((\text{corad } C)^\perp \cap D^*)^{\perp(D)} = \sigma(\text{corad } C)$, it follows that $\sigma(\text{corad } C) \supset (\text{rad } D^*)^{\perp(D)} = \text{corad } D$.

LEMMA 2.3.13 Let C be a finite dimensional k-coalgebra and assume that the coradical corad C is separable. If D is a k-subcoalgebra of C, then a projection from D to $D \cap (\text{corad } C)$ can be extended to a projection from C to corad C.

Proof Denote by π a projection from D to $D \cap (\text{corad } C)$. Let $\alpha = \pi' \circ i$ be the composite of the canonical embedding $i: \text{corad } C \to C$ and the canonical k-coalgebra morphism $\pi': C \to C/\text{Ker } \pi = E$. Then we have $\text{Ker } \alpha = \{0\}$ and $\text{Im } \alpha \subset \text{corad } E$. On the other hand, Lemma

2.3.12 shows that $\operatorname{Im}\alpha \supset \operatorname{corad} E$, which implies that $\operatorname{Im}\alpha = \operatorname{corad} E$. Since the dual k-algebra morphism $\alpha^* : E^* \to (\operatorname{corad} C)^*$ of the k-coalgebra morphism α is surjective, and because $(\operatorname{corad} C)^*$ is a separable k-algebra, Theorem 1.4.9 implies that there exists a k-algebra morphism $\pi''^* : (\operatorname{corad} C)^* \to E^*$ such that $\alpha^* \circ \pi''^*$ is the identity map on $(\operatorname{corad} C)^*$. Therefore there exists a k-coalgebra morphism $\pi'' : E \to \operatorname{corad} C$ such that $\pi'' \circ \alpha$ is the identity map on $\operatorname{corad} C$. Setting $\tilde{\pi} = \pi'' \circ \pi'$, we see that $\tilde{\pi}$ is a projection from C to $\operatorname{corad} C$ and that $\tilde{\pi}$ is an extension of π.

***Proof of Theorem* 2.3.11** Let (F, π) be a pair consisting of a k-subcoalgebra F of the k-coalgebra C and the projection π from F to $F \cap (\operatorname{corad} C)$. Let $\mathscr{F}$ denote the set of all such pairs. Since the pair consisting of $\operatorname{corad} C$ and the identity map is an element of $\mathscr{F}$, the set $\mathscr{F}$ is not empty. For $(F, \pi), (F', \pi) \in \mathscr{F}$, the definition

$$(F', \pi') \leqq (F, \pi) \Leftrightarrow F' \subset F \quad \text{and} \quad \pi|_{F'} = \pi'$$

makes $\mathscr{F}$ an ordered set. The set $\mathscr{F}$ satisfies the hypothesis of Zorn's lemma, so there exists a maximal element, say (F_0, π_0). Now it suffices to show that $F_0 = C$. Suppose that $F_0 \subsetneqq C$. Then there exists $x \in C$ such that $x \notin F_0$. Let D be the k-subcoalgebra of C generated by x. Then we have $\operatorname{Im}(\pi_0|_{F_0 \cap D}) = D \cap (\operatorname{corad} C) = \operatorname{corad} D$; and since D is finite dimensional, applying Lemma 2.3.13 to $F_0 \cap D \subset D$ and $\pi_0|_{F_0 \cap D}$, we see that there exists a projection π from D to $D \cap (\operatorname{corad} C)$ which is an extension of $\pi_0|_{F_0 \cap D}$. Since the pairs (D, π) and (F_0, π_0) have projections onto their coradicals which coincide on their intersection, they can be extended to a projection π' from $F_0 + D$ to $(F_0 + D) \cap (\operatorname{corad} C)$. Thus we have $(F_0 + D, \pi') \gneqq (F_0, \pi_0)$, which contradicts the maximality of (F_0, π_0). This forces $F_0 = C$.

3.5 The correspondence between algebras and coalgebras

As defined in 2.3, there is a correspondence between k-algebras and k-coalgebras

$$F : \mathbf{Cog}_k \to \mathbf{Alg}_k, \qquad C \mapsto C^*,$$
$$G : \mathbf{Alg}_k \to \mathbf{Cog}_k, \qquad A \mapsto A^\circ.$$

Now, if $\sigma : C \to D$ is a k-coalgebra morphism, then its dual k-linear map $\sigma^* : D^* \to C^*$ is a k-algebra morphism, and if $\rho : A \to B$ is a k-algebra morphism, then the restriction of its dual k-linear map ρ^* to B°, $\rho^\circ : B^\circ \to A^\circ$, is a k-coalgebra morphism. From this fact, we see that F and G are contravariant functors from the category of k-coalgebras to the category of k-algebras, and vice versa. Moreover, the following theorem shows that they are adjoint.

THEOREM 2.3.14 Let A be a k-algebra and let C be a k-coalgebra. If we define maps

$$\mathbf{Alg}_k\,(A, C^*) \underset{\varphi}{\overset{\Phi}{\rightleftarrows}} \mathbf{Cog}_k\,(C, A^\circ)$$

by $\Phi : \rho \mapsto \rho^\circ \circ \lambda_C$ and $\Psi : \sigma \mapsto \sigma^* \circ \lambda_A$, then Φ and Ψ are inverses and give a bijective correspondence between the two sets.

Proof Since $\Psi \circ \Phi(\rho) = \Psi(\rho^\circ \circ \lambda_C) = (\rho^\circ \circ \lambda_C)^* \circ \lambda_A = \lambda_C{}^* \circ \rho^{\circ *} \circ \lambda_A$, for $a \in A$ and $c \in C$, we have

$$\begin{aligned} \langle \lambda_C{}^* \circ \rho^{\circ *} \circ \lambda_A(a), c \rangle &= \langle \rho^{\circ *} \circ \lambda_A(a), \lambda_C(c) \rangle \\ &= \langle a, \rho^\circ(c) \rangle \\ &= \langle \rho(a), c \rangle. \end{aligned}$$

Therefore, $\Psi \circ \Phi(\rho) = \rho$. Similarly, we obtain $\Phi \circ \Psi(\sigma) = \sigma$.

We know that if C is a k-coalgebra, C° is a k-subalgebra of the dual k-algebra C^* of C. Thus, if H is a k-Hopf algebra, H° inherits a k-Hopf algebra structure. We call H° the **dual k-Hopf algebra** of H. The correspondence given by $H \mapsto H^\circ$ is a contravariant functor from the category $\mathbf{Hopf}_k$ to itself, and it is adjoint to itself. Namely, if H, K are k-Hopf algebras, the correspondence of Theorem 2.3.14 gives rise to a bijective correspondence

$$\mathbf{Hopf}_k(H, K^\circ) \cong \mathbf{Hopf}_k(K, H^\circ).$$

Next, we investigate the relationship between k-algebras and k-coalgebras which correspond by the functor defined above.

THEOREM 2.3.15 Given $\sigma \in \mathbf{Cog}_k\,(C, D)$,

$$\sigma \text{ is a surjection} \Leftrightarrow \sigma^* \text{ is an injection};$$

$$\sigma \text{ is an injection} \Leftrightarrow \sigma \text{ is a surjection}.$$

The proof is straightforward.

THEOREM 2.3.16 Given $\sigma \in \mathbf{Alg}_k(A, B)$,

(i) σ is a surjection $\Rightarrow \sigma^\circ$ is an injection.

(ii) σ° is an injection, B is proper, and A is weak-reflexive $\Rightarrow \sigma$ is a surjection and B is reflexive.

(iii) σ is an injection, B is proper, and A is weak-reflexive $\Rightarrow \sigma^\circ$ is a surjection and A is reflexive.

(iv) σ° is a surjection and A is proper $\Rightarrow \sigma$ is an injection.

Proof (i) This follows readily from the fact that σ° is the restriction of σ^* to B°.

(ii) Since σ° is injective and λ_A is surjective, it follows that $\sigma^{\circ *} \circ \lambda_A = \lambda_B \circ \sigma$ is surjective. Thus λ_B is surjective. Meanwhile, by hypothesis, λ_B is injective. Hence B is reflexive. Since λ_B is bijective, σ is a surjection.

(iii) Since σ and λ_B are injective, it follows that $\lambda_B \circ \sigma = \sigma^{\circ *} \circ \lambda_A$ is injective. Thus λ_A is injective and A is proper; and, by hypothesis, λ_A is surjective. Hence λ_A is bijective. Therefore, since $\sigma^{\circ *}$ is injective, σ° is surjective.

(iv) Since σ° is surjective and λ_A is injective, it follows that $\sigma^{\circ *} \circ \lambda_A = \lambda_B \circ \sigma$ is injective. Consequently σ is an injection.

Remark In Theorem 2.3.16 (ii), (iii), and (iv), we cannot leave out any of the conditions of the hypothesis concerning A or B. For instance, let C be the k-coalgebra defined in Example 2.2. If we let $B = C^*$ be its dual k-algebra, then $B \cong k[[x]]$. Let A denote its k-subalgebra $k[x]$. If $\sigma : A \to B$ is the canonical embedding, then $\sigma^\circ : B^\circ \cong C \to A^\circ$ is an injection but not a surjection. On the other hand, B is proper, but A is not weak-reflexive. In summary, if we omit the condition that A is weak-reflexive in (ii), then the implication in (ii) does not necessarily hold. Moreover, if we let $A = k$ and let $k(x) = B$ be the field of rational functions over k in one variable, then $B^\circ = \{0\}$ and B is not proper but A is proper and weak-reflexive. In this situation, $\sigma : A \to B$ is injective and $\sigma^\circ : B^\circ = \{0\} \to A^\circ$ is not surjective. Therefore, if we omit the condition that B is proper in (iii), the result in (iii) does not necessarily hold.

COROLLARY 2.3.17 (i) If C is a k-coalgebra, C is cocommutative $\Leftrightarrow C^*$ is commutative.

(ii) If A is a k-algebra, then A is commutative $\Rightarrow A^\circ$ is cocommutative.

Conversely, A° is cocommutative and A is proper $\Rightarrow A$ is commutative.

Remark In Corollary 2.3.17 (ii), if A is not proper, it does not necessarily follow that A is commutative even if A° is cocommutative. For instance, let k be a finite field and let G be an infinite simple group (e.g. the inductive limit $\varinjlim_n A_n$ of alternating groups A_n with respect to the canonical embeddings $A_n \subset A_{n+1}$). Letting $A = kG$, we see that the k-algebra A is not commutative, and since $R_k(G) = k$ by the following Exercise 2.5, $A^\circ = R_k(A_m) \cap A^*$ is cocommutative.

EXERCISE 2.5 Let k be a finite field and let G be a group that does not have any proper subgroup of finite index. Then $R_k(G) = k$.

THEOREM 2.3.18 The dual k-algebra C^* of a k-coalgebra C is proper.

Proof By Corollary 2.2.14, an element $c \in C$ generates a finite dimensional k-subcoalgebra in C. Thus C is the sum of its finite dimensional k-subcoalgebras $C_\lambda (\lambda \in \Lambda)$, so that we can write $C = \sum_{\lambda \in \Lambda} C_\lambda$. Then

$$\mathfrak{a}_\lambda = C_\lambda{}^\perp = \{f \in C^*; \langle f, C_\lambda \rangle = 0\}$$

is an ideal of C^*; and since $C^*/\mathfrak{a}_\lambda \cong C_\lambda{}^*$, it follows that $\mathfrak{a}_\lambda$ is finite codimensional. Since $\bigcap_\lambda \mathfrak{a}_\lambda = \bigcap_\lambda C_\lambda{}^\perp = \left(\sum_\lambda C_\lambda\right)^\perp = \{0\}$, Theorem 2.2.17 implies that C^* is proper

THEOREM 2.3.19 A finitely generated commutative k-algebra is proper.

Proof Let $a \neq 0$ be an element of a finitely generated commutative k-algebra A. It suffices to show that there exists a finite codimensional ideal of A that does not contain a.

If we set $\mathfrak{a} = \{x \in A;\ xa = 0\}$, then $\mathfrak{a}$ is a proper ideal of A. Let $\mathfrak{m}$ be a maximal ideal containing $\mathfrak{a}$, and let $A_{\mathfrak{m}}$ be the quotient ring of A by $S = A - \mathfrak{m}$. Then $A_{\mathfrak{m}}$ is finitely generated k-algebra with a unique maximal ideal $\mathfrak{m}A_{\mathfrak{m}} = \text{rad}\,(A_{\mathfrak{m}})$. Therefore, by Krull's intersection theorem (Theorem 1.5.9 (ii)), $\bigcap_n (\mathfrak{m}A_{\mathfrak{m}})^n = \{0\}$. Thus there exists a natural number n such that $a \notin \mathfrak{m}^n$. On the other hand, $A/\mathfrak{m}$ is an algebraic extension field of k of finite degree (cf. Theorems 1.5.4, 1.5.5). In other words, $\mathfrak{m}$ is a finite codimensional ideal. Since $\mathfrak{m}^n$ is a finite codimensional ideal because of the following lemma, it follows that A is proper.

LEMMA 2.3.20 Given a commutative k-algebra A, let $\mathfrak{a}$ be a finite codimensional ideal of A and let M be a finitely generated A-module. Then $\mathfrak{a}M$ is a finite codimensional A-submodule of M.

Proof It suffices to show that $M/\mathfrak{a}M \cong A/\mathfrak{a} \otimes_A M$ is of finite dimension. Since M is a finitely generated A-module, for some natural number n, there exists an exact sequence $\underbrace{A \oplus \cdots \oplus A}_{n \text{ times}} \to M \to 0$. Thus

$$A/\mathfrak{a} \oplus \cdots \oplus A/\mathfrak{a} \to A/\mathfrak{a} \otimes_A M \to 0$$

is also exact (cf. Exercise 1.10). On the other hand, since $\mathfrak{a}$ is finite codimensional, it follows that $A/\mathfrak{a} \otimes_A M$ is finite dimensional.

COROLLARY 2.3.21 If $\mathfrak{a}$ is a finite codimensional ideal of a finitely generated k-algebra A, then for any natural number n, $\mathfrak{a}^n$ is a finite codimensional ideal.

4 Irreducible bialgebras

A k-coalgebra or a k-bialgebra that has only one simple k-subcoalgebra is said to be **irreducible**. In this section, we investigate some properties of irreducible k-subcoalgebras of k-bialgebras and coradical filtrations on irreducible k-bialgebras. In particular, a cocommutative k-bialgebra can be decomposed into the direct sum of its irreducible k-subcoalgebras. Such properties will be applied extensively in Chapters 4 and 5.

4.1 Filtered bialgebras

In Chapter 1, §2.2, we defined filtered algebras and graded algebras. Here, we first define filtered k-coalgebras, which are dual to the notion of filtered k-algebras, and then proceed to define filtered k-bialgebras which have both these structures.

Let $I = \{0, 1, 2, \ldots\}$ be the set of all non-negative integers. Given a k-coalgebra C, if a family $\{C_i\}_{i \in I}$ of k-linear subspaces of C satisfies the conditions

$$C_i \subset C_{i+1} (i \in I), \quad C = \bigcup_{i \in I} C_i,$$

$$\Delta C_n \subset \sum_{i=0}^{n} C_i \otimes C_{n-i} \quad (n \in I),$$

then C is called a **filtered k-coalgebra**, and $\{C_i\}_{i \in I}$ is said to be a **filtration** on C. By definition, C_i $(i \in I)$ are k-subcoalgebras of C. If there exists a family of k-linear subspaces $\{C_{(i)}\}_{i \in I}$ of C such that

$$C = \coprod_{i \in I} C_{(i)}, \quad \varepsilon(C_{(n)}) = 0 \qquad (n \neq 0),$$

$$\Delta C_{(n)} \subset \sum_{i=0}^{n} C_{(i)} \otimes C_{(n-i)} \qquad (n \in I),$$

then C is called a **graded k-coalgebra**. If we set $C_n = \coprod_{i \leqq n} C_{(i)}$, then $\{C_n\}_{n \in I}$ becomes a filtration on C, so we obtain a filtered k-coalgebra. If in particular, $C_{(0)} \cong k$ (and hence there is only one group-like element in C) and $C_{(1)} = P(C)$, C is said to be a **strictly graded k-coalgebra**. A strictly graded k-coalgebra is a pointed irreducible k-coalgebra. If C is a filtered k-coalgebra with filtration $\{C_i\}_{i \in I}$, then setting $C_{(i)} = C_i / C_{i-1}$ $(i \in I$; where we set $C_{-1} = \{0\})$, we automatically obtain a graded k-coalgebra gr $C = \coprod_{i \in I} C_{(i)}$. This is called the **associated graded k-coalgebra of the filtered k-coalgebra** C.

Let B be a k-bialgebra. If there exists a family of k-linear subspaces $\{B_i\}_{i \in I}$ of B such that $\{B_i\}_{i \in I}$ is simultaneously a k-algebra filtration and a k-coalgebra filtration, then B is called a **filtered k-bialgebra**. Moreover, if there exists a family of k-linear subspaces $\{B_{(i)}\}_{i \in I}$ of B with respect to which B is a graded k-algebra as a k-algebra, and is a graded k-coalgebra when regarded as a k-coalgebra, then B is said to be a **graded k-bialgebra**. Specifically, given a k-Hopf algebra B, suppose that $\{B_i\}_{i\in I}$ is a k-bialgebra filtration on B which satisfies $S(B_i)\subset B_i$ $(i\in I)$. Then B is called a **filtered k-Hopf algebra**.

Now let C, D be filtered k-coalgebras with filtrations $\{C_i\}_{i\in I}$ and $\{D_i\}_{i\in I}$ respectively. A k-coalgebra morphism $\sigma: C\to D$ such that $\sigma(C_i)\subset D_i$ $(i\in I)$ is said to be a **filtered** k-coalgebra morphism. Filtered **k-bialgebra morphisms** and **filtered k-Hopf algebra morphisms** may be defined similarly.

EXAMPLE 2.10 The tensor k-algebra $T(V)$ and the symmetric k-algebra $S(V)$ over a k-linear space V (cf. Chapter 1, §2.2) together with the ith homogeneous component $T_{(i)}(V)=V\otimes\cdots\otimes V$ (i-fold tensor product) and its image $S_{(i)}(V)$ in $S(V)$ respectively are graded k-algebras. Now, if V_1, V_2 are two k-linear spaces, then $S(V_1\oplus V_2)$ is isomorphic to $S(V_1)\otimes S(V_2)$. Let

$$\Delta: S(V)\to S(V)\otimes S(V)\cong S(V\oplus V),$$
$$\varepsilon: S(V)\to S(0)=k,$$
$$S: S(V)\to S(V)$$

respectively be k-algebra morphisms induced by the k-linear maps

$$x\mapsto x+x \quad (V\to V\oplus V),$$
$$x\mapsto 0 \quad (V\to k),$$

and

$$x\mapsto -x \quad (V\to V).$$

These maps define a graded k-Hopf algebra structure on $S(V)$. Further, the universal enveloping k-bialgebra $U(L)$ of a k-Lie algebra L (cf. Example 2.5) together with the image $U_i(L)$ of $T_i(V)$ in $U(L)$ admits a filtered k-bialgebra structure. In particular, the k-bialgebra $C=\coprod_{i\in I} kc_i$ in Example 2.6 becomes a graded k-bialgebra when we set $C_{(i)}=kc_i$ $(i\in I)$.

Given a k-coalgebra C, let D be a k-subcoalgebra, and set $D_i=\sqcap^{i+1}D$ $(i\in I)$ and $D_\infty=\bigcup_{i\in I} D_i$. The family $\{D_i\}_{i\in I}$ of k-submodules of D is a filtration on D_∞ by the next theorem to be introduced, so that

D_∞ becomes a filtered k-coalgebra. If in particular D is conilpotent, i.e. if $D_\infty = C$, then $\{D_i\}_{i\in I}$ is a filtration on C and C becomes a filtered k-coalgebra. If we let D be the coradical corad C of C, then D is conilpotent, and we obtain a filtration $\{D_i\}_{i\in I}$ on C. We call this the **coradical filtration** on C.

THEOREM 2.4.1 Given a k-coalgebra C, let D be a k-subcoalgebra. If we set $D_i = \sqcap^{i+1} D$, then $\Delta D_n \subset \sum_{i=0}^{n} D_i \otimes D_{n-i}$.

Proof Since $\sqcap^{n+1} D = (\sqcap^i D) \sqcap (\sqcap^{n-i+1} D)$, we have $\Delta(\sqcap^{n+1} D) \subseteq C \otimes (\sqcap^{n-i+1} D) + (\sqcap^i D) \otimes C$ for each i $(1 \leqq i \leqq n)$. The fact that $\sqcap^{n+1} D$ is a k-subcoalgebra implies that the above inclusion holds even for $i = 0$ and $i = n+1$. Thus it suffices to show that given a k-linear space V and a family $\{V_i\}_{i\in I}$ of k-linear subspaces such that $V_0 \subset V_1 \subset V_2 \subset \cdots$, the equality

$$\bigcap_{i=0}^{n} (V \otimes V_{n-i} + V_i \otimes V) = \sum_{i=1}^{n} V_i \otimes V_{n+1-i}$$

holds. We can set $V = \bigcup_{i\in I} V_i$. For each V_j, take one complementary subspace of V_{j-1} in V_j and denote it by W_j. Then

$$V_1 = W_1, \quad V_j = \coprod_{i=1}^{j} W_i, \quad V \otimes V = \coprod_{i,j} W_i \otimes W_j.$$

and

$$V \otimes V_{n-i} + V_i \otimes V = \coprod_{\substack{r \quad i \text{ or} \\ s \leqq n-i}} (W_r \otimes W_s).$$

Thus

$$\bigcap_{i=0}^{n} \coprod_{\substack{r<i \text{ or} \\ s \leqq n-i}} (W_r \otimes W_s) = \coprod_{r+s\leqq n+1} W_r \otimes W_s$$

$$= \sum_{i=1}^{n} V_i \otimes V_{n+1-i}.$$

4.2 Cofree k-coalgebras

Next we define cofree k-coalgebras over a k-linear space V and cocommutative cofree k-coalgebras over V which are respectively dual to tensor k-algebras and symmetric k-algebras over V. When a pair (C, π) consisting of a k-coalgebra C and a k-linear map $\pi : C \to V$ satisfies the following property (P), (C, π) or simply C is called a **cofree k-coalgebra** over V.

(P) Given an arbitrary k-coalgebra D and a k-linear map $f : D \to V$, there exists a unique k-coalgebra morphism $F : D \to C$ such that $f = \pi \circ F$. Namely, the map

$$\mathbf{Cog}_k\,(D, C) \to \mathbf{Mod}_k\,(D, V)$$

given by $F \mapsto \pi \circ F = f$ is bijective.

If such a pair (C, π) exists, then it is unique up to isomorphism. A cocommutative k-coalgebra (C, π) which is obtained by replacing C and D in the above definition by cocommutative k-coalgebras is called a **cocommutative cofree k-coalgebra** over V.

THEOREM 2.4.2 There exists a cofree k-coalgebra over a k-linear space V.

Proof Let $T(V^*)$ be the tensor k-algebra over the dual k-linear space V^* of V, and let $i : V^* \to T(V^*)$ be the canonical embedding. Let π be the following composite map:

$$T(V^*)^\circ \to T(V^*)^* \xrightarrow{i^*} V^{**}.$$

Then $(T(V^*)^\circ, \pi)$ is a cofree k-coalgebra over V^{**}. In fact, if X, Y are k-linear spaces, then since $\mathbf{Mod}_k(X, Y^*) \cong \mathbf{Mod}_k(Y, X^*)$, for any k-coalgebra D, we have

$$\begin{aligned}\mathbf{Mod}_k(D, V^{**}) \cong \mathbf{Mod}_k(V^*, D^*) &\cong \mathbf{Alg}_k(T(V^*), D^*) \\ &\cong \mathbf{Cog}_k(D, T(V^*)^\circ)\end{aligned}$$

(cf. Chapter 1, §2.2, and Theorem 2.3.14.). Moreover, the morphism $F \in \mathbf{Cog}_k\,(D, T(V^*)^\circ)$ which corresponds to $f \in \mathbf{Mod}_k\,(D, V^{**})$ satisfies the condition $f = \pi \circ F$. Hence the proof of this theorem would be complete if we verify the following lemma.

LEMMA 2.4.3 Let (C, π) be the cofree k-coalgebra over a k-linear space V and let $W \subset V$ be a k-linear subspace of V. If we let D be the sum of all k-subcoalgebras E of C which satisfy $\pi(E) \subset W$ and if we let ρ be the restriction of π to D, then (D, ρ) is the cofree k-coalgebra over W.

Proof Let M be a k-coalgebra and $f: M \to W$ a k-linear map. Then f can be regarded as a k-linear map $M \to V$, so there exists a k-coalgebra morphism $F: M \to C$ satisfying $f = \pi \circ F$. Since $\pi(F(M)) \subset W$, we get $F(M) \subset D$, so that F is a k-coalgebra morphism from M to D.

THEOREM 2.4.4 There exists a cocommutative cofree k-coalgebra over a k-linear space V. We denote it by $C(V)$.

Proof Let $(\bar{C}, \bar{\pi})$ be the cofree k-coalgebra over V. If we let C be the sum of all cocommutative k-subcoalgebras of $\bar{C}$ and let π be the restriction of $\bar{\pi}$ to C, it can be shown that (C, π) is a cocommutative cofree k-coalgebra over V.

Now let V_1, V_2 be k-linear spaces and let (C_i, π_i) be cocommutative cofree k-coalgebras over V_i $(i = 1, 2)$. If we define a k-linear map $\rho: C_1 \otimes C_2 \to V_1 \otimes V_2$ by

$$\rho(c \otimes d) = \pi_1(c)\varepsilon(d) + \pi_2(d)\varepsilon(c), \quad c \otimes d \in C_1 \otimes C_2,$$

then $(C_1 \otimes C_2, \rho)$ becomes a cocommutative cofree k-coalgebra over $V_1 \oplus V_2$. Defining k-linear maps

$$\mu: C(V) \otimes C(V) \cong C(V \oplus V) \to C(V),$$
$$\eta: C(\{0\}) \cong k \to C(V),$$
$$S: C(V) \to C(V)$$

to be the k-coalgebra morphisms respectively induced by

$$(v, w) \mapsto v + w \ (V \oplus V \to V),$$
$$0 \mapsto 0 \ (\{0\} \to V),$$

and

$$v \mapsto -v \ (V \to V),$$

then these maps determine a k-algebra structure on $C(V)$. The

k-coalgebra $C(V)$ together with this k-algebra structure becomes a commutative as well as cocommutative k-Hopf algebra.

EXERCISE 2.6 Prove that the maps μ, η, S above are indeed k-algebra structure maps of $C(V)$.

Remark Given a k-algebra A, we can employ the same method to define a k-bialgebra structure on the cocommutative cofree k-coalgebra $C(A)$ over the k-linear space A in such a way that $\pi : C(A) \to A$ becomes a k-algebra morphism.

4.3 Irreducible k-coalgebras and irreducible components of k-coalgebras

As already defined, a k-coalgebra which has only one simple k-subcoalgebra is said to be 'irreducible'. This is to say that the intersection of two arbitrary non-zero k-subcoalgebras is not 0.

THEOREM 2.4.5 Given a k-coalgebra C, let $C = \sum_{\lambda \in \Lambda} C_\lambda$, where each C_λ $(\lambda \in \Lambda)$ is a family of k-subcoalgebras of C.

(i) A simple k-subcoalgebra of C is contained in some C_λ.

(ii) C is irreducible $\Leftrightarrow$ each C_λ is irreducible and $\bigcap_\lambda C_\lambda \neq \{0\}$.

(iii) C is pointed $\Leftrightarrow$ each C_λ is pointed.

Proof Let D be a simple k-subcoalgebra of C. Since $\dim D < \infty$, we can pick a finite family $\{C_{\lambda_i}\}_{1 \leqq i \leqq n}$ such that $D \subset \sum_{i=1}^{n} C_{\lambda_i}$. By induction on n, it suffices to show that $D \subset C_\lambda + C_{\lambda'}$ implies either $D \subset C_\lambda$ or $D \subset C_{\lambda'}$. If $D \not\subset C_\lambda$, then, since D is simple, we have $D \cap C_\lambda = \{0\}$. Thus there exists $f \in C^*$ such that $f|_D = \varepsilon_D$ and $f|_{C_\lambda} = 0$. On the other hand, since $D \subset C_\lambda + C_{\lambda'}$, it follows that $\Delta(D) \subset C_\lambda \otimes C_\lambda + C_{\lambda'} \otimes C_{\lambda'}$. Thus, for $d \in D$, we get

$$\sum_{(d)} d_{(1)} \langle f, d_{(2)} \rangle = \sum_{(d)} d_{(1)} \varepsilon(d_{(2)}) = d \in C_{\lambda'},$$

which implies that $D \subset C_{\lambda'}$.

(ii) Let C be irreducible and denote by R its unique simple k-subcoalgebra. By (i), $R \subset C_\lambda$ for all $\lambda \in \Lambda$. Thus C_λ is irreducible and $\bigcap_\lambda C_\lambda \neq \{0\}$. Conversely, suppose that each C_λ is irreducible and $\bigcap_\lambda C_\lambda \neq \{0\}$. If we let R be the simple k-subcoalgebra contained in $\bigcap_\lambda C_\lambda$, then R is the only simple k-subcoalgebra of each C_λ. Since a simple k-subcoalgebra of C is contained in some C_λ by (i), it follows that R is the only simple k-subcoalgebra of C.

The proof of (iii) follows from (i).

COROLLARY 2.4.6 Let $C = \sum_{\lambda \in \Lambda} C_\lambda$ be as in the above theorem. C is pointed irreducible $\Leftrightarrow$ each C_λ is pointed irreducible and $\bigcap_{\lambda \in \Lambda} C_\lambda \neq \{0\}$.

If C is a k-coalgebra, then a maximal irreducible k-subcoalgebra of C is called an **irreducible component** (IC) of C. An irreducible component that is pointed is called a **pointed irreducible component** (PIC).

THEOREM 2.4.7 Let C be a k-coalgebra.

(i) An arbitrary irreducible k-subcoalgebra E of C is contained in an irreducible component of C.

(ii) A sum of distinct irreducible components is a direct sum.

(iii) If C is cocommutative, then C is the direct sum of its irreducible components.

Proof (i) Let F be the sum of irreducible k-subcoalgebras containing E. Then F is an irreducible k-subcoalgebra (cf. Theorem 2.4.5 (ii)). By definition, F is the unique maximal irreducible k-subcoalgebra containing E.

(ii) Let $\{C_\lambda\}_{\lambda \in \Lambda}$ be a family of distinct irreducible components. Suppose $C_\beta \cap \sum_{\lambda \neq \beta} C_\lambda \neq \{0\}$. If R is the unique simple k-subcoalgebra of C_β, then $R \subset \sum_{\lambda \neq \beta} C_\lambda$. Theorem 2.4.5 (i) implies that there exists $\gamma \neq \beta$ such that $R \subset C_\gamma$. Thus $C_\beta \cap C_\gamma \neq \{0\}$. Hence $C_\beta + C_\gamma$ is irreducible (cf. Theorem 2.4.5 (ii)), and, since $C_\beta + C_\gamma$ are irreducible components,

$C_\beta = C_\gamma = C_\beta + C_\gamma$. This contradicts the choice of $\{C_\lambda\}_{\lambda \in \Lambda}$. Therefore $\sum_{\lambda \in \Lambda} C_\lambda$ is a direct sum.

(iii) By (i) and (ii), it suffices to show that C is a sum of irreducible k-subcoalgebras. For $x \in C$, let C_x be the k-subcoalgebra of C generated by x. We need only prove (iii) for C_x. Now, since C_x is finite dimensional, we may assume that C is finite dimensional. Then C^* becomes a finite dimensional commutative k-algebra, and C^* can be expressed as a direct sum $C^* = A_1 \oplus \cdots \oplus A_n$ where the A_i are local k-algebras. Therefore

$$C \cong C^{**} \cong A_1{}^* \oplus \cdots \oplus A_n{}^*$$

and each $A_i{}^*$ is an irreducible k-coalgebra.

COROLLARY 2.4.8 Let C be a k-coalgebra.

(i) A sum of distinct simple k-subcoalgebras is a direct sum.

(ii) C is irreducible $\Leftrightarrow$ any element of C is contained in some irreducible k-subcoalgebra.

(iii) C is pointed irreducible $\Leftrightarrow$ any element of C is contained in some pointed irreducible k-subcoalgebra.

Proof (i) Since a simple k-subcoalgebra is irreducible, it is contained in some irreducible component. Moreover, distinct simple k-subcoalgebras are contained in distinct irreducible components. Since the sum of distinct irreducible components is a direct sum, a sum of distinct simple k-subcoalgebras must also be a direct sum.

(ii) The direction ($\Rightarrow$) is straightforward. As for ($\Leftarrow$), suppose that C is not irreducible. Then C contains two simple k-subcoalgebras D and E. By (i), $D + E$ is a direct sum. Let $d \in D$ and $e \in E$ be non-zero elements. If we let F be the k-subcoalgebra of C generated by $d + e \neq 0$, then, by hypothesis, F is contained in an irreducible k-subcoalgebra. Therefore F must be irreducible. If R is the unique simple k-subcoalgebra of F, then either $R = D$ or $R = E$ (cf. Theorem 2.4.5 (i)). If $R = D$, then d, $d + e \in F$, so that $e \in F$. Thus we obtain D, $E \subseteq F$, which contradicts the fact that F is irreducible. The proof of (iii) follows readily from (ii).

LEMMA 2.4.9 Let M, N be k-subcoalgebras of a k-coalgebra C, and let L be a simple k-subcoalgebra of C. If $L \subset N \sqcap M$, then $L \subset M$ or $L \subset N$.

Proof Suppose that $L \not\subset M$. Since L is simple, $L \cap M = \{0\}$. Thus there exists $f \in C^*$ such that $f|_L = \varepsilon_L$, $f|_M = 0$. On the other hand, if $l \in L \subset N \sqcap M$, then $\Delta(l) \in C \otimes M + N \otimes C$. Therefore

$$l = \sum_{(l)} l_{(1)}\varepsilon(l_{(2)}) = \sum l_{(1)}\langle f, l_{(2)}\rangle \subset N\langle f, C\rangle \subset N,$$

in other words, $L \subset N$.

LEMMA 2.4.10 Let D be an irreducible k-subcoalgebra of a k-coalgebra C and set $D_n = \sqcap^{n+1} D$. Then $D_\infty = \bigcup_{n=0}^{\infty} D_n$ is an irreducible k-coalgebra and D_∞ is the irreducible component of C containing D.

Proof Let L be a simple k-subcoalgebra of D_∞. Since L is finite dimensional, there is a natural number n for which $L \subset D_n$. Applying Lemma 2.4.9 to $D_n = D \sqcap D_{n-1}$, we obtain $L \subset D$ or $L \subset D_{n-1}$. Induction on n shows that $L \subset D$. Since D is irreducible, L is the unique simple k-subcoalgebra of D. Thus D_∞ is also irreducible. Let E be the irreducible component of C containing D. Then, since D_∞ is irreducible, $D_\infty \subset E$. On the other hand, if L is the unique simple k-subcoalgebra of D (and hence of E), we have $L = \operatorname{corad} E$. Denoting the operation $\sqcap$ in E by $\sqcap_E$, we obtain $E = \bigcup_{n=0}^{\infty} \sqcap_E{}^n L \subset \bigcup_{n=0}^{\infty} \sqcap_E{}^n D \subset D_\infty$ thanks to Theorem 2.3.9 (ii). Therefore $E = D_\infty$.

THEOREM 2.4.11 Let C, D be k-coalgebras and let $f: C \to D$ be a k-coalgebra morphism. If C is pointed irreducible, then f is injective $\Leftrightarrow f|_{P(C)}$ is injective; namely $\operatorname{Ker} f \cap P(C) = \{0\}$.

We need some lemmas in order to prove this theorem. If g is the only group-like element of C, then $C_0 = kg$ is the coradical of C. Let $\{C_n\}_{n \in I}$ $(I = \{0, 1, 2, \ldots\}, C_n = \sqcap^{n+1} kg)$ be the coradical filtration on C.

LEMMA 2.4.12 If C is a pointed irreducible k-coalgebra, then

$$C_1 = kg \sqcap kg = kg \oplus P(C).$$

Proof Since $P(C) \subset \text{Ker}\,\varepsilon$ and $\varepsilon(g) = 1$, it follows that $kg + P(C)$ is a direct sum. For $c \in C_1$, set $d = c - \varepsilon(c)g$. Then $d \in C_1 \cap \text{Ker}\,\varepsilon$. We will show that $d \in P(C)$. Since $d \in C_1$, $\Delta d \in C \otimes kg + kg \otimes C$. Thus, setting $\Delta d = d_1 \otimes g + g \otimes d_2$, we obtain $\varepsilon(d) = (\varepsilon \otimes \varepsilon)\Delta d = \varepsilon(d_1) + \varepsilon(d_2) = 0$. In turn, since $(1 \otimes \varepsilon)\Delta d = d = (\varepsilon \otimes 1)\Delta d$, it follows that $d_1 \varepsilon(g) + g\varepsilon(d_2) = d = \varepsilon(d_1)g + \varepsilon(g)d_2$. Thus

$$\begin{aligned}\Delta d = d_1 \otimes g + g \otimes d_2 &= (d - g\varepsilon(d_2)) \otimes g + g \otimes (d - g\varepsilon(d_1))\\ &= d \otimes g + g \otimes d - (\varepsilon(d_1) + \varepsilon(d_2))g \otimes g\\ &= d \otimes g + g \otimes d,\end{aligned}$$

in other words, $d \in P(C)$ and $c = \varepsilon(c)g + d \in kg \oplus P(C)$.

LEMMA 2.4.13 Set $C_n^+ = C_n \cap \text{Ker}\,\varepsilon$. If $c \in C_n^+$, then $\Delta c = g \otimes c + c \otimes g + y$, $y \in \sum_{i=1}^{n-1} C_i^+ \otimes C_{n-i}^+ \subset C_{n-1}^+ \otimes C_{n-1}^+$.

Proof Writing $y = \Delta c - g \otimes c - c \otimes g$, it follows that $(1 \otimes \varepsilon)y = (1 \otimes \varepsilon)\Delta c - c\varepsilon(g) = c - c = 0$. Thus $y \in C \otimes (\text{Ker}\,\varepsilon)$. Similarly, $y \in (\text{Ker}\,\varepsilon) \otimes C$, and hence $y \in (\text{Ker}\,\varepsilon) \otimes (\text{Ker}\,\varepsilon)$. Meanwhile, $\Delta c \in \sum_{i=0}^{n} C_i \otimes C_{n-1}$, so $y \in \left(\sum_{i=0}^{n} C_i \otimes C_{n-i}\right) \cap (\text{Ker}\,\varepsilon \otimes \text{Ker}\,\varepsilon) = \sum_{i=0}^{n} C_i^+ \otimes C_{n-i}^+$. Since $C_0^+ = \{0\}$, we obtain $y \in \sum_{i=1}^{n-1} C_i^+ \otimes C_{n-1}^+ \subset C_{n-1}^+ \otimes C_{n-1}^+$.

Proof of Theorem **2.4.11** The implication $(\Rightarrow)$ is obvious. $(\Leftarrow)$ It suffices to show that $f|_{C_i}$ is injective for each i. The fact that f is a k-coalgebra morphism implies that $\varepsilon f(g) = \varepsilon(g) = 1$ and $\varepsilon f(P(C)) = \varepsilon(P(C)) = 0$. Therefore $f(kg) + f(P(C))$ is a direct sum. By Lemma 2.4.12, $C_1 = kg \oplus P(C)$. Thus, if $f|_{P(C)}$ is injective, then $f|_{C_1}$ must also be injective. By induction on n, it suffices to show that $f|_{C_n}$ injective implies $f|_{C_{n+1}}$ injective. If $x \in C_{n+1}^+ = C_{n+1} \cap \text{Ker}\,\varepsilon$, then, by Lemma 2.4.13, $\Delta x = g \otimes x + x \otimes g + y$ for $y \in C_n^+ \otimes C_n^+$. Thus $\Delta f(x)$

$= (f \otimes f)\Delta(x) = f(g) \otimes f(x) + f(x) \otimes f(g) + (f \otimes f)(y)$. Now, if $x \in \operatorname{Ker} f$, then $(f \otimes f)(y) = 0$, and since $f|_{C_n}$ is injective, $y = 0$. Thus $x \in P(C) \cap \operatorname{Ker} f = \{0\}$. This shows that $f|_{C_{n+1}}$ is also injective.

COROLLARY 2.4.14 Given a coideal I of a pointed irreducible k-coalgebra C, $I \cap P(C) = \{0\}$ implies $I = \{0\}$.

Proof Apply Theorem 2.4.11 to the canonical k-coalgebra morphism $\pi : C \to C/I$. By hypothesis, $\pi|_{P(C)}$ is injective, so π is also injective. Thus $\operatorname{Ker} \pi = I = \{0\}$.

COROLLARY 2.4.15 Given k-coalgebras C and D, let $f, g : C \to D$ be k-coalgebra morphisms. If C is pointed irreducible, then

$$f = g \Leftrightarrow \operatorname{Im}(f - g) \cap P(D) = \{0\}.$$

COROLLARY 2.4.16 Given a pointed irreducible k-coalgebra C, let C^* be its dual k-algebra. If a k-subalgebra A of C^* satisfies $A^{\perp(C)} \cap P(C) = \{0\}$, then A is dense in C^*. Namely, $A^{\perp(C)} = \{0\}$.

COROLLARY 2.4.17 Let $D = \coprod_{i \in I} D_{(i)}$ be a strictly graded k-coalgebra and let $\pi : D \to D_{(1)}$ be the canonical projection.

(i) If f, g are k-coalgebra morphisms from C to D, then $\pi f = \pi g$ implies $f = g$.

(ii) Given a graded k-coalgebra C, if $f, g : C \to D$ are graded k-coalgebra morphisms, then $f|_{C_{(1)}} = g|_{C_{(1)}}$ implies $f = g$.

THEOREM 2.4.18 Given graded k-bialgebras C and D, suppose that $C_{(0)} = D_{(0)} = k$ and that D is a strictly graded k-bialgebra. Then a graded k-coalgebra morphism f from C to D is a graded k-bialgebra morphism.

Proof Since $f(C_{(0)}) \subset D_{(0)}$ and 1 is the only group-like element of D, we get $f(1) = 1$. On the other hand, if $(C \otimes C)_{(n)} = \sum_{i+j=n} C_{(i)} \otimes C_{(j)}$, then $C \otimes C$ becomes a graded k-coalgebra. Moreover, if $x \in (C \otimes C)_{(1)} = C_{(1)} \otimes k + k \otimes C_{(1)}$, then x can be written in the form

$x = 1 \otimes c + d \otimes 1$. Thus $f \circ \mu_C(x) = f(c) + f(d) = \mu_D \circ (f \otimes f)(x)$. Therefore $f \circ \mu_C$ coincides with $\mu_D(f \otimes f)$ on $(C \otimes C)_{(1)}$. By Corollary 2.4.17, we get $f \circ \mu_C = \mu_D \circ (f \otimes f)$. Thus f is a k-bialgebra morphism.

COROLLARY 2.4.19 Given pointed irreducible k-bialgebras C and D, let $f : C \to D$ be a filtered k-coalgebra morphism with respect to the coradical filtrations on C and D. Then $\operatorname{gr} f : \operatorname{gr} C \to \operatorname{gr} D$ is a graded k-bialgebra morphism.

THEOREM 2.4.20 Given k-coalgebras C and E, where C is irreducible, let R be the unique simple k-subcoalgebra of C. If $f : C \to E$ is a surjective k-coalgebra morphism, then

(i) If $F \neq \{0\}$ is a k-subcoalgebra of E, then $F \cap f(R) \neq \{0\}$.

(ii) $f(R)$ contains all simple k-subcoalgebras of E.

(iii) E is irreducible $\Leftrightarrow f(R)$ is irreducible.

(iv) If R is cocommutative, then E is irreducible and $f(R)$ is its unique simple k-subcoalgebra.

Proof (i) Given $d \neq 0$ in F, pick $c \in C$ such that $f(c) = d$. Let Y be the k-subcoalgebra of C generated by c. Then Y is finite dimensional irreducible, R is its unique simple k-subcoalgebra, and $F \cap f(Y) \neq \{0\}$ is a k-subcoalgebra of $f(Y)$. Thus, by replacing C by Y, E by $f(Y)$, and F by $f(Y)$, we may assume that C and E are finite dimensional. The dual k-algebra morphism $f^* : E^* \to C^*$ of $f : C \to E$ is injective, and $M = R^\perp$ is the unique maximal ideal of C^*. Thus M is a nilpotent ideal and $N = M \cap E^*$ is also a nilpotent ideal. Hence $N \subset \operatorname{rad} E^*$, and any maximal ideal of E^* contains N. Thus any simple k-subcoalgebra of E is contained in $N^{\perp(E)}$. On the other hand, $N^{\perp(E)} = (E^* \cap R^\perp)^{\perp(E)} = f(R)$ and F contains a simple k-subcoalgebra, so we obtain $F \cap f(R) \neq \{0\}$. The implications (i) $\Rightarrow$ (ii) $\Rightarrow$ (iii) are clear.

(iv) It suffices to show that $f(R)$ is simple. Since R is a finite dimensional cocommutative k-coalgebra, R^* is a finite dimensional commutative simple k-algebra. Thus R^* is an extension field of k of finite degree, and all k-subalgebras of R^* are simple. Therefore, all factor k-coalgebras of R are simple, so in particular $f(R)$ is also simple.

COROLLARY 2.4.21 The image of a pointed irreducible k-coalgebra under a k-coalgebra morphism is pointed irreducible.

THEOREM 2.4.22 Given irreducible k-coalgebras C and E let R and S respectively be the unique simple k-subcoalgebras of C and E.

(i) If $X \neq \{0\}$ is a k-subcoalgebra of $C \otimes E$, then $X \cap (R \otimes S) \neq \{0\}$.

(ii) $R \otimes S$ contains all simple k-subcoalgebras of $C \otimes E$.

(iii) $C \otimes E$ is irreducible $\Leftrightarrow R \otimes S$ is irreducible.

(iv) If C is pointed irreducible, then $R \otimes S$ is a simple k-subcoalgebra of $C \otimes E$. Thus $C \otimes E$ is pointed irreducible.

Proof Since (i)$\Rightarrow$(ii)$\Rightarrow$(iii)$\Rightarrow$(iv) is clear, we need only prove (i). Let $\sum_{i=1}^{n} c_i \otimes e_i \neq 0$ be an element of X. Let C', E' be k-subcoalgebras of C, E respectively generated by $\{c_i\}_{1 \leqq i \leqq n}$ and $\{e_i\}_{1 \leqq i \leqq n}$. Then $R \subset C'$, $S \subset E'$, C', E' are irreducible and $X \cap (C' \otimes E') \neq \{0\}$. Therefore, by replacing C by C' and E by E', we may assume that C and E are finite dimensional. Thus we obtain $(C \otimes E)^* \cong C^* \otimes E^*$ as k-algebras, and $M = R^{\perp}, N = S^{\perp}$ are respectively the unique maximal ideals of C^* and E^*. Since both M and N are nilpotent ideals, $P = M \otimes E^* + C^* \otimes N$ is a nilpotent ideal of $C^* \otimes E^*$. Therefore we have $P \subset \operatorname{rad}(C^* \otimes E^*)$. Hence all maximal ideals of $C^* \otimes E^*$ contain P. Thus all simple k-subcoalgebras of $C \otimes E$ are contained in $P^{\perp(C \otimes E)} = R \otimes S$. In particular, $X \cap (R \otimes S) \neq \{0\}$.

4.4 Irreducible k-bialgebras and the irreducible components of k-bialgebras

A k-bialgebra which is irreducible as a k-coalgebra is said to be an **irreducible k-bialgebra**. Here, we will discuss some properties of irreducible k-bialgebras and the decomposition of k-bialgebras into their irreducible components.

LEMMA 2.4.23 Given a k-coalgebra C, let D be a k-subcoalgebra. Then $\sqcap^n D \otimes \sqcap^m D \subset \sqcap^{n+m-1}(D \otimes D)$.

Proof We proceed by induction on $n+m$. The case $n+m \leqq 2$ is clear. So suppose that $n+m \geqq 2$ and set $\sqcap^{n+1}D = D_n$. By Theorem 2.4.1 and the inductive hypothesis, we have

$$\begin{aligned}\Delta(D_{n-1}\otimes D_{m-1}) &\subset \Delta D_{n-1}\otimes\Delta D_{m-1}\\ &\subset \left(\sum_{i=0}^{n-1} D_i\otimes D_{n-i+1}\right)\otimes\left(\sum_{j=0}^{m-1} D_j\otimes D_{m-j+1}\right)\\ &\subset (D\otimes D)\otimes(C\otimes C) + C\otimes C\otimes(\sqcap^{n+m-2}D\otimes D).\end{aligned}$$

Therefore $D_{n-1}\otimes D_{m-1}\subset \sqcap^{n+m-1}(D\otimes D)$.

THEOREM 2.4.24 Let B be an irreducible k-bialgebra.

(i) B is pointed, and its coradical is k.

(ii) If we set $B_i = \sqcap^{i+1}k$, then $B = \bigcup_{i=0}^{\infty} B_i$ and $B_iB_j \subset B_{i+j}$.

(iii) B is a k-Hopf algebra, and $S(B_i)\subset B_i$.

Therefore an irreducible k-bialgebra is a filtered k-Hopf algebra with respect to its coradical filtration.

Proof (i) Since B is a k-bialgebra, kl is a simple k-subcoalgebra, and thus B is a pointed k-coalgebra.

(ii) Since the multiplication map $\mu: B\otimes B \to B$ is a k-coalgebra morphism, $\mu((B\otimes B)_{i+j})\subset B_{i+j}$ (cf. Exercise 2.4 (iii)). Now Lemma 2.4.23 implies that $B_i\otimes B_j \subset (B\otimes B)_{i+j}$. Therefore $B_iB_j \subseteq B_{i+j}$. By Theorem 2.3.9 (ii), it follows readily that $B = \bigcup_{i=0}^{\infty} B_i$.

(iii) The next lemma shows that B has antipode S. Moreover, B^{op} is pointed and $(B_i)^{\text{op}} = (B^{\text{op}})_i$. On the other hand, since S is an anti-k-coalgebra morphism, we have $S(B_i) = S(B^{\text{op}})_i \subset B_i$.

LEMMA 2.4.25 Given a pointed irreducible k-coalgebra C, let g be the unique group-like element of C. If A is a k-algebra, then $f\in \mathbf{Mod}_k(C, A)$ is an invertible element with respect to convolution $\Leftrightarrow f(g)$ is invertible in A. In particular, if B is an irreducible k-bialgebra, then there exists an inverse S of 1_B.

Proof If f has an inverse h, then

$$f(g)h(g) = (f*h)(g) = \varepsilon(g)1 = (h*f)(g) = h(g)f(g).$$

Since $\varepsilon(g) = 1$, $h(g)$ is the inverse of $f(g)$.

Conversely, assume $a = f(g)$ is an invertible element of A and let b be the inverse of a. Let $x \in \mathbf{Mod}_k(C, A)$ be defined by $x(c) = f(c) - \varepsilon(c)a$ for $c \in C$. Since $x(g) = 0$, $R = kg \subset \operatorname{Ker} x$. Set $C_n = \sqcap^{n+1} R$. We will show that $C_{n-1} \subset \operatorname{Ker} x^n$ by induction on n. When $n = 1$, $R \subset \operatorname{Ker} x$ holds. Suppose that the inclusion holds up to $n - 1$. Since $x^n = x * x^{n-1} = \mu(x \otimes x^{n-1})\Delta$, we have

$$\begin{aligned}\operatorname{Ker} x^n &\supset \{z \in C;\ \Delta z \in R \otimes C + C \otimes \operatorname{Ker} x^{n-1}\}\\ &= R \sqcap (\operatorname{Ker} x^{n-1}).\end{aligned}$$

By the inductive hypothesis, we have $\operatorname{Ker} x^{n-1} \supset C_{n-2}$. Thus, indeed, $\operatorname{Ker} x^n \supset C_{n-1}$.

From the fact that $C_n \subset C_{n+1}$ and $C = \bigcup_{n=0}^{\infty} C_n$, it follows that if $c \in C$, there exists a natural number n for which $c \in C_{n-1}$. Thus $c \in \operatorname{Ker} x^n$. Hence we can define $h \in \mathbf{Mod}_k(C, A)$ by

$$h(c) = b\varepsilon(c) - b^2 x(c) + b^3 x^2(c) - \cdots, \quad c \in C.$$

For this h,

$$f * h = (a\varepsilon + x)(b\varepsilon - b^2 x + b^3 x^2 - \cdots) = \eta \circ \varepsilon.$$

Similarly, $h * f = \eta \circ \varepsilon$. Therefore h is the inverse of f.

If, in particular, B is a pointed irreducible k-bialgebra, then $1_B \in \mathbf{Mod}_k(B, B)$, and $1_B(g) = g$ is an invertible element of B. Thus the inverse S of 1_B exists.

COROLLARY 2.4.26 Given a k-coalgebra C which is the direct sum of its pointed irreducible k-subcoalgebras $C_\lambda (\lambda \in \Lambda)$, let g_λ be the unique group-like element of C_λ, and let A be a k-algebra. Then $f \in \mathbf{Mod}_k(C, A)$ is invertible $\Leftrightarrow f(g_\lambda)$ is an invertible element of A for each λ.

THEOREM 2.4.27 Let B be a k-bialgebra and let $G(B)$ be the semigroup consisting of all group-like elements of B. For $g \in G(B)$, let B_g be the irreducible component of B which contains kg.

(i) If $g, h \in G(B)$, then $B_g B_h \subset B_{gh}$. Therefore, if 1 is the identity element of $G(B)$, then B_1 is a k-sub-bialgebra of B.

(ii) If B is a k-Hopf algebra, then B_1 is a k-sub-Hopf algebra.

Proof (i) Since kg is a simple k-subcoalgebra of B_g, it follows that B_g is pointed irreducible. By Theorem 2.4.22 (iv), $B_g \otimes B_h$ is pointed irreducible, and has a one-dimensional simple k-subcoalgebra $k(g \otimes h)$. Since the multiplication map μ is a k-coalgebra morphism, Corollary 2.4.21 shows that $B_g B_h$ is pointed irreducible and contains gh. Therefore $B_g B_h \subset B_{gh}$,

(ii) Let S be the antipode of B. Then $S:(B_1)^{\mathrm{op}} \to B$ is a k-coalgebra morphism, and, since $(B_1)^{\mathrm{op}}$ is pointed irreducible, $S((B_1)^{\mathrm{op}}) \subset B_1$. Thus $S(B_1) \subseteq B_1$.

COROLLARY 2.4.28 If B is a cocommutative pointed k-bialgebra, then

$$B = \coprod_{g \in G(B)} B_g \quad (k\text{-coalgebra direct sum}).$$

Proof By Theorem 2.4.7 (iii), B is the direct sum of its irreducible components. Thus it suffices to show that the B_g for $g \in G(B)$ constitute a complete set of irreducible components of B. If E is an irreducible component of B, then $G(E) = \{g\}$. Therefore E and B_g are irreducible components containing g. Hence $E = B_g$.

COROLLARY 2.4.29 If B is a cocommutative pointed k-bialgebra, the following conditions are equivalent.

(i) B is a k-Hopf algebra.

(ii) $G(B)$ is a group.

(iii) Each element of $G(B)$ is invertible in B.

Proof (i) $\Rightarrow$ (ii) If B is a k-Hopf algebra, given $g \in G(B)$, $S(g)$ is the inverse of g, and hence $G(B)$ becomes a group. The implication (ii) $\Rightarrow$ (iii) is clear.

(iii) $\Rightarrow$ (i) By Corollary 2.4.28, we can write $B = \coprod_{g \in G(B)} B_g$. By Lemma 2.4.25, $1_B \in \mathbf{Mod}_k(B, B)$ is invertible, and its inverse S is the antipode of B.

COROLLARY 2.4.30 Given a cocommutative pointed k-bialgebra B, set $B = \coprod_{g \in G(B)} B_g$. If we define maps $\pi_g : B_g \to kg$ by $\pi_g(h) = \varepsilon(h)g$, then

$$\pi = \coprod_{g \in G(B)} \pi_g : B \to kG$$

is a k-bialgebra morphism.

5 Irreducible cocommutative bialgebras

If k has characteristic 0, then an irreducible cocommutative k-bialgebra is isomorphic to the universal enveloping k-bialgebra of a k-Lie algebra. Moreover, from this we can conclude that the dual k-algebra of an irreducible cocommutative k-bialgebra (regarded as a k-coalgebra) is reduced. If k has characteristic $p > 0$, this property does not generally hold. We will investigate the conditions under which such properties do hold.

5.1 The bialgebra $B(V)$

We showed in §4.2 that a cofree cocommutative k-coalgebra $C(V)$ over a k-linear space V admits a commutative k-Hopf algebra structure. Let the irreducible component of the k-Hopf algebra $C(V)$ which contains the identity element 1 be denoted by $B(V)$. Then $B(V)$ is a pointed irreducible cocommutative as well as commutative k-Hopf algebra. We now proceed to discuss some of the properties of $B(V)$.

THEOREM 2.5.1 Let C_0 be the coradical of a pointed irreducible cocommutative k-coalgebra C, and set $C^+ = \operatorname{Ker} \varepsilon \cong C/C_0$. If V is a k-linear space, the bijection $\mathbf{Cog}_k\,(C, C(V)) \cong \mathbf{Mod}_k\,(C, V)$ induces the map

$$\mathbf{Cog}_k\,(C, B(V)) \cong \mathbf{Mod}_k\,(C^+, V).$$

Proof Let D be an arbitrary cocommutative k-coalgebra and let $f : C \to D$ be a k-coalgebra morphism. Given a group-like element g of D, let D_g denote the irreducible component of D containing g. Then we have $f(C) \subset D_g \Leftrightarrow f(C_0) \subset kg$. In fact, if $f(C) \subset D_g$, then $f(C_0) \subset kg$ (cf. Theorem 2.4.20). Conversely, let $f(C_0) \subset kg$. Then, if $\{C_n\}_{n \in I}$ is the coradical filtration on C, $f(C_n) \subset \sqcap^{n+1} kg$ (cf. Exercise 2.4). On the other hand, since $C = \bigcup_{n \in I} C_n$ (cf. Theorem 2.3.9 (ii)) and $\bigcup_{n \in I} (\sqcap^{n+1} kg) = D_g$ (cf. Theorem 2.4.10), we get $f(C) \subset D_g$. Applying this fact to $D = B(V)$ and $g = 1$, we obtain

$$\begin{aligned}\mathbf{Cog}_k(C, B(V)) &\cong \{f \in \mathbf{Cog}_k(C, C(V)); f(C_0) = k1\} \\ &\cong \operatorname{Ker}(\mathbf{Mod}_k(C, V) \to \mathbf{Mod}_k(C_0, V)) \\ &\cong \mathbf{Mod}_k(C/C_0, V) \cong \mathbf{Mod}_k(C^+, V).\end{aligned}$$

It is clear that the above correspondence is induced by the isomorphism $\mathbf{Cog}_k(C, C(V)) \cong \mathbf{Mod}_k(C, V)$.

The structure of $B(V)$

The case $\dim V = 1$. Let $B = \coprod_{i=0}^{\infty} kc_i$ be the k-coalgebra defined in Example 2.2. Then, $B_n = \coprod_{i=0}^{n} kc_i$ is a k-subcoalgebra of B, and $\{B_n\}_{n \in I}$ is the coradical filtration on B. Define a k-linear map $\pi : B \to k$ by $\pi(c_n) = \delta_{1_n}$. Let C be an arbitrary k-coalgebra and let $\{C_n\}_{n \geq 0}$ be the coradical filtration on C. Given $\sigma \in \mathbf{Cog}_k(C, B)$ and setting $\pi \circ \sigma \in \mathbf{Mod}_k(C, k)$, we obtain $\pi \circ \sigma(C_0) = 0$. Therefore $\pi \circ \sigma$ induces a k-linear map $g : C/C_0 \to k$. This gives rise to a map

$$\Phi : \mathbf{Cog}_k(C, B) \to \mathbf{Mod}_k(C/C_0, k),$$

which turns out to be a surjection. In fact, $g \in \mathbf{Mod}_k(C/C_0, k)$ defines a morphism $f \in \mathbf{Mod}_k(C, k)$ which is zero on C_0. If we define a k-linear map σ from C to B by

$$\sigma(c) = \sum_{n=0}^{\infty} f^n(c) c_n, \qquad c \in C,$$

since $f^n|_{C_{n-1}} = 0$, a sum as above with respect to an arbitrary c must be a finite sum. Thus $\sigma(c)$ is a well-defined element of B. Now we have

$$(\sigma \otimes \sigma)\Delta(c) = \sum_{i=0}^{n} f^i(c_{(1)}) f^{n-i}(c_{(2)}) c_i \otimes c_{n-i},$$

$$\Delta\sigma(c) = \sum_{i=0}^{\infty} f^n(c) c_i \otimes c_{n-i}.$$

On the other hand, since $f^n(c) = \sum_{i=0}^{n} f^i(c_{(1)})f^{n-i}(c_{(2)})$, we obtain $(\sigma \otimes \sigma) \circ \Delta = \Delta \circ \sigma$. Moreover, $\varepsilon(\sigma(c)) = f^\circ(c) = \varepsilon(c)$. Thus $\sigma \in \mathbf{Cog}_k(C, B)$. Since $\pi \circ \sigma(c) = f^1(c) = g(c)$, $\Phi(\sigma) = g$. Therefore, Φ is a bijection, and (B, π) is the cocommutative cofree pointed irreducible k-coalgebra over the one-dimensional k-linear space k. The k-algebra structure of $B = B(k)$ is determined by the equality $\pi \circ \mu = \pi \otimes \varepsilon + \varepsilon \otimes \pi = \zeta$. Since $\zeta(c_i \otimes c_j) = \delta_{(0,1)(i,j)} + \delta_{(1,0)(i,j)}$,

$$\mu(c_i \otimes c_j) = \sum_{\substack{i_1 + \ldots + i_r = i \\ j_1 + \ldots + j_r = j}} \zeta(c_{i_1} \otimes c_{j_1}) \cdots \zeta(c_{i_r} \otimes c_{j_r}) c_r.$$

Hence

$$\mu(c_i \otimes c_j) = \binom{i+j}{i} c_{i+j}.$$

Therefore B is precisely the k-bialgebra defined in Example 2.6. The set $\{c_i\}_{i \geqq 0}$ is said to be the **standard basis** for $B = B(k)$.

The general case Let V be a k-linear space and let $\{v_\lambda\}_{\lambda \in \Lambda}$ be one of its bases. Set $I = \{0, 1, 2, \ldots\}$ and write $I^{(\Lambda)} = \{f : \Lambda \to I$, a map; $f(\lambda) = 0$ except for a finite number of λs$\}$.

Given $f, g \in I^{(\Lambda)}$, define $f + g, fg \in I^{(\Lambda)}$ by

$$(f + g)(\lambda) = f(\lambda) + g(\lambda), \quad (fg)(\lambda) = f(\lambda)g(\lambda).$$

Furthermore, set $\operatorname{Supp} f = \{\lambda \in \Lambda;\ f(\lambda) \neq 0\}$, $|f| = \sum_{\lambda \in \Lambda} f(\lambda)$

and

$$\binom{f}{g} = \prod_{\lambda \in \Lambda} \binom{f(\lambda)}{g(\lambda)},$$

where we set $\binom{f(\lambda)}{g(\lambda)} = 0$ if $f(\lambda) < g(\lambda)$. Now, $kv_\lambda \subset V$ is a one-dimensional k-linear subspace of V and $B(kv_\lambda) \subset B(V)$. Let $\{v_{\lambda(i)}\}_{i \geqq 0}$ be the standard basis for $B(kv_\lambda)$. Define and fix a total ordering on Λ, and regard Λ as a totally ordered set. For $f \in I^{(\Lambda)}$, set

$$v_{(f)} = \prod_{\lambda \in \Lambda} v_{\lambda(f(\lambda))} \in B(V), \quad v_{\lambda(0)} = 1.$$

Then $\{v_{(f)};\ f\in I^{(\Lambda)}\}$ becomes a basis for $B(V)$. Moreover,

$$\Delta(v_{(f)}) = \sum_{g+h=f} v_{(g)}\otimes v_{(h)}, \qquad \varepsilon(v_{(f)}) = \delta_{0,|f|},$$

$$v_{(f)}v_{(g)} = \binom{f+g}{f} v_{(f+g)}, \qquad v_{(0)} = 1.$$

If we let $B(V)_{(i)}$ be a k-linear subspace of $B(V)$ generated by $\{v_{(f)};|f| = i\}$, then $B(V) = \coprod_{i=0}^{\infty} B(V)_{(i)}$, and we see that $B(V)$ becomes a strictly graded k-Hopf algebra. Further, if we set $B(V)_n = \coprod_{i=0}^{n} B(V)_{(i)}$, then $\{B(V)_n\}_{n\geqq 0}$ is the coradical filtration on $B(V)$. In fact, since we may assume without loss of generality that $\dim V<\infty$, if we set $\Lambda = \{\lambda_1,\ldots,\lambda_n\}$, then the map

$$\varphi : B(kv_{\lambda_1})\otimes\cdots\otimes B(kv_{\lambda_n})\to B(V)$$

defined by $x_1\otimes\cdots\otimes x_n\mapsto x_1x_2\cdots x_n$ turns out to be an isomorphism of graded k-bialgebras. Since $\{v_{\lambda_i(j)}\}_{j\geqq 0}$ is a basis for $B(kv_{\lambda_i})$, the set $\{v_{(f)}\}_{f\in I^{(\Lambda)}}$ becomes a basis for $B(V)$. The rest is straightforward.

THEOREM 2.5.2 A pointed irreducible cocommutative k-coalgebra is isomorphic to a k-subcoalgebra of some $B(V)$.

Proof Let C be a pointed irreducible cocommutative k-coalgebra. Set $V = P(C)$, let $i : V\to C^+ = \operatorname{Ker}\varepsilon_C$ be the canonical embedding, and pick a k-linear map $f : C^+\to V$ such that $f\circ i = 1_V$. By Theorem 2.5.1, there exists a k-coalgebra morphism $F : C\to B(V)$ which satisfies $\pi\circ F|_{C^+} = f$. Since $F|_{P(C)} = \pi\circ F|_{P(C)} = f|_{P(C)}$, the restriction of F to $P(C)$ is injective. Thus F is injective by Theorem 2.4.11. Therefore C is isomorphic to a k-subcoalgebra of $B(V)$.

5.2 Irreducible cocommutative k-Hopf algebras (in case k has characteristic $= 0$)

THEOREM 2.5.3 If k is a field of characteristic 0, an irreducible cocommutative k-Hopf algebra H is isomorphic to the universal enveloping k-Hopf algebra $U(P(H))$ of the k-Lie algebra $P(H)$.

Proof We first show that $H \cong U(P(H))$ as k-coalgebras. Let $\{v_\lambda\}_{\lambda \in \Lambda}$ be a basis for $P(H)$. Define an ordering on Λ and regard Λ as a totally ordered set. If we write $v_\lambda{}^{(i)} = \frac{1}{i!} v_\lambda{}^i$, then we obtain

$$\Delta(v_\lambda{}^{(n)}) = \sum_{i=0}^{n} v_\lambda{}^{(i)} \otimes v_\lambda{}^{(n-i)}, \quad \varepsilon(v_\lambda{}^{(n)}) = \delta_{0n}.$$

Now set

$$v^{(f)} = \prod_{\lambda \in \Lambda} \frac{v_\lambda{}^{f(\lambda)}}{f(\lambda)!}, \qquad f \in I^{(\Lambda)}, \qquad I = \{0, 1, 2, \ldots\},$$

and define a k-linear map

$$\chi : B(P(H)) \to H$$

by $v_{(f)} \mapsto v^{(f)}$. Then χ is a k-coalgebra morphism, and is moreover the identity map on $P(H)$. Therefore, by Corollary 2.4.17 (ii), χ is a k-coalgebra isomorphism. Since $\{v_{(f)}; f \in I^{(\Lambda)}\}$ is a basis for $B(P(H))$, it follows that $\{v^{(f)}; f \in I^{(\Lambda)}\}$ is a basis for H. On the other hand, by defining a k-coalgebra morphism

$$\sigma : B(P(H)) \to U(P(H))$$

by $v_{(f)} \mapsto v^{(f)}$, σ is the identity map on $P(H)$, so σ is injective. Moreover, since $\{v^{(f)}; f \in I^{(\Lambda)}\}$ generates $U(P(H))$, σ is surjective. Therefore, indeed, $H \cong U(P(H))$ as k-coalgebras. Now we will verify that the correspondence is also a k-algebra isomorphism. Let $\rho : P(H) \to U(P(H))$ be the canonical embedding. Then the definition of universal enveloping k-algebras ensures the existence and uniqueness of a k-algebra morphism $\nu : U(P(H)) \to H$ such that $\nu \circ \rho$ is the canonical embedding from $P(H)$ to H. Here, we have $\nu(v^{(f)}) = v^{(f)}$. Thus v is a k-algebra isomorphism. Since v is the inverse of the k-coalgebra morphism $\sigma \circ \chi^{-1}$, it is also a k-bialgebra isomorphism, thereby making H isomorphic to $U(P(H))$ as k-Hopf algebras.

COROLLARY 2.5.4 If k is a field of characteristic 0, then a commutative k-Hopf algebra is reduced.

Proof We may assume that k is an algebraically closed field and that H is finitely generated. Then H becomes a finitely generated commutative k-algebra, and thus H is proper (cf. Theorem 2.3.19). Namely, the canonical k-algebra morphism $\lambda_H : H \to H^{\circ *}$ is injective. Hence it suffices to show that $H^{\circ *}$ is reduced. Since H° is a pointed cocommutative k-Hopf algebra (cf. Theorem 2.3.3), it follows from Corollary 2.4.28 and Theorem 2.5.3 that H° is the coalgebra direct sum of $U(P(H^\circ))g$ for $g \in G(H^\circ)$. Hence its dual k-algebra $H^{\circ *}$ is isomorphic to the k-algebra direct product of copies of k-algebras isomorphic to $U(P(H^\circ))^*$. The next lemma shows that $U(P(H^\circ))^*$ is isomorphic to the power series ring over k, and hence is an integral domain. Consequently, $H^{\circ *}$ is reduced.

LEMMA 2.5.5 $B(V)^* \cong k[[X_\lambda]]_{\lambda \in \Lambda}$ is an integral domain.

Proof Let $f, g \in B(V)^*$ be non-zero elements. Then there exists x, $y \in B(V)$ such that $\langle f, x \rangle \neq 0$ and $\langle g, y \rangle \neq 0$. Pick a finite dimensional k-linear space V_1 of V such that $x, y \in B(V_1) \subseteq B(V)$, and let $i^* : B(V)^* \to B(V_1)^*$ be the dual k-algebra morphism of the canonical embedding $i : B(V_1) \to B(V)$. Then, $i^*f \neq 0$, $i^*g \neq 0$, and $B(V_1)^*$ is an integral domain, so we obtain $i^*f \cdot i^*g = 0$, which forces $fg \neq 0$.

5.3 Irreducible cocommutative k-Hopf algebras (when k has characteristic $p > 0$)

In this section, we assume that k is a perfect field of characteristic $p > 0$. Let $p : k \to k$ be the pth power map from k to k given by $\alpha \mapsto \alpha^p$. Moreover, given a commutative k-algebra A, the k-linear map

$$\mathscr{F} : A \to p_* A \quad \text{given by} \quad a \mapsto a^p, \qquad a \in A,$$

is called the **Frobenius map** of A. We will define a map which is dual to this map on k-coalgebras. Let the tensor k-algebra and the symmetric k-algebra both over a k-linear space V be denoted by $T(V)$ and $S(V)$ respectively, and let $T_{(n)}(V)$ and $S_{(n)}(V)$ denote their respective nth homogeneous components. For an element σ of the symmetric group S_n on n letters, we define an action of σ on $T_{(n)}(V)$ by

$$\sigma : x_1 \otimes \cdots \otimes x_n \mapsto x_{\sigma(1)} \otimes \cdots \otimes x_{\sigma(n)}, \quad x_i \in V \quad (1 \leqq i \leqq n),$$

and set

$$\mathscr{S}_{(n)}(V) = \{t \in T_{(n)}(V); \quad \sigma(t) = t \ \forall \sigma \in S_n\}.$$

Then $\mathscr{S}_{(n)}(V)$ is a k-linear subspace of $T_{(n)}(V)$. Now let

$$i : \mathscr{S}_{(n)}(V) \to T_{(n)}(V), \quad \pi : T_{(n)}(V) \to S_{(n)}(V)$$

respectively be the canonical embedding and the canonical projection.

THEOREM 2.5.6 (i) If we define a k-linear map $F : V \to p_* S_{(p)}(V)$ by

$$F : y \mapsto y \otimes \cdots \otimes y \text{ (p-fold tensor product)}, \ y \in V,$$

then there exists a k-linear map $V : p_* \mathscr{S}_{(p)}(V) \to V$ which makes the following diagram commutative.

$$\begin{array}{ccc} p_* \mathscr{S}_{(p)}(V) & \xrightarrow{\ p_*(i)\ } & p_* T_{(p)}(V) \\ {\scriptstyle V}\downarrow & & \downarrow{\scriptstyle p_*(\pi)} \\ V & \xrightarrow{\ F\ } & p_* S_{(p)}(V) \end{array}$$

(ii) Given an arbitrary $y \in \mathscr{S}_{(p)}(V)$, there exists a unique $V(y)$ which satisfies $\pi \circ i(y) = F \circ V(y)$.

Proof Since F is injective, $V(y)$ is unique provided it exists. Let $X = \{v_\lambda ; \lambda \in \Lambda\}$ be a basis for V, and set $P = \{1, 2, \ldots, p\}$. For each map $f \in X^P$ from P into X, write $[f] = f(1) \otimes \cdots \otimes f(p) \in T_{(p)}(V)$. Fix a total ordering relation on X, and let Z denote the set of all $f \in X^P$ which satisfies $f(1) \leqq f(2) \leqq \cdots \leqq f(p)$. Now set

$$\text{Orb}\,[f] = \{\sigma[f]; \sigma \in S_p\}$$

and

$$\text{Sym}\,[f] = \sum_{y \in \text{Orb}[f]} y.$$

Then $\{[f]; f \in X^P\}$ is a basis for $T_{(p)}(V)$. If we denote the number of elements of Orb $[f]$ by $|\text{Orb}\,[f]|$ for $f \in X^P$, then $|\text{Orb}\,[f]| = pm$ ($m \geqq 1$) when f is not a constant function, and $|\text{Orb}\,[f]| = 1$ when f is a constant function. Moreover, only one element of Z is contained in Orb $[f]$. The set $\{\text{Sym}\,[f]; f \in Z\}$ turns out to be a basis for $\mathscr{S}_{(p)}(V)$.

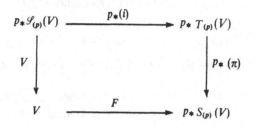

Now, if we define V for $f \in Z$ by $V(\mathrm{Sym}[f]) = 0$ when f is not a constant function, and by $V(\mathrm{Sym}[f]) = f(1) \otimes 1$ when f is a constant function, then this map induces a k-linear map

$$V : p_* \mathcal{S}_{(p)}(V) \to V.$$

Now we assert that this map satisfies the required conditions of the theorem. We see that $F \circ V$ is a k-linear map, and if f is a constant function, then we have $\mathrm{Sym}[f] = f(1) \otimes \cdots \otimes f(1) = \otimes^p f(1)$, so

$$\pi \circ i(\mathrm{Sym}[f]) = \otimes^p f(1) = F(f(1) \otimes 1) = F(V(\mathrm{Sym}[f])).$$

Now, if f is not a constant function, we set $\mathrm{Orb}[f] = \{[g_1], \ldots, [g_{pm}]\}$ for $g_i \in X^p$ $(1 \leqq i \leqq pm)$. Since $\pi(g_1) = \cdots = \pi(g_{pm})$, we get

$$\pi \circ i(\mathrm{Sym}[f]) = pm\pi(g_1) = 0 = F(0) = F(V(\mathrm{Sym}[f])).$$

Therefore V satisfies the required conditions of the theorem.

Let C be a cocommutative k-coalgebra. Since $\Delta_{p-1}(C) \subset \mathcal{S}_{(p)}(C)$, if we set $\mathcal{V} = V \circ \Delta_{p-1}$, then we obtain a k-linear map $\mathcal{V} : p_* C \to C$. When we want to specify which C is meant, we denote $\mathcal{V}$ by $\mathcal{V}_C$.

LEMMA 2.5.7 Let C be a cocommutative k-coalgebra and let A be a commutative k-algebra. Write the pth power map of $f \in \mathbf{Mod}_k(C, A)$ in the form $f^p = f * \cdots * f$ (p times), where $*$ is the convolution defined on $\mathbf{Mod}_k(C, A)$.

(i) $f^p(c) = \mathcal{F}(f(\mathcal{V}(c)))$, $\quad c \in C$.

(ii) Let B be a dense subset of C^*. Given $c \in C$, there exists a unique $\mathcal{V}(c)$ which satisfies $f^p(c) = \mathcal{F}(f(\mathcal{V}(c)))$ for any $f \in B$.

Proof First of all, let $A = S(C)$ and let $\rho \in \mathbf{Mod}_k(C, A)$ be the canonical embedding. For $c \in C$, $\mathcal{V}(c)$ is the unique element which satisfies $\rho^p(c) = \mathcal{F}(\rho(\mathcal{V}(c)))$. Indeed,

$$\mathcal{F}(\rho(\mathcal{V}(c))) = F \circ V(\Delta_{p-1}(c)) = \pi \circ i(\Delta_{p-1}(c)) = \rho^p(c).$$

Proof of (i): The map $f \in \mathbf{Mod}_k(C, A)$ can be extended uniquely to $\bar{f} \in \mathbf{Alg}_k(S(C), A)$. Define a map

$$\varphi : \mathbf{Mod}_k(C, S(C)) \to \mathbf{Mod}_k(C, A)$$

by $\varphi(g) = \bar{f} \circ g$. Then φ is a k-algebra morphism and $\varphi(\rho) = f$. Thus $\varphi(\rho^p) = \varphi(\rho)^p = f^p$. Hence, for any element $c \in C$, we obtain

$$f^p(c) = \bar{f} \circ \rho^p(c) = \bar{f} \circ \mathscr{F}(\rho(\mathscr{V}(c))) = \mathscr{F}(\bar{f} \circ \rho(\mathscr{V}(c)))$$
$$= \mathscr{F}(f(\mathscr{V}(c))).$$

Proof of (ii): Let $x \in C$. Given an arbitrary $f \in B$, if we assume that $f^p(c) = \mathscr{F}(f(x))$ holds, then it follows that $\mathscr{F}(f(\mathscr{V}(c))) = \mathscr{F}(f(x))$. Meanwhile, if we assume $\mathscr{V}(c) \neq x$, then there exists $f \in B$ which satisfies $f(\mathscr{V}(c)) \neq f(x)$. Thus $\mathscr{F}(f(\mathscr{V}(c))) \neq \mathscr{F}(f(x)) = f^p(c)$, which is a contradiction. Consequently, we must have $\mathscr{V}(c) = x$.

LEMMA 2.5.8 Suppose that a sequence $\{c_0, c_1, \ldots, c_n\}$ of elements of a cocommutative k-coalgebra C satisfies $\Delta(c_n) = \sum_{i=0}^{n} c_i \otimes c_{n-i}$. Then

$$\mathscr{V}(c_n) = \begin{cases} 0 & \text{if } n \text{ and } p \text{ are relatively prime.} \\ c_{n/p} & \text{if } n \text{ is divisible by } p. \end{cases}$$

Proof It follows readily that $V = \coprod_{i=0}^{n} kc_i$ is a k-subcoalgebra of C, and the dual k-algebra V^* of V is isomorphic to $A = k[X]/(X^{n+1})$. Let $x^i \in A$ be the equivalence class containing X^i. Then $\langle x^i, c_j \rangle = \delta_{ij}$ $(1 \leqq i, j \leqq n)$. Thus we get

$$\left\langle \left(\sum_{i=0}^{n} \alpha_i x^i \right)^p, c_j \right\rangle = \begin{cases} \alpha_r{}^p = \mathscr{F}(\alpha_r) = \mathscr{F}\left(\sum_{i=0}^{n} \alpha_i x^i \right)(c_r) & \text{when } j = pr. \\ 0 = \mathscr{F}(0) & \text{when } j \text{ is not divisible by } p. \end{cases}$$

Therefore an application of Lemma 2.5.7 yields the desired result.

We obtain the next theorem by applying the above to the cocommutative irreducible k-Hopf algebra $B(V)$ and its standard basis $\{v_\lambda\}_{\lambda \in \Lambda}$. Given $\lambda \in \Lambda$, define $\delta_\lambda \in I^{(\Lambda)}$ by $\delta_\lambda(\mu) = \delta_{\lambda\mu}$. Then $v_{(\delta_\lambda)} = v_\lambda \in V = B(V)_{(1)}$. For $f \in I^{(\Lambda)}$, set $\operatorname{Supp} f = \{\lambda \in \Lambda;\ f(\lambda) \neq 0\}$.

THEOREM 2.5.9 (i) If $f, g \in I^{(\Lambda)}$ and $\operatorname{Supp} f \cap \operatorname{Supp} g = \emptyset$, then $v_{(f)} v_{(g)} = v_{(f+g)}$.

(ii) If $f \in I^{(\Lambda)}$, then f can be written uniquely in the form $f = n_1\delta_{\lambda_1} + \cdots + n_r\delta_{\lambda_r} (\lambda_i \in \Lambda,\ n_i \in I)$.

(iii) $\mathscr{V}(v_{(f)}) = \begin{cases} 0 & \text{if } p \nmid f. \\ v_{(g)} & \text{if } p \mid f,\ \ f = pg. \end{cases}$

The notation $p|f$ (resp. $p\nmid f$) stands for '$f(\lambda)(\lambda \in \Lambda)$ is divisible by p' (resp. 'there exists λ for which $f(\lambda)$ is not divisible by p'). Moreover, by $f = pg$, we mean that $pg(\lambda) = f(\lambda)\ (\lambda \in \Lambda)$.

Now, let H be a cocommutative irreducible k-Hopf algebra, $f: H \to B(V)$ an injective k-coalgebra morphism, and C the k-subcoalgebra $\operatorname{Im} f$ of $B(V)$. In Lemmas 2.5.10 to 2.5.13, we will assume that H, C, and V are as defined above. These lemmas are used in the proof of Theorem 2.5.14.

LEMMA 2.5.10 Let $c \in C_n$ and $d \in C_m$.

(i) There exists $e \in C_{n+m}$ which satisfies $e - cd \in B(V)_{n+m-1}$.

(ii) If $\mathscr{V}(c) = 0$, we can pick e in (i) such that $\mathscr{V}(e) = 0$.

Proof Since $f: H \to C$ is an isomorphism of filtered k-coalgebras, there exist unique $x \in H_n$ and $y \in H_m$ which respectively satisfy $f(x) = c$ and $f(y) = d$.

(i) We write $f(xy) = e$. Corollary 2.4.19 shows that $\operatorname{gr} f: \operatorname{gr} H \to \operatorname{gr} B(V)$ is a k-bialgebra morphism. Thus $f(xy) - f(x)f(y) \in B(V)_{n+m-1}$.

(ii) We have $0 = \mathscr{V}(c) = \mathscr{V}(f(x)) = f \circ \mathscr{V}_H(x)$, and since f is injective, $\mathscr{V}_H(x) = 0$. Thus $\mathscr{V}(e) = f \circ \mathscr{V}_H(xy) = f(\mathscr{V}_H(x)\mathscr{V}_H(y)) = 0$.

LEMMA 2.5.11 Let $P(C) = V$. Write $c \in C_n$ in the form $c = \sum_i \alpha_i v_{(f_i)} + d$, where $|f_i| = n$ and $d \in B(V)_{n-1}$. Given any $\lambda \in \Lambda$, there exists $\tilde{c}$ such that

$$\tilde{c} = \sum_i \alpha_i f_i(\lambda) v_{(f_i)} + e \in C_n,\quad e \in B(V)_{n-1},\quad \mathscr{V}(\tilde{c}) = 0.$$

Proof Pick $h \in B(V)^*$, such that $\langle h, v_{(\delta_\lambda)} \rangle = 1$, and $\langle h, v_{(f)} \rangle = 0$ for

$f \neq \delta_\lambda$. If we set $h \rightharpoonup c = \sum_{(c)} c_{(1)} \langle h, c_{(2)} \rangle$, $c \in B(V)$, then

$$h \rightharpoonup v_{(f)} = \begin{cases} v_{(f - \delta_\lambda)} & \text{if } \lambda \in \operatorname{Supp} f. \\ 0 & \text{if } \lambda \notin \operatorname{Supp} f. \end{cases}$$

Thus $h \rightharpoonup c = \sum_i \alpha_i' v_{(f_i - \delta_\lambda)} + \tilde{d} \in C_{n-1}$. Now, if $f_i(\lambda) > 0$, then $\alpha_i' = \alpha_i$, and if $f_i(\lambda) = 0$, then $\alpha_i' = 0$ and $\tilde{d} \in B(V)_{n-2}$. Meanwhile, since $P(C) = V$, we get $v_\lambda \in C$, and Lemma 2.5.10 (i) shows that there exists $\tilde{c} \in C_{(n-1)+1}$ which satisfies

$$(h \rightharpoonup c) v_\lambda - \tilde{c} \in B(V)_{n-1}.$$

If $f_i(\lambda) > 0$, then since $v_{(f_i - \delta_\lambda)} v_{\delta_\lambda} = f_i(\lambda) v_{(f_i)}$, we obtain

$$(h \rightharpoonup c) v_\lambda = \sum_i \alpha_i f_i(\lambda) v_{(f_i)} + \tilde{d} v_\lambda.$$

Since $\tilde{d} \in B(V)_{n-2}$ in the above equality, we have $\tilde{d} v_\lambda \in B(V)_{n-1}$ and $\tilde{c}$ is a desired element. Further, $\tilde{c}$ can be chosen to satisfy $\mathscr{V}(\tilde{c}) = 0$ by Lemma 2.5.10 (ii).

LEMMA 2.5.12 Let $P(C) = V$. Given $c \in C_n$, set $c = \sum_i \alpha_i v_{(f_i)} + d$ where $|f_i| = n$ and $d \in B(V)_{n-1}$. Then there exists $c' \in C_n$ such that

$$c' = \sum_{p | f_i} \alpha_i v_{(f_i)} + d', \quad d' \in B(V)_{n-1}, \quad \mathscr{V}(c) = \mathscr{V}(c').$$

In particular, if $p \nmid f_i$ for all i, then $c' \in B(V)_{n-1}$.

Proof If $p \nmid f_1$, then there exists $\lambda \in \Lambda$ such that $f_1(\lambda) \not\equiv 0 \pmod p$. By Lemma 2.5.11, there exists an element $\tilde{c}$ such that

$$\tilde{c} = \sum_i \alpha_i f_i(\lambda) v_{(f_i)} + e \in C_n, \quad e \in B(V)_{n-1}, \quad \mathscr{V}(\tilde{c}) = 0.$$

For $c - \tilde{c}/f_1(\lambda) \in C_n$, we have

$$\sum_i \alpha_i (1 - f_i(\lambda)/f_1(\lambda)) v_{(f_i)} + (d - e/f_1(\lambda)).$$

Here, we have $d - e/f_1(\lambda) \in B(V)_{n-1}$. In the case $p | f_i$, we have $f_i(\lambda) \equiv 0 \pmod p$, so the coefficient $\alpha_i(1 - f_i(\lambda)/f_1(\lambda))$ of $v_{(f_i)}$ equals α_i. On the other hand, since $\mathscr{V}(\tilde{c})=0$, we get $\mathscr{V}(c-\tilde{c}/f_1(\lambda))=\mathscr{V}(c)$. Thus we may eliminate the term involving $v_{(f_1)}$. We obtain the c' desired in the lemma by similarly eliminating the terms which satisfy $p\nmid f_i$.

LEMMA 2.5.13 Let $P(C)=V$. For $c\in C_n\cap\mathscr{V}(C)$, write $c=\sum v_{(f_i)}\otimes\alpha_i+d$ where $|f_i|=n$, $\alpha_i\in k$ and $d\in B(V)_{n-1}$. Then there exists $\tilde{c}\in C$ which satisfies

$$\tilde{c}=\sum\bar{\alpha}_i^{\,p}v_{(pf_i)}+e\in C_{pn},\quad e\in B(V)_{pn-1},\quad \mathscr{V}(\tilde{c})=c.$$

Proof Pick $\tilde{c}\in C$ such that $\mathscr{V}(\tilde{c})=c$ and $\tilde{c}\in C_r$, where we let r be the least possible number for this to hold. Write $\tilde{c}=\sum\beta_j v_{(g_j)}+z$ where $|g_j|=r$, $z\in B(V)_{r-1}$. Then, by Lemma 2.5.12, we may write $g_j=ph_j$ for each j. Further, since r is the least possible number for $\tilde{c}$ to be in C_r, we may assume $\beta_j\neq 0$. The fact that $|g_j|=r$ implies that $|h_j|=s$ for each j where $r=ps$. Lemma 2.5.8 shows that $\mathscr{V}(\tilde{c})=\sum\beta_j^{1/p}v_{(h_j)}+\mathscr{V}(z)$. Now $\mathscr{V}(z)\in B(V)_{s-1}$. Therefore $\tilde{c}$ satisfies the requirements of the lemma.

THEOREM 2.5.14 Let H be a cocommutative irreducible k-Hopf algebra and $f:H\to B(V)$ a k-coalgebra morphism, and set $C=\operatorname{Im} f$. If $P(C)=V$, then f is surjective $\Leftrightarrow\mathscr{V}:p_*C\to C$ is surjective.

Proof If $C=B(V)$, for $x=\sum_i\alpha_i v_{(f_i)}\in B(V)$, set $y=\sum_i\alpha_i^{1/p}v_{(pf_i)}$. Then $\mathscr{V}(y)=x$. Thus $\mathscr{V}$ is a surjection. Conversely, if $\mathscr{V}$ is a surjection, we will prove that $C_n=B(V)_n$ by induction on n. Since $C_0\neq\{0\}$, it follows that $C_0=k=B(V)_0$. Meanwhile, since $P(C)=V$, we obtain

$$B(V)_1=k\oplus V\subset C_0+P(C)=C_1\subset B(V)_1.$$

Let $n\geqq 2$ and assume that $C_{n-1}=B(V)_{n-1}$. We will show that $C_n=B(V)_n$. It suffices to show that $v_{(f)}\in C$ for any $f\in I^{(\Lambda)}$ such that $|f|=n$. When Supp f has more than two elements, we can write $f=n_1\delta_{\lambda_1}+\cdots+n_r\delta_{\lambda_r}$ $\left(r>1,\ \sum_{i=1}^{r}n_i=n\right)$. Since $0\leqq n_i<n$, we get $v_{(n_i\delta_\lambda)}\in C$ by induction. Since $\operatorname{gr} f:\operatorname{gr} H\to\operatorname{gr} B(V)$ is a k-bialgebra morphism, there exists $c\in C_n$ which satisfies the condition

$$c-v_{(n_1\delta_{\lambda_1})}\cdots v_{(n_r\delta_{\lambda_r})}\in B(V)_{n-1}.$$

Since $B(V)_{n-1} = C_{n-1}$, we have $v_{(f)} \in C$. Therefore, we need only prove the case $r = 1$, namely that $v_{(n\delta_\lambda)} \in C$. If $p \nmid n$, then $v_{(\delta_\lambda)}$, $v_{((n-1)\delta_\lambda)} \in C$, so Lemma 2.5.10 (i) implies that there exists $c \in C_n$ which satisfies the condition

$$c - v_{(\delta_\lambda)} v_{((n-1)\delta_\lambda)} \in B(V)_{n-1}.$$

We obtain $v_{(\delta_\lambda)} v_{((n-1)\delta_\lambda)} = n v_{(n\delta_\lambda)}$. Since $n \neq 0$ and $B(V)_{n-1} = C_{n-1}$, we get $v_{(n\delta_\lambda)} \in C_n$. When $p | n$, set $n = pm$ ($n \geqq 2$, $m \geqq 1$). Then, by the inductive hypothesis, $v_{(m\delta_\lambda)} \in C_m$ for any λ. Since $\mathscr{V}$ is a surjection, by Lemma 2.5.13, there exists $\tilde{c}$ such that

$$\tilde{c} = v_{(pm\delta_\lambda)} + e, \quad e \in B(V)_{n-1}, \quad \mathscr{V}(\tilde{c}) = v_{(m\delta_\lambda)}.$$

Since $B(V)_{n-1} = C_{n-1}$ and $pm = n$, we obtain $v_{(n\delta_\lambda)} \in C_n$. We thus conclude that $v_{(n\delta_\lambda)} \in C_n$ in any case.

COROLLARY 2.5.15 Let V be the k-linear space $P(H)$ of all primitive elements of a cocommutative irreducible k-Hopf algebra H. Then $H \cong B(V)$ as k-coalgebras $\Leftrightarrow \mathscr{V}_H : p_* H \to H$ is surjective.

Proof By Theorem 2.5.2, H is isomorphic to a k-subcoalgebra of some $B(V)$ as k-coalgebras. If we let $f : H \to B(V)$ be its embedding into $B(V)$ and set $\operatorname{Im} f = C$, then $V = P(C)$. Now, if $\mathscr{V}_H$ is a surjection, then $\mathscr{V}_C$ is also a surjection, and by Theorem 2.5.14 we have $C = B(V)$. Thus $H \cong B(V)$ as k-coalgebras. Conversely, if $H \cong B(V)$ as k-coalgebras, then $\mathscr{V}_{B(V)}$ is a surjection, so $\mathscr{V}_H$ is also surjective.

COROLLARY 2.5.16 For a cocommutative irreducible k-Hopf algebra H, the following conditions are equivalent.

(i) $H^* \cong k[[X_1, \ldots, X_n]]$ (the power series ring in n variables over k).

(ii) $H \cong B(V)$ as k-coalgebras, where $V = P(H)$.

(iii) H^* is an integral domain.

(iv) H^* is reduced.

Proof (i)$\Leftrightarrow$(ii): If $H \cong B(V)$ as k-coalgebras, then $H^* \cong k[[X_1, \ldots, X_n]]$. Conversely, if we let V be an n-dimensional k-linear space, then $B(V)^* \cong k[[X_1, \ldots, X_n]]$. If $H^* \cong k[[X_1, \ldots, X_n]]$, Theorem 2.2.16 as well as the remark following it imply that $B(V)$ and H are coreflexive. Therefore we obtain

$$H \cong H^{*\circ} \cong k[[X_1, \ldots, X_n]]^\circ \cong B(V)^{*\circ} \cong B(V)$$

as k-coalgebras.

The implications (ii)$\Rightarrow$(iii)$\Rightarrow$(iv) are trivial. Hence it is enough to show (iv)$\Rightarrow$(ii). Suppose that H is not isomorphic to $B(V)$ as k-coalgebras. Then Corollary 2.5.15 would imply that $\mathscr{V}_H : p_*H \to H$ is not surjective. Thus there exists $f \in H^*$ such that $f \neq 0$ and $\operatorname{Im} \mathscr{V}_H \subset \operatorname{Ker} f$. Now for any $h \in H$, we have $f^p(h) = \mathscr{F}(f(\mathscr{V}_H(h))) = 0$. In other words, $f^p = 0$. Consequently H^* is not reduced.

Remark If k has characteristic 0, then the conditions of Corollary 2.5.16 hold for any H. Generally speaking, if k is not a perfect field, conditions (i) and (ii) of Corollary 2.5.16 are equivalent. Furthermore, the following conditions are equivalent to (i).

(ii)′ $H \otimes_k k^{1/p} \cong B(V) \otimes_k k^{1/p} = B(V \otimes_k k^{1/p})$ as k-coalgebras.

(iii)′ $\mathbf{Mod}_{k^{1/p}}(H \otimes_k k^{1/p}, k^{1/p})$ is an integral domain.

(iv)′ $\mathbf{Mod}_{k^{1/p}}(H \otimes_k k^{1/p}, k^{1/p})$ is reduced.

5.4 Birkhoff–Witt k-bialgebras

Let C be a k-coalgebra and set $I = \{0, 1, 2, \ldots\}$. A sequence $\{d_i\}_{i \in I}$ of elements of C which satisfies the conditions

$$d_0 = 1, \quad d_1 = d, \quad \Delta d_n = \sum_{i=0}^{n} d_i \otimes d_{n-i} \quad (i \in I)$$

is said to be an **∞-sequence of divided powers lying over** d.

Let B be a pointed irreducible cocommutative k-bialgebra. When an ∞-sequence of divided powers exists over any element $x \in P(B)$, B is said to be a **Birkhoff–Witt k-bialgebra**. A cocommutative cofree k-coalgebra $B(V)$ over a k-linear space V is a Birkhoff–Witt k-bialgebra. In fact, for $P(B(V)) = V$, let $\{v_\lambda\}_{\lambda \in \Lambda}$ be a basis for V. If we let $\{v_{\lambda(i)}\}_{i \in I}$ be the standard basis for $B(kv_\lambda)$, then $\{v_{\lambda(i)}\}_{i \in I}$ is an ∞-sequence of divided powers lying over v_λ. In general, for $\alpha, \beta \in k$, we write

$$(\alpha v_\lambda + \beta v_\mu)_{(n)} = \sum_{i=0}^{n} \alpha^i \beta^{n-i} v_{\lambda(i)} v_{\mu(n-i)}.$$

Then $\{(\alpha v_\lambda + \beta v_\mu)_{(i)}\}_{i \in I}$ is an ∞-sequence of divided powers lying over $\alpha v_\lambda + \beta v_\mu$. Therefore an ∞-sequence of divided powers exists over any element of V.

Conversely, it is a standard result that a Birkhoff–Witt k-bialgebra is isomorphic as a k-coalgebra to the cocommutative cofree k-coalgebra $B(V)$ over some k-linear space V. However, we will not present its proof here.

Remark Let L be a semi-simple Lie algebra over the field of complex numbers $\mathbb{C}$, and let Φ be a root system of L. Let $L_{\mathbb{Z}}$ be the $\mathbb{Z}$-Lie algebra spanned by a Chevalley basis $\{H_1, \ldots, H_l, X_\alpha, \alpha \in \Phi\}$ of L. Then

$$X_\alpha^{(n)} = \frac{1}{n!} X_\alpha^{\ n} \qquad (\alpha \in \Phi,\ n \in I),$$

$$H_i^{(n)} = \frac{1}{n!} H_i(H_i - 1) \cdots (H_i - n + 1) \qquad (1 \leqq i \leqq l,\ n \in I)$$

form ∞-sequences of divided powers in the universal enveloping $\mathbb{C}$-bialgebra $U(L)$ of L. Let $U_{\mathbb{Z}}$ be the $\mathbb{Z}$-subalgebra of $U(L)$ generated by $X_\alpha^{(n)}$, $H_i^{(n)}$ $(\alpha \in \Phi,\ 1 \leqq i \leqq l,\ n \in I)$. If we set $U_k = U_{\mathbb{Z}} \otimes_{\mathbb{Z}} k$ for an arbitrary field k, then U_k is a Birkhoff–Witt k-bialgebra. If k has characteristic 0, then U_k is isomorphic to the universal enveloping k-bialgebra $U(L_k)$ of the k-Lie algebra $L_k = L_{\mathbb{Z}} \otimes_{\mathbb{Z}} k$. However, if the characteristic of k is prime, U_k is not isomorphic to $U(L_k)$.

If we let $H(L_k)$ denote the dual k-Hopf algebra $(U_k)^\circ$ of U_k, then $H(L_k)$ is a commutative k-Hopf algebra. In the case when k is an algebraically closed field, by the definition due to Borel–Chevalley, $G(k) = \mathbf{M}_k(H(L_k), k)$ becomes a connected as well as a simply connected semi-simple affine algebraic k-group.

3

Hopf algebras and representations of groups

1 Comodules and bimodules

In this section we introduce comodules over a k-coalgebra C which are dual to modules over a k-algebra and analyze some of its properties. It turns out that comodules correspond to modules with certain finiteness conditions, i.e. rational modules. We also define bimodules, which come equipped with structures of both modules and comodules and then prove their structure theorem (cf. Theorem 3.1.8). This theorem will be utilized in Chapter 4. Throughout this chapter, we let k again denote a field.

1.1 Comodules

Let C be a k-coalgebra. If we are given a k-linear space M and a k-linear map

$$\psi : M \to M \otimes C$$

which makes the following diagrams

$$\begin{array}{ccc} M\otimes k & \xleftarrow{1\otimes\varepsilon} & M\otimes C \\ & \nwarrow{\scriptstyle\sim} & \uparrow{\scriptstyle\psi} \\ & & M \end{array} \qquad \begin{array}{ccc} M\otimes C\otimes C & \xleftarrow{\psi\otimes 1} & M\otimes C \\ {\scriptstyle 1\otimes\Delta}\uparrow & & \uparrow{\scriptstyle\psi} \\ M\otimes C & \xleftarrow{\psi} & M \end{array}$$

commutative, then the pair (M, ψ) is called a **right C-comodule** and ψ is said to be its **structure map**. A **left C-comodule** can be defined similarly.

Just as a k-algebra A becomes a right A-module or a left A-module when the multiplication map μ is taken to be the structure map, likewise, a k-coalgebra C together with its comultiplication map Δ as the structure map can be viewed as a left or a right C-comodule.

Notation As a matter of notation, we employ the following symbolism with regard to a right C-comodule M and its structure map ψ.

$$\psi(m) = \sum_{(m)} m_{(0)} \otimes m_{(1)}, \qquad m, m_{(0)} \in M, \quad m_{(1)} \in C.$$

$$(\psi \otimes 1)\psi(m) = (1 \otimes \Delta)\psi(m) = \sum_{(m)} m_{(0)} \otimes m_{(1)} \otimes m_{(2)}, \quad m_{(2)} \in C.$$

Due to the property of right C-comodules, we can write

$$m = \sum_{(m)} m_{(0)} \varepsilon(m_{(1)}), \quad \sum_{(m)} \psi(m_{(0)}) \otimes m_{(1)} = \sum_{(m)} m_{(0)} \otimes \Delta(m_{(1)}).$$

Similarly, if M is a left C-comodule, we write

$$\psi(m) = \sum_{(m)} m_{(-1)} \otimes m_{(0)}, \qquad m, m_{(0)} \in M, \quad m_{(-1)} \in C,$$

$$(\psi \otimes 1)\psi(m) = \sum_{(m)} m_{(-2)} \otimes m_{(-1)} \otimes m_{(0)}, \quad m_{(-2)} \in C.$$

When a k-linear subspace N of a right C-comodule M satisfies $\psi(N) \subset N \otimes C$, N becomes a right C-comodule with the restriction $\psi|_N$ of ψ to N as the structure map. Such an N is called a **right C-subcomodule** of M. Now let M, N be right C-comodules and ψ_M, ψ_N the structure maps of M, N respectively. If a k-linear map $f: M \to N$ satisfies $\psi_N \circ f = (f \otimes 1) \circ \psi_M$, namely if the diagram

$$\begin{array}{ccc} M & \xrightarrow{\quad f \quad} & N \\ {\scriptstyle \psi_M}\downarrow & & \downarrow{\scriptstyle \psi_N} \\ M \otimes C & \xrightarrow{\quad f \otimes 1 \quad} & N \otimes C \end{array}$$

commutes, then f is said to be a **C-comodule morphism**. The category consisting of right C-comodules and their morphisms is called the category of right C-comodules and denoted by $\mathbf{Com}_C$. For right C-comodules M, N, the set of all C-comodule morphisms from M to N is written $\mathbf{Com}_C(M, N)$.

THEOREM 3.1.1 Let $f : M \to N$ be a right C-comodule morphism. Then

$$\operatorname{Ker} f = \{m \in M;\ f(m) = 0\},$$
$$\operatorname{Im} f = \{f(m) \in N;\ m \in M\}$$

are right C-subcomodules of M, N respectively.

Proof Let ψ_M, ψ_N be structure maps of M, N respectively. Then we have

$$\psi_M(\operatorname{Ker} f) \subset \operatorname{Ker}(f \otimes 1) = (\operatorname{Ker} f) \otimes C,$$
$$\psi_N(\operatorname{Im} f) = (f \otimes 1)\,\psi_M(M) \subset (\operatorname{Im} f) \otimes C,$$

and hence $\operatorname{Ker} f, \operatorname{Im} f$ are respectively right C-subcomodules of M, N.

EXERCISE 3.1 Let N be a right C-subcomodule of a right C-comodule M. Show that there is a unique right C-comodule structure on the quotient space M/N which makes the canonical k-linear map $\pi : M \to M/N$ a right C-comodule morphism.

1.2 Rational modules

Let G be a semigroup with identity element e, and let $M_V(G)$ denote the set of all maps from G to a k-linear space V. For $f, g \in M_V(G), x \in G$ and $c \in k$, by defining

$$(f + g)(x) = f(x) + g(x), \qquad (cf)(x) = cf(x),$$

$M_V(G)$ becomes a k-linear space. Moreover, for $x, y \in G$, by defining

$$(xf)(y) = f(yx), \qquad (fx)(y) = f(xy),$$

$M_V(G)$ becomes a two-sided G-module. Now define a k-linear map $\zeta : V \otimes M_k(G) \to M_V(G)$ by

$$\zeta(v \otimes f)(x) = f(x)v, \qquad v \in V, \quad f \in M_k(G), \quad x \in G.$$

Then ζ is an isomorphism of k-linear spaces. Setting

$$R_V(G) = \{f \in M_V(G);\ \dim\ kGf < \infty\},$$

$R_V(G)$ becomes a left G-submodule of $M_V(G)$.

Given a left G-module V, let $\rho : G \times V \to V$ be its structure map. If each $v \in V$ generates a finite dimensional left G-submodule, then V is called a **locally finite left G-module**. Given a locally finite left G-module V, if we define a k-linear map $\rho^* : V \to M_V(G)$ by

$$\rho^*(v)(x) = \rho(x, v) \qquad x \in G,\ v \in V,$$

then we have $\rho^*(V) \subset R_V(G)$, and the k-linear map

$$\psi_\rho = \zeta^{-1} \circ \rho^* : V \to V \otimes R_k(G)$$

defines a right $R_k(G)$-comodule structure on V. In fact, if we write

$$\rho(x, v_i) = \sum_j \rho_{ji}(x) v_j$$

with respect to a basis $\{v_i\}_{i \in I}$ of V, the fact that V is locally finite implies that the sum is a finite sum, and hence the expression above is well-defined. Now, since $\rho^*(v_i)(x) = \zeta\left(\sum_j v_j \otimes \rho_{ji}\right)(x)$, we obtain

$$\psi_\rho(v_i) = \sum_j v_j \otimes \rho_{ji};$$

and from $\Delta\rho_{ij} = \sum_l \rho_{il} \otimes \rho_{lj}$, we get $(\psi_\rho \otimes 1)\psi_\rho = (1 \otimes \Delta)\psi_\rho$ and also $(1 \otimes \varepsilon)\psi_\rho(v_i) = v_i \otimes 1$. Conversely, suppose that V is a right $R_k(G)$-comodule and let $\psi : V \to V \otimes R_k(G)$ be its structure map. For

$$\psi(v) = \sum_{(v)} v_{(0)} \otimes v_{(1)},$$

defining

$$\rho_\psi(x, v) = \sum_{(v)} v_{(1)}(x) v_{(0)},$$

V becomes a locally finite left G-module. We have just shown that there exists a one-to-one correspondence,

$$\text{locally finite left } G\text{-modules} \leftrightarrow \text{right } R_k(G)\text{-comodules}, \tag{3.1}$$

given by $\rho \mapsto \psi_\rho$, $\psi \mapsto \rho_\psi$.

Given a k-algebra A, let A_m be the underlying semigroup of A via multiplication on A. Applying (3.1) to $G = A_m$, we obtain the one-to-one correspondence

$$\text{locally finite } A_m\text{-modules} \leftrightarrow \text{right } R_k(A_m)\text{-comodules}. \tag{3.2}$$

If, in particular, V is a locally finite left A-module, then we have

$\rho^*(V) \subset \zeta(V \otimes A^\circ)$, and hence V corresponds to a right A°-comodule. Conversely, for a right A°-comodule V, since $A^\circ = R_k(A_m) \cap A^*$, V also becomes a right $R_k(A_m)$-comodule. Further, the locally finite left A_m-module which corresponds to V is an A-module. Therefore, from the correspondence (3.2), we obtain the following correspondence.

$$\text{locally finite left } A\text{-modules} \leftrightarrow \text{right } A^\circ\text{-comodules.} \tag{3.3}$$

The dual k-coalgebra A° of a k-algebra A is a right A°-comodule with structure map given by its comultiplication map $\mu^\circ : A^\circ \to A^\circ \otimes A^\circ$; and the structure map of the locally finite left A-module obtained from the correspondence (3.3) is given by

$$\rho_{\mu^\circ}(x, f) = \sum_{(f)} \langle f_{(2)}, x \rangle f_{(1)}, \quad f \in A^\circ, x \in A.$$

We denote this action by $x \rightharpoonup f$. Namely, we have

$$\langle x \rightharpoonup f, y \rangle = \langle f, yx \rangle. \tag{3.4}$$

This action turns out to be the dual of the process of regarding A as a right A-module via multiplication. Generally speaking, A^* becomes a left A-module through this action, albeit not necessarily locally finite. Similarly, A° or A^* can be endowed with right A-module structures by setting

$$\langle f \leftharpoonup x, y \rangle = \langle f, xy \rangle. \tag{3.5}$$

Now let H be a k-Hopf algebra and let S be its antipode. We can define left or right H-module structures on H^* given for $f \in H^*$ and x, $y \in H$ by

$$\langle x \rightharpoonup f, y \rangle = \langle f, S(x)y \rangle, \tag{3.6}$$

$$\langle f \leftharpoonup x, y \rangle = \langle f, yS(x) \rangle, \tag{3.7}$$

respectively.

Moreover, applying (3.3) to the dual k-algebra C^* of a k-coalgebra C, we obtain the correspondence

$$\text{locally finite left } C^*\text{-modules} \leftrightarrow \text{right } C^{*\circ}\text{-comodules.} \tag{3.8}$$

Given a right C-comodule V, V becomes a right $C^{*\circ}$-comodule by the canonical embedding $\lambda_C : C \to C^{*\circ}$, and the locally finite left C^*-module to which it corresponds by (3.8) is called a **rational left C^*-module**. In general, a locally finite left C^*-module does not necessarily correspond to a right C-comodule. By definition, we also get the correspondence

$$\text{rational left } C^*\text{-modules} \leftrightarrow \text{right } C\text{-comodules.} \tag{3.9}$$

When we regard the comultiplication map $\Delta : C \to C \otimes C$ as the structure map, C becomes a right C-comodule, and the structure map of the rational left C^*-module to which it corresponds is given by

$$\rho_\Delta(f, x) = \sum_{(x)} \langle x_{(2)}, f \rangle x_{(1)}, \quad x \in C,\ f \in C^*.$$

We denote this action by $f \rightharpoonup x$. Namely,

$$\langle f \rightharpoonup x, g \rangle = \langle x, gf \rangle, \quad g \in C^*. \tag{3.10}$$

Similarly, the rational right C^*-module structure of C can be defined by

$$\langle x \leftharpoonup f, g \rangle = \langle x, fg \rangle, \quad g \in C^*. \tag{3.11}$$

Now let A be a dense k-subalgebra of C^*. In other words, let

$$A^{\perp(C)} = \{x \in C; \langle f, x \rangle = 0 \quad \forall f \in A\} = \{0\}.$$

Given the canonical embedding $i : A \to C^*$, let $i^\circ : C^{*\circ} \to A^\circ$ be its dual k-coalgebra morphism, and construct the k-coalgebra morphism $\psi = i^\circ \circ \lambda_C : C \to A^\circ$. Now, for $x \in C$, we obtain

$$\psi(x) : f \mapsto \langle f, x \rangle, \quad f \in A,$$

and since A is dense in C^*, ψ is an injection. Therefore the right C-comodule V becomes a right A°-comodule via $\psi : C \to A^\circ$. Furthermore, by correspondence (3.3), it also admits a structure of a locally finite left A-module. Such an A-module is called a **rational left A-module**. By definition, we obtain the correspondences:

$$\begin{aligned}\text{rational left } A\text{-modules} &\leftrightarrow \text{rational left } C^*\text{-modules} \\ &\leftrightarrow \text{right } C\text{-comodules}\end{aligned} \tag{3.12}$$

If A is dense in C^*, then the operator domain of action of rational left A-modules can be extended uniquely from A to C^*, and rational left

A-modules can be regarded as rational left C^*-modules. Thus, giving a rational C^*-module is equivalent to giving a rational A-module.

THEOREM 3.1.2. Let C^* be the dual k-algebra of a k-coalgebra C. Given a rational left C^*-module V, let $\psi : V \to V \otimes C$ be the structure map of the right C-comodule to which it corresponds. Then we have

$$\begin{aligned} V^{C^*} &= \{v \in V;\ fv = \langle f, 1\rangle v \quad \forall\, f \in C^*\} \\ &= \{v \in V;\ \psi(v) = v \otimes 1\}. \end{aligned}$$

Proof If $\psi(v) = v \otimes 1$, then it follows readily that $fv = \langle f, 1\rangle v$. Conversely, let $fv = \langle f, 1\rangle v$ for $f \in C^*$. If we pick a basis $\{v_\lambda\}_{\lambda \in \Lambda}$ of V and set $\psi(v) = \sum_{\lambda \in \Lambda} v_\lambda \otimes a_\lambda$, $v = \sum_{\lambda \in \Lambda} \alpha_\lambda v_\lambda$, then

$$fv = \sum_\lambda v_\lambda \langle f, a_\lambda\rangle = \sum_\lambda v_\lambda \alpha_\lambda \langle f, 1\rangle.$$

Thus $\langle f, a_\lambda\rangle = \langle f, \alpha_\lambda 1\rangle$ for all $f \in C^*$. Therefore $a_\lambda = \alpha_\lambda 1$, and hence $\psi(v) = \sum_\lambda v_\lambda \otimes \alpha_\lambda 1 = v \otimes 1$.

THEOREM 3.1.3 For a k-coalgebra C, the following conditions are equivalent.

(i) C is coreflexive, i.e. $C \cong C^{*\circ}$

(ii) All finite dimensional C^*-modules are rational.

Proof The implication (i)$\Rightarrow$(ii) is straightforward. Conversely, we assume (ii) and show that each $c \in C^{*\circ}$ is contained in C. Let M be the k-subcoalgebra of $C^{*\circ}$ generated by c. Then M is finite dimensional and can be regarded as a right $C^{*\circ}$-comodule with structure map Δ. Thus, by correspondence (3.8), M has a left $C^{*\circ}$-module structure, and by the hypothesis of (ii), M is moreover a rational left C^*-module. Now by (3.12), M is a right C-comodule, and since the correspondence is one-to-one, the right C-comodule structure coincides with its right $C^{*\circ}$-comodule structure. Hence, if we write $\Delta(c) = \sum_{(c)} c_{(1)} \otimes c_{(2)}$, then the k-linear subspace of $C^{*\circ}$ generated by $\{c_{(2)}\}$ is contained in C. Therefore $c = (\varepsilon \otimes 1)\Delta(c) = \sum_{(c)} \varepsilon(c_{(1)}) c_{(2)} \in C$.

Remark Let H be a k-Hopf algebra and let H° be the dual k-Hopf algebra of H. The group $G(H^\circ)$ of all group-like elements of H° coincides with $\mathbf{Alg}_k(H, k)$, where the multiplication is defined via convolution (cf. Theorem 2.1.5). In fact, for $f \in H^*$, $x, y \in H$ we have

$$\langle \Delta f - f \otimes f, x \otimes y \rangle = f(xy) - f(x)f(y),$$

and hence $\Delta f = f \otimes f \Leftrightarrow f(xy) = f(x)f(y)$. In particular, if k is an algebraically closed field and H is a reduced commutative finitely generated k-algebra, $kG(H^\circ)$ is dense in H^* (cf. Lemma 4.2.10). Furthermore, a rational $kG(H^\circ)$-module can be identified with a rational H^*-module. This fact will be applied to the representations of algebraic groups in Chapter 4.

Given a k-coalgebra C, let M be a right C-comodule with structure map $\psi : M \to M \otimes C$. Fix a basis $\{m_\lambda\}_{\lambda \in \Lambda}$ of M and let the subspace of C spanned by

$$\{c_\lambda \in C;\ \psi(m) = \sum_\lambda m_\lambda \otimes c_\lambda \quad \forall m \in M\}$$

be denoted by $C(M)$. Then $C(M)$ does not depend on the choice of a basis for M, and from the equality $(\psi \otimes 1)\psi = (1 \otimes \Delta)\psi$, we see that $C(M)$ is a k-subcoalgebra of C. Moreover, the following implications hold.

M is one-dimensional (resp. finite dimensional) $\Rightarrow C(M)$ is one-dimensional (resp. finite dimensional).

$$M_1 \subset M_2 \Rightarrow C(M_1) \subset C(M_2).$$

$$M = M_1 + M_2 \Rightarrow C(M) = C(M_1) + C(M_2).$$

We also have the following.

THEOREM 3.1.4 (i) If M is a simple right C-comodule, then $C(M)$ is a simple k-subcoalgebra.

(ii) If D is a simple k-subcoalgebra of C, then there exists a simple right C-comodule M which satisfies $C(M) = D$.

Proof (i) Let $\{m_\lambda\}_{\lambda \in \Lambda}$ be a basis for M. If we write $\psi(m_\lambda) = \sum_{\lambda'} m_{\lambda'} \otimes c_{\lambda'\lambda}$, then $C(M)$ is a k-linear space spanned by $\{c_{\lambda'\lambda}\}_{\lambda, \lambda' \in \Lambda}$. For

$f \in C^*$, we have

$$fm_\lambda = \sum_{\lambda'} m_{\lambda'} \langle f, c_{\lambda'\lambda} \rangle = 0 \quad \forall \lambda \in \Lambda \Rightarrow \langle f, c_{\lambda'\lambda} \rangle = 0 \quad \forall \lambda', \lambda \in \Lambda.$$

Thus $(\operatorname{ann} M)^{\perp(C)} \supset C(M)$. Regarding M as a left C^*-module, we have $M \cong C^*/\operatorname{ann} M$, which shows that $\operatorname{ann} M$ is a maximal ideal of C^*. Now Theorem 2.3.4 implies that $(\operatorname{ann} M)^{\perp(C)}$ is a simple k-subcoalgebra of C, and hence $(\operatorname{ann} M)^{\perp(C)} = C(M)$. Consequently, $C(M)$ is a simple k-subcoalgebra of C.

(ii) For a minimal right coideal M of D, we have $\Delta M \subset M \otimes D$. Thus $C(M) \subset D$. Since D is a simple k-subcoalgebra, $D = C(M)$.

COROLLARY 3.1.5 The following conditions are equivalent for a k-coalgebra C.

(i) All rational simple left C^*-modules are one-dimensional.

(ii) All minimal right coideals of C are one-dimensional.

(iii) C is a pointed k-coalgebra, namely, all simple k-subcoalgebras are one-dimensional.

Proof With regard to a C-comodule M, the proof follows readily by Theorem 3.1.4 and the fact that $\dim M = 1 \Leftrightarrow \dim C(M) = 1$.

COROLLARY 3.1.6 Given a k-coalgebra C,

(i) C is the sum of all k-subcoalgebras $C(M)$, where M ranges over all finite dimensional right C-comodules.

(ii) corad C is the sum of all k-subcoalgebras $C(M)$ where M ranges over all simple right C-comodules.

Proof If $c \in C$, write $\Delta c = \sum_i c_i \otimes d_i$ in such a way that $\{c_i\}$ is linearly independent over k. Let M denote the k-linear subspace of C spanned by $\{c_i\}$. Then M becomes a finite dimensional right C-comodule with respect to the structure map Δ. Now, for all i, we have $d_i \in C(M)$, and hence $c = (\varepsilon \otimes 1)\Delta c = \sum_i \varepsilon(c_i) d_i \in C(M)$. Therefore we get (i). The proof of (ii) follows readily from (i) and Theorem 3.1.4.

THEOREM 3.1.7 Let C be a k-coalgebra. The following conditions are equivalent.

(i) All rational left C^*-modules are completely reducible.
(ii) C is co-semi-simple.

Proof The implication (i)$\Rightarrow$(ii) is clear by Corollary 3.1.6.

(ii)$\Rightarrow$(i) Let M be a rational left C^*-module. In order to prove that M is completely reducible, it is enough to show that each $m \in M$ is contained in a simple left C^*-submodule of M. Since m is contained in a finite dimensional left C^*-submodule of M, we may assume M is finite dimensional. Hence $C(M)$ can also be considered finite dimensional. Since C is co-semi-simple, $C(M)$ is contained in a direct sum D of a finite number of simple k-subcoalgebras of C. Now, since D^* is a finite dimensional semi-simple k-algebra and M is a right D-comodule, we have that M is a rational left D^*-module and is completely reducible as a D^*-module. Meanwhile, the k-algebra morphism $C^* \to D^*$ is surjective, so M is also completely reducible as a C^*-module.

1.3 **Bimodules**

Given a k-bialgebra K, let M be a right K-module with structure map $\varphi_M : M \otimes K \to M$, and set $\varphi(m \otimes x) = mx$. If we define a k-linear map $\varphi_{M \otimes K} : (M \otimes K) \otimes K \to M \otimes K$ by $\varphi_{M \otimes K} = (\varphi_M \otimes \mu)(1 \otimes \tau \otimes 1)(1 \otimes \Delta)$, namely, if we define

$$\varphi_{M \otimes K}(m \otimes x \otimes y) = \sum_{(y)} m y_{(1)} \otimes x y_{(2)}, \quad m \otimes x \in M \otimes K, \quad y \in K, \tag{3.13}$$

then $M \otimes K$ becomes a right K-module with structure map $\varphi_{M \otimes K}$. Dually, let M be a right K-comodule with structure map $\psi_M : M \to M \otimes K$. If we define a k-linear map $\psi_{M \otimes K} : M \otimes K \to (M \otimes K) \otimes K$ by $\psi_{M \otimes K} = (1 \otimes \mu)(1 \otimes \tau \otimes 1)(\psi_M \otimes \Delta)$, namely by

$$\psi_{M \otimes K}(m \otimes x) = \sum_{(m)(x)} m_{(0)} \otimes x_{(1)} \otimes m_{(1)} x_{(2)}, \quad m \otimes x \in M \otimes K, \tag{3.14}$$

then $M \otimes K$ becomes a right K-comodule with structure map $\psi_{M \otimes K}$. For a given left K-module or left K-comodule M, we can similarly define structures of a left K-module or a left K-comodule on $K \otimes M$.

Now suppose M is simultaneously a right K-module and a right

K-comodule. Let φ_M, ψ_M respectively be its structure maps, and regard $M \otimes K$ as a right K-module and right K-comodule via (3.13) and (3.14). Then we obtain

$$\begin{aligned}&\varphi_M \text{ is a right } K\text{-comodule morphism}\\ &\quad\Leftrightarrow \psi_M \text{ is a right K-module morphism.} \qquad (3.15)\end{aligned}$$

In fact, $(\varphi_M \otimes 1) \circ \psi_{M \otimes K} = \psi_M \circ \varphi_M$ and $\varphi_{M \otimes N}(\psi_M \otimes 1) = \psi_M \circ \varphi_M$ are both given by

$$(\varphi_M \otimes \mu)(1 \otimes \tau \otimes 1)(\psi_M \otimes \Delta) = \psi_M \circ \varphi_M,$$

namely by

$$\psi_M(mx) = \sum_{(m)\,(x)} m_{(0)} x_{(1)} \otimes m_{(1)} x_{(2)} \tag{3.16}$$

for $m \in M$, $x \in K$. If M is both a right K-module and a right K-comodule and satisfies one of the equivalent conditions (3.15) or (3.16), M is said to be a **right K-bimodule**. A **left K-bimodule** can be defined similarly.

Given K-bimodules M, N, a k-linear map $f: M \to N$ which is simultaneously a K-module morphism and a K-comodule morphism is said to be a **K-bimodule morphism**.

THEOREM 3.1.8 (Structure theorem of bimodules.) Given a k-Hopf algebra K, let M be a right K-bimodule and let φ, ψ be its structure maps when M is viewed as a right K-module and right K-comodule respectively. Then

$$N = \{m \in M;\ \psi(m) = m \otimes 1\}$$

is a right K-subcomodule of M, and $N \otimes K$ turns out to be a right K-subcomodule of $M \otimes K$. Moreover, $N \otimes K$ is a right K-module with structure map the k-linear map $\varphi_{N \otimes K}: N \otimes K \otimes K \to N \otimes K$ given by

$$\varphi_{N \otimes K}(n \otimes x \otimes y) = n \otimes xy, \quad n \in N, \quad x, y \in K;$$

and $N \otimes K$ becomes a right K-bimodule with respect to these two structures. As right K-bimodules, we have the isomorphism

$$M \cong N \otimes K.$$

Proof Define a k-linear map $\alpha : N \otimes K \to M$ by $\alpha(n \otimes x) = nx$ for $n \in N$, $x \in K$. Then

$$(\alpha \otimes 1)\psi_{N\otimes K}(n \otimes x) = \sum_{(n)(x)} n_{(0)}x_{(1)} \otimes n_{(1)}x_{(2)}$$
$$= \sum_{(x)} nx_{(1)} \otimes x_{(2)}$$
$$= \psi(n)x = \psi(nx) = \psi(\alpha(n \otimes x)).$$

Hence, α is a right K-comodule morphism. Define a k-linear map $\beta : M \to N \otimes K$ for $m \in M$ by

$$\beta(m) = \sum_{(m)} m_{(0)}S(m_{(1)}) \otimes m_{(2)}.$$

We will show that β is a right K-comodule morphism from M to $N \otimes K$ and is the inverse of α. We have

$$\psi\left(\sum_{(m)} m_{(0)}S(m_{(1)})\right) = \sum_{(m)} m_{(0)}S(m_{(3)}) \otimes m_{(1)}S(m_{(2)})$$
$$= \sum_{(m)} m_{(0)}S(m_{(2)}) \otimes \varepsilon(m_{(1)}) = \sum m_{(0)}S(m_{(1)}) \otimes 1.$$

Hence $\sum_{(m)} m_{(0)}S(m_{(1)}) \in N$ and $\beta(M) \subset N \otimes K$. Moreover,

$$(\beta \otimes 1)\psi(m) = \sum_{(m)} m_{(0)}S(m_{(1)}) \otimes m_{(2)} \otimes m_{(3)} = (1 \otimes \Delta)\beta(m),$$

which shows that β is a right K-comodule morphism. We also have

$$(\alpha \circ \beta)(m) = \sum_{(m)} \alpha(m_{(0)}S(m_{(1)}) \otimes m_{(2)})$$
$$= \sum_{(m)} m_{(0)}S(m_{(1)})m_{(2)} = m,$$
$$(\beta \circ \alpha)(n \otimes x) = \beta(nx) = \sum nx_{(1)}S(x_{(2)}) \otimes x_{(3)} = n \otimes x.$$

This shows that α and β are each other's inverses. Therefore α is an isomorphism of right K-comodules. That α is a right K-module morphism is clear. Now, since

$$(\psi \otimes 1)\psi(mx) = \sum_{(m)(x)} m_{(0)}x_{(1)} \otimes m_{(1)}x_{(2)} \otimes m_{(2)}x_{(3)}$$
$$= (1 \otimes \Delta)\psi(mx),$$

we have

$$\begin{aligned}\beta(mx) &= \sum_{(m)(x)} m_{(0)}x_{(1)}S(m_{(1)}x_{(2)}) \otimes m_{(2)}x_{(3)} \\ &= \sum_{(m)(x)} m_{(0)}x_{(1)}S(x_{(2)})S(m_{(1)}) \otimes m_{(2)}x_{(3)} \\ &= \sum_{(m)} m_{(0)}S(m_{(1)}) \otimes m_{(2)}x = \beta(m)x.\end{aligned}$$

Hence β is also a right K-module morphism. Thus we conclude that α is an isomorphism of right K-bimodules.

EXAMPLE 3.1 Given a cocommutative pointed k-Hopf algebra H, let $G = G(H)$ be the group of all group-like elements of H. Let $\pi: H \to kG$ be a k-bialgebra morphism as in Corollary 2.4.30. Let the restriction of the multiplication map μ of H to $H \otimes kG$ be denoted by

$$\varphi: H \otimes kG \to H$$

and let

$$\psi: H \to H \otimes kG$$

be the k-linear map $(1 \otimes \pi)\Delta$. Then φ, ψ are structure maps of H considered as a right kG-module and as a right kG-comodule respectively. Therefore H becomes a right kG-bimodule. Since we have

$$H_1 = \{h \in H; \psi(h) = h \otimes 1\}, H_1 g = H_g,$$

it follows from Theorem 3.1.8 that the map

$$\alpha: H_1 \otimes kG \to H$$

which is induced by the multiplication map of H is an isomorphism of right kG-bimodules.

EXAMPLE 3.2 Let H be a k-Hopf algebra and let M be the sum of all finite dimensional left ideals of H^*. Set

$M' = \{f \in H^*; \operatorname{Ker} f \supset I, I$ is a finite codimensional left coideal of $H\}$.

We claim that $M = M'$ and that M becomes a rational left H^*-module. Therefore it has the structure of a right H-comodule.

Meanwhile, M is a right H-module by the action $\leftharpoonup$ which is given by

$$(f \leftharpoonup x)(y) = f(yS(x)),\ f \in M,\ x,\ y \in H.$$

With respect to these two structures, M becomes a right H-bimodule. In fact, M is a locally finite left H^*-module by definition. Moreover if $\mathfrak{a}$ is a finite dimensional left ideal of H^*, then $\mathfrak{a}^{\perp(H)} = \{x \in H; \langle f, x \rangle = 0\ \forall f \in \mathfrak{a}\}$ is a finite codimensional left coideal of H, so $\mathfrak{a} \subset M'$. Hence $M \subset M'$. Conversely, let J be a finite codimensional left coideal of H. Regarding H as a left H-comodule, J is a left H-subcomodule of H. By (3.9), J, H are rational right H^*-modules, and from the exact sequence of right H^*-modules

$$0 \to J \to H \to H/J \to 0,$$

we automatically obtain the exact sequence of left H^*-modules

$$0 \leftarrow J^* \leftarrow H^* \leftarrow (H/J)^* \leftarrow 0.$$

Here, the action of H^* on H^* is given by the dual of the action $\leftharpoonup$ of H^* on H, namely by

$$\langle x \leftharpoonup f, g \rangle = \langle x, fg \rangle,\ x \in H\ \ f, g \in H^*.$$

Now, $(H/J)^* \cong J^\perp$ is a rational left H^*-module which is contained in M when regarded as a submodule of H^*, and hence $M' \subset M$. Therefore we conclude that $M' = M$. In order to demonstrate that M is a right H-bimodule, we make use of the following lemma.

LEMMA 3.1.9 Given a k-Hopf algebra H, $f, g \in H^*, a \in H$, we have

$$f(g \leftharpoonup a) = \sum_{(a)} ((a_{(2)} \rightharpoonup f)g) \leftharpoonup a_{(1)}.$$

Proof For $x \in H$,

$$\begin{aligned}
\sum_{(a)} \langle (a_{(2)} \rightharpoonup f)g \leftharpoonup a_{(1)}, x \rangle &= \sum_{(a)} \langle (a_{(2)} \rightharpoonup f)g, xS(a_{(1)}) \rangle \\
&= \sum_{(a)(x)} \langle a_{(2)} \rightharpoonup f, x_{(1)}S(a_{(1)})_{(1)} \rangle \langle g, x_{(2)}S(a_{(1)})_{(2)} \rangle \\
&= \sum_{(a)(x)} \langle a_{(3)} \rightharpoonup f, x_{(1)}S(a_{(2)}) \rangle \langle g, x_{(2)}S(a_{(1)}) \rangle \\
&= \sum_{(x)} \langle f, x_{(1)} \rangle \langle g, x_{(2)}S(a) \rangle = \langle f(g \leftharpoonup a), x \rangle,
\end{aligned}$$

which gives the required equality.

Now let ψ be the structure map of M regarded as a right H-comodule. For $g \in M$, set $\psi(g) = \sum g_{(0)} \otimes g_{(1)}$. Then, for any $f \in H^*$, we have $fg = \sum \langle f, g_{(1)} \rangle g_{(0)}$. Hence thanks to Lemma 3.1.9, we obtain

$$\begin{aligned} f(g \leftharpoonup a) &= \sum_{(a)} ((a_{(2)} \rightharpoonup f)g) \leftharpoonup a_{(1)} \\ &= \sum_{(a)(g)} \langle a_{(2)} \rightharpoonup f, g_{(1)} \rangle g_{(0)} \leftharpoonup a_{(1)} \\ &= \sum_{(a)(g)} \langle f, g_{(1)} a_{(2)} \rangle (g_{(0)} \leftharpoonup a_{(1)}). \end{aligned}$$

This gives $\psi(g \leftharpoonup a) = \sum_{(a)(g)} (g_{(0)} \leftharpoonup a_{(1)}) \otimes g_{(1)} a_{(2)} = \psi(g) \leftharpoonup a$, and we see that M is indeed a right H-bimodule.

Set $M^H = \{f \in M; \psi(f) = f \otimes 1\}$. Then, by Theorem 3.1.8,

$$\alpha : M^H \otimes H \to M, \qquad f \otimes x \mapsto f \leftharpoonup x$$

becomes an isomorphism of right H-bimodules.

EXAMPLE 3.3 Given a commutative k-bialgebra H, let K be a k-sub-bialgebra of H and let $i : K \to H$ be the canonical embedding. Suppose there exists a k-bialgebra projection $q : H \to K$, i.e. a k-bialgebra morphism q whose restriction to K is the identity map. Moreover, assume that H has a right H-comodule structure with structure map $\psi' : H \to H \otimes H$ (e.g. Δ). Then H becomes a right K-module as well as a right K-comodule with the respective structure maps given by

$$\begin{aligned} \varphi &= \mu(1 \otimes i) : H \otimes K \to H, \\ \psi &= (1 \otimes q)\psi' : H \to H \otimes K. \end{aligned}$$

Furthermore, H becomes a right K-bimodule with respect to these structures. Therefore, by Theorem 3.1.8, we obtain the right K-bimodule isomorphism $H^K \otimes K \cong H$.

EXERCISE 3.2 Verify that H is a right K-bimodule with respect to the above mentioned structure maps.

2 Bimodules and bialgebras

In this section, we begin by defining modules, comodules and bimodules over a k-bialgebra. We then proceed to define algebraic systems which admit one of the above structures while at the same time admitting a k-algebra, k-coalgebra, or k-bialgebra structure. We also present some examples which will become useful in the following section. As a result of these definitions, we obtain a construction for Hopf algebras which turns out to be an analogue of a semi-direct product of groups or rings.

2.1 K-module k-algebras and K-comodule k-coalgebras

Let K be a k-bialgebra and A a k-algebra. Further, assume that A is a right K-module with structure map $\varphi : A \otimes K \to A$, written $\varphi(a \otimes x) = a \leftharpoonup x$. We see that $A \otimes A$, k both have right K-module structures when we define

$$(a \otimes b) \leftharpoonup x = \sum (a \leftharpoonup x_{(1)}) \otimes (b \leftharpoonup x_{(2)}),$$
$$1 \leftharpoonup x = \varepsilon(x)1, \quad a, b \in A, \quad x \in K.$$

Suppose that, with respect to the above K-module structures, the structure maps $\mu_A : A \otimes A \to A$, $\eta_A : k \to A$ of the k-algebra A turn out to be right K-module morphisms, namely that

$$(ab) \leftharpoonup x = \sum (a \leftharpoonup x_{(1)})(b \leftharpoonup x_{(2)})$$
$$1 \leftharpoonup x = \varepsilon(x)1.$$

Then we call A a **right K-module k-algebra**. A **left K-module k-algebra** may be defined similarly. Dually, given a k-coalgebra C, let C also be a right K-comodule with structure map $\psi : C \to C \otimes K$. For $c \in C$ write $\psi(c) = \sum_{(c)} c_{(0)} \otimes c_{(1)}$. Now $C \otimes C$, k become right K-comodules if we define

$$\psi_{C \otimes C}(c \otimes d) = \sum c_{(0)} \otimes d_{(0)} \otimes c_{(1)} d_{(1)}, \quad c, d \in C,$$
$$\psi_k(\alpha) = 1 \otimes \eta_K(\alpha), \quad \alpha \in k.$$

When the structure maps $\Delta_C : C \to C \otimes C$, $\varepsilon_C : C \to k$ of the k-coalgebra C are right K-comodule morphisms with respect to the above right

K-comodule structures, namely when the equalities

$$(\Delta_C \otimes 1)\psi = (1 \otimes 1 \otimes \mu_K)(1 \otimes \tau \otimes 1)(\psi \otimes \psi)\Delta_C,$$
$$(\varepsilon_C \otimes 1)\psi = (1 \otimes \eta_K)\varepsilon_C$$

are satisfied, we call C a **right K-comodule k-coalgebra**. If A is a right K-module k-algebra, then its dual k-coalgebra A° is a right K°-comodule k-coalgebra, and if C is a right K-comodule k-coalgebra, then C° becomes a right K°-module k-algebra. We can similarly define a **left K-comodule k-coalgebra**.

2.2 K-comodule k-algebras and K-module k-coalgebras

Let K be a k-bialgebra. If a k-algebra A is also a right K-comodule and the structure maps $\mu_A : A \otimes A \to A, \eta_A : k \to A$ of the k-algebra A are K-comodule morphisms, namely when

$$\psi(ab) = \sum_{(a)(b)} a_{(0)}b_{(0)} \otimes a_{(1)}b_{(1)}, \quad \psi(1) = 1 \otimes 1,$$

then A is said to be a **right K-comodule k-algebra**. For instance K is a right K-comodule k-algebra when $\Delta_K : K \to K \otimes K$ is taken as its right K-comodule structure map. A **left K-comodule k-algebra** can be defined similarly. Dually, if a k-coalgebra C is also a right K-module and the structure maps $\Delta_C : C \to C \otimes C, \varepsilon_C : C \to k$ of C are K-module morphisms, namely when we have

$$\sum_{(c)(x)} c_{(0)}x_{(1)} \otimes c_{(1)}x_{(2)} = \sum (cx)_{(0)} \otimes (cx)_{(1)},$$
$$\varepsilon(cx) = \varepsilon(c)\varepsilon(x),$$

then C is said to be a **right K-module k-coalgebra**. With respect to its multiplication map $\mu_K : K \otimes K \to K$, K is a right K-module and further a right K-module k-coalgebra. If A is a K-comodule k-algebra, then A° is a K°-module k-coalgebra.

EXAMPLE 3.4 Let H be a k-bialgebra and H^* its dual k-algebra. For $a, b \in H$, $f \in H^*$, set

$$\langle a \rightharpoonup f, b\rangle = \langle f, ba\rangle, \ \langle f \leftharpoonup a, b\rangle = \langle f, ab\rangle.$$

Then H^* becomes a left or right H-module k-algebra with respect to the actions $\rightharpoonup$, $\leftharpoonup$ respectively. If, in particular, H is a k-Hopf algebra, set

$$\langle a \rightharpoonup f, b\rangle = \langle f, S(a)b\rangle,$$
$$\langle f \leftharpoonup a, b\rangle = \langle f, bS(a)\rangle.$$

Then H^* becomes a left or right H-module k-algebra with respect to $\rightharpoonup$, $\leftharpoonup$ respectively. (cf. (3.4), (3.5), (3.6), (3.7).)

EXAMPLE 3.5 Let B be a k-bialgebra and B^* its dual k-algebra. For $f, g \in B^*$, $a \in B$, set

$$\langle f \rightharpoonup a, g\rangle = \langle a, gf\rangle, \quad \langle a \leftharpoonup f, g\rangle = \langle a, fg\rangle,$$

i.e. define $f \rightharpoonup a = \sum_{(a)} a_{(1)}\langle a_{(2)}, f\rangle$, $a \leftharpoonup f = \sum_{(a)} \langle a_{(1)}, f\rangle a_{(2)}$. Then B becomes a left or right B^*-module with respect to $\rightharpoonup$, $\leftharpoonup$ respectively. Moreover, if B is commutative, then B becomes a left or right B^*-module k-algebra.

2.3 K-module k-bialgebras and K-comodule k-bialgebras

Let K, B be k-bialgebras. Further, assume B is also a right K-module with structure map $\varphi : B \otimes K \to B$ with respect to which B is a K-module k-algebra as well as a K-module k-coalgebra. Namely, suppose that μ_B, η_B, Δ_B, ε_B are K-module morphisms, i.e. setting $\varphi(b \otimes x) = b \leftharpoonup x$ for $x \in K$, $b \in B$, we have

$$(ab) \leftharpoonup x = \sum_{(x)} (a \leftharpoonup x_{(1)})(b \leftharpoonup x_{(2)}), \quad 1 \leftharpoonup x = \varepsilon(x)1,$$

$$\Delta(a \leftharpoonup x) = \sum_{(a)(x)} (a_{(1)} \leftharpoonup x_{(1)}) \otimes (a_{(2)} \leftharpoonup x_{(2)}),$$

$$\varepsilon(a \leftharpoonup x) = \varepsilon(a)\varepsilon(x).$$

In these circumstances, B is said to be a **right K-module k-bialgebra**. Dually, if a k-bialgebra B is also a right K-comodule with structure map $\psi : B \to B \otimes K$ such that with respect to ψ it is a right K-comodule k-algebra as well as a right K-comodule k-coalgebra, i.e. if μ_B, η_B, Δ_B, ε_B are K-comodule morphisms, then B is said to be a **right K-comodule k-bialgebra**. A **left K-module** (resp. **left K-comodule**)

k-**bialgebra** may be defined similarly. If B is a right K-module (resp. right K-comodule) k-bialgebra, then B° is a right K°-comodule (resp. right K°-module) k-bialgebra. A k-Hopf algebra which is also a right K-module (resp. right K-comodule) k-bialgebra when regarded as a k-bialgebra is called a **right K-module** (resp. **right K-comodule**) k-**Hopf algebra**.

EXAMPLE 3.6 Let G, S be groups and let Aut (S) be the group of all automorphisms of S. Let $K = kG, L = kS$ respectively be the group k-bialgebras of G, S. Given a group morphism $\rho : G \to \text{Aut}(S)$, if we define a k-linear map $\varphi : K \otimes L \to L$ by

$$\varphi(x \otimes a) = \rho(x)a = x \rightharpoonup a, \qquad x \in G, a \in S,$$

then L becomes a left K-module k-bialgebra.

EXAMPLE 3.7 Let M, N be k-Lie algebras. Let $\text{Der}_k(N)$ be the k-Lie algebra consisting of all k-derivations from N to N, and let $K = U(M), L = U(N)$ be the universal enveloping k-algebras of M, N respectively. Suppose that we are given a k-Lie algebra morphism $\rho : M \to \text{Der}_k(N)$. If we define a k-linear map $\varphi : K \otimes L \to L$ by

$$\varphi(x \otimes a) = \rho(x)a = x \rightharpoonup a, \qquad x \in M, a \in N,$$

then L becomes a left K-module k-bialgebra.

EXAMPLE 3.8 Let H be a k-Hopf algebra, and define a k-linear map $\psi : H \to H \otimes H$ by

$$\psi(x) = \sum_{(x)} x_{(1)} S(x_{(3)}) \otimes x_{(2)}, \qquad x \in H.$$

Then H is a left H-comodule, and if H is commutative, then H is a left H-comodule k-Hopf algebra. Now let H be a commutative k-Hopf algebra. Setting $G = G(H^\circ) = \mathbf{Alg}_k(H, k)$, $L = P(H^\circ)$, the dual k-linear map $\psi^\circ : H^\circ \otimes H^\circ \to H^\circ$ of ψ induces the maps

$$G(\psi) : G \times G \to G, \quad (a, b) \mapsto aba^{-1},$$
$$L(\psi) : L \times L \to L, \quad (a, b) \mapsto [a, b] = ab - ba.$$

The automorphism of G given by $b \mapsto aba^{-1}$ for $b \in G$ and the derivation of L given by $b \mapsto [a, b]$ for $b \in L$ are respectively called **inner automorphism** and **inner derivation** (cf. Example 1.8).

EXERCISE 3.3 Verify what is stated in Example 3.7.

We define a k-linear map $\psi': H \to H \otimes H$ by

$$\psi'(x) = \sum_{(x)} x_{(2)} \otimes S(x_{(1)}) x_{(3)}, \qquad x \in H.$$

Show that H becomes a right H-comodule and that if H is commutative, it becomes a right H-comodule k-Hopf algebra.

EXAMPLE 3.9 Given a cocommutative k-Hopf algebra H, set $G = G(H)$ and let H_1 be the irreducible component containing the identity element 1. Taking the k-linear map

$$\varphi : kG \otimes H_1 \to H_1, \qquad g \otimes h \mapsto ghg^{-1}, \qquad h \in H_1, \qquad g \in G$$

as the structure map, H_1 becomes a left kG-module k-bialgebra.

THEOREM 3.2.1 Let H, K be k-Hopf algebras. If H is a left K-module k-Hopf algebra, then the antipode S of H is K-linear, namely

$$S(x \rightharpoonup a) = x \rightharpoonup S(a), \qquad x \in K, a \in H.$$

Proof Since $\varepsilon(a)1 = \sum_{(a)} a_{(1)} S(a_{(2)})$, we obtain

$$\begin{aligned}\varepsilon(x)\varepsilon(a)1 &= x \rightharpoonup \varepsilon(a)1 = x \rightharpoonup \sum_{(a)} a_{(1)} S(a_{(2)}) \\ &= \sum_{(a)(x)} (x_{(1)} \rightharpoonup a_{(1)})(x_{(2)} \rightharpoonup S(a_{(2)})).\end{aligned}$$

Thus

$$\begin{aligned}S(x \rightharpoonup a) &= \sum_{(a)(x)} S(x_{(1)} \rightharpoonup a_{(1)}) \varepsilon(x_{(2)} \rightharpoonup a_{(2)}) \\ &= \sum_{(x)(a)} S(x_{(1)} \rightharpoonup a_{(1)}) \varepsilon(x_{(2)}) \varepsilon(a_{(2)}) \\ &= \sum_{(x)(a)} S(x_{(1)} \rightharpoonup a_{(1)})(x_{(2)} \rightharpoonup a_{(2)})(x_{(3)} \rightharpoonup S(a_{(3)})) \\ &= \sum_{(x)(a)} \varepsilon(x_{(1)} \rightharpoonup a_{(1)})(x_{(2)} \rightharpoonup S(a_{(2)})) \\ &= \sum_{(x)(a)} \varepsilon(x_{(1)}) x_{(2)} \rightharpoonup \varepsilon(a_{(1)}) S(a_{(2)}) \\ &= x \rightharpoonup S(a).\end{aligned}$$

2.4 Semi-direct product of cocommutative k-Hopf algebras

Let K, L be cocommutative k-Hopf algebras. Suppose L is a left K-module k-Hopf algebra with structure map $\varphi : K \otimes L \to L$, and write $\varphi(x \otimes a) = x \rightharpoonup a$ for $x \in K$, $a \in L$. We then define a cocommutative k-Hopf algebra $L \sharp K$ as follows and call it the **semi-direct product** of K and L.

(1) As a k-linear space, $L \sharp K$ coincides with $L \otimes K$, and we denote $a \otimes x$ by $a \sharp x$.

(2) We define the k-algebra structure map by

$$\mu_{L \sharp K} = (\mu_L \otimes \mu_K)(1 \otimes \varphi \otimes 1 \otimes 1)(1 \otimes 1 \otimes \tau \otimes 1)(1 \otimes \Delta \otimes 1 \otimes 1),$$
$$\eta_{L \sharp K} = \eta_L \otimes \eta_K.$$

In other words, for $a, b \in L$, $x, y \in K$, we define

$$(a \sharp x)(b \sharp y) = \sum_{(x)} a(x_{(1)} \rightharpoonup b) \sharp x_{(2)} y,$$
$$1 = 1 \sharp 1.$$

(Verify that $\mu_{L \sharp K}$, $\eta_{L \sharp K}$ are indeed k-algebra structure maps.)

(3) The k-coalgebra structure is defined via the k-coalgebra tensor product of L and K. Namely, for $a \in L$, $x \in K$, we define

$$\Delta(a \sharp x) = \sum_{(a)(x)} (a_{(1)} \sharp x_{(1)}) \otimes (a_{(2)} \sharp x_{(2)}),$$
$$\varepsilon(a \sharp x) = \varepsilon(a)\varepsilon(x).$$

(Verify that Δ, ε are k-algebra morphisms.)

(4) We define the antipode $S_{L \sharp K}$ by

$$S_{L \sharp K} = (S_L \otimes S_K)(\varphi \otimes 1)(S_K \otimes 1 \otimes 1)(\tau \otimes 1)(1 \otimes \Delta).$$

Namely, for $a \in L$, $x \in K$, we define

$$S_{L \sharp K}(a \sharp x) = S_K(x_{(1)}) \rightharpoonup S_L(a) \sharp S_K(x_{(2)}).$$

(Verify that $\mu_{L \sharp K}(1 \otimes S_{L \sharp K})\Delta_{L \sharp K} = \eta_{L \sharp K} \circ \varepsilon_{L \sharp K}$. cf. Theorem 3.2.1.)

EXAMPLE 3.10 Let G, S be groups and let $K = kG, L = kS$ be group k-Hopf algebras of G, S respectively. These turn out to be cocommutative k-Hopf algebras. Furthermore, as was shown in Example 3.6, when given a group morphism $\rho : G \to \mathrm{Aut}\,(S)$, L becomes a left

K-module k-Hopf algebra. In this situation, the semi-direct product $L \sharp K$ of L and K becomes a cocommutative k-Hopf algebra, and for $a, b \in S$, $x, y \in G$, we have

$$(a \sharp x)(b \sharp y) = a\rho(x)b \sharp xy.$$

Denoting the semi-direct product of S and G with respect to ρ by $S \times_\rho G$, it turns out that $L \sharp K$ is isomorphic to the group k-Hopf algebra $k(S \times_\rho G)$ of $S \times_\rho G$.

EXAMPLE 3.11 Let M, N be k-Lie algebras, and let $K = U(M)$, $L = U(N)$ be the universal enveloping k-Hopf algebras of M, N respectively. Then K, L are cocommutative k-Hopf algebras. Moreover, as was shown in Example 3.7, when given a k-Lie algebra morphism $\rho : M \to \mathrm{Der}_k(N)$, L becomes a left K-module k-Hopf algebra. Now the semi-direct product $L \sharp K$ of L and K is a cocommutative k-Hopf algebra and is isomorphic to the universal enveloping k-Hopf algebra $U(N \oplus_\rho M)$ of the semi-direct sum $N \oplus_\rho M$ of N and M with respect to ρ. (cf. Exercise 1.18.)

2.5 The co-semi-direct product of cocommutative k-Hopf algebras

We will define co-semi-direct products which are dual to semi-direct products. Let K, L be commutative k-Hopf algebras. Further, suppose L is a left K-comodule k-Hopf algebra with a structure map $\psi : L \to K \otimes L$. Now we define a commutative k-Hopf algebra $L \flat K$ as follows, and call it the **co-semi-direct product** of L and K.

(1) As a k-linear space, $L \flat K$ coincides with $L \otimes K$; for $a \in L$, $x \in K$, we denote $a \otimes x$ by $a \flat x$.

(2) The k-algebra structure of $L \flat K$ is defined by the k-algebra tensor product of L and K. Namely, given $a, b \in L$, $x, y \in K$, we define

$$(a \flat x)(b \flat y) = ab \flat xy.$$

(3) We define the k-coalgebra structure map of $L \flat K$ as follows.

$$\Delta_{L \flat K} = (1 \otimes \mu_K \otimes 1 \otimes 1)(1 \otimes 1 \otimes \tau \otimes 1)(1 \otimes \psi \otimes 1 \otimes 1)(\Delta_L \otimes \Delta_K),$$
$$\varepsilon_{L \flat K} = \varepsilon_L \otimes \varepsilon_K.$$

(4) The antipode $S_{L \flat K}$ is defined as follows.

$$S_{L \flat K} = (1 \otimes \mu_K)(\tau \otimes 1)(S_K \otimes 1 \otimes 1)(\psi \otimes 1)(S_L \otimes S_K).$$

EXERCISE 3.4 Prove that the structure maps defined in (2), (3), (4) above satisfy the axioms of a k-Hopf algebra.

EXERCISE 3.5 If K, L are commutative k-Hopf algebras and L is also a left K-comodule k-Hopf algebra, then K°, L° are cocommutative k-Hopf algebras and L° is also a left K°-module k-Hopf algebra. Moreover, we have $(L \flat K)^\circ \cong L^\circ \sharp K^\circ$. Now show that

$$G((L \flat K)^\circ) \cong G(L^\circ)\cdot G(K^\circ) \quad \text{(semi-direct product)},$$
$$P((L \flat K)^\circ) = P(L^\circ)\otimes 1 + 1 \otimes P(K^\circ)$$
$$\cong P(L^\circ)\oplus P(K^\circ) \quad \text{(semi-direct sum)}.$$

3 Integrals for Hopf algebras

It is a standard result that the set $\mathscr{R}(G)$ of all real-valued continuous representative functions on a compact topological group G becomes a Hopf algebra over the real number field $\mathbb{R}$ (cf. Chapter 2, §2.2), and we can define a Haar measure on G which is invariant by the left (or right) translation $x \mapsto gx$ (or $x \mapsto xg$) where g is an element of G, and we see that an integral over G is a linear form on $\mathscr{R}(G)$ which is invariant by such transformations. Here we define integrals for a k-Hopf algebra to be linear forms on the k-Hopf algebra with invariant properties similar to those of topological groups. In this section, we consider when such integrals exist and investigate their properties. It will be shown that the complete reducibility of representations of finite groups or compact groups and the existence of integrals are interrelated.

3.1 The definition of integrals and some examples

Let H be a k-Hopf algebra. We call $\sigma \in H^*$ a **left integral** (or simply an **integral**) for H if for any $f \in H^*$, σ satisfies

$$f\sigma = \langle f, 1\rangle\sigma. \tag{3.17}$$

We let $\mathscr{I}_H$ be the set of all left integrals for H. It is easily seen that $\mathscr{I}_H$ is a k-linear space as well as a left ideal of H^*. Moreover, we have

$$\mathscr{I}_H = \{\sigma \in H^*;\, (1\otimes\sigma)\circ\Delta = \eta\circ\sigma\}$$
$$= \{\sigma \in H^*;\, \sigma \text{ is a left } H\text{-comodule morphism}\}.$$

Indeed, for $f \in H^*$ and $x \in H$,

$$\langle (1\otimes\sigma)\Delta x, f\rangle = \sum_{(x)} \langle f, x_{(1)}\rangle \sigma(x_{(2)}) = \langle f\sigma, x\rangle,$$

$$\langle \eta\circ\sigma(x), f\rangle = \langle \sigma, x\rangle\langle 1, f\rangle = \langle f, 1\rangle\sigma(x),$$

so that we obtain $(1\otimes\sigma)\Delta = \eta\circ\sigma \Leftrightarrow f\sigma = \langle f, 1\rangle\sigma$ for any $f \in H^*$. Further, M^H defined in Example 3.2 is the same as $\mathscr{I}_H$. In fact, if $\sigma \in \mathscr{I}_H$, we see that $\sigma \in M$ since $k\sigma$ is a one-dimensional left ideal of H^*. On the other hand, we have

$$f\sigma = \langle f, 1\rangle\sigma \text{ for any } f \in H^* \Leftrightarrow \psi(\sigma) = \sigma\otimes 1,$$

and hence we get $M^H = \mathscr{I}_H$. The fact that $M^H \otimes H$ is isomorphic to M via the correspondence $f\otimes x \mapsto f \leftharpoonup x$ as right H-comodules and $f \leftharpoonup x = S(x) \rightharpoonup f$ implies that S is injective if $\mathscr{I}_H \neq \{0\}$. Since $M^H = \mathscr{I}_H$, we obtain the following theorem.

THEOREM 3.3.1 For a k-Hopf algebra H, the following conditions are equivalent.

(i) H has a non-zero left integral, namely, $\mathscr{I}_H \neq \{0\}$.
(ii) H^* contains a non-zero left ideal of finite dimension.
(iii) H contains a proper left coideal of finite codimension.

EXAMPLE 3.12 Let G be a group and $H = kG$ the group k-Hopf algebra, and let $e \in G$ be the identity element. If we define $\sigma \in H^*$ by $\sigma(e) = 1$ and $\sigma(x) = 0$ for $x \in G$, $x \neq e$, then σ is a left integral for H. Indeed, for $x \in G$ and for any $f \in H^*$, we have

$$\begin{aligned}\langle f\sigma, x\rangle &= \langle f\otimes\sigma, x\otimes x\rangle = \langle f, x\rangle\langle\sigma, x\rangle\\ &= \langle f, 1\rangle\sigma(x).\end{aligned}$$

EXAMPLE 3.13 Let G be a finite group and $H = (kG)^*$ the dual k-Hopf algebra of the group k-Hopf algebra kG. Identifying H^* with kG, we see that $\sigma = \sum_{g\in G} g \in H^*$ is an integral for H. In fact, for $x \in H$ and for $f = \sum_{i=1}^{n} a_i g_i$, $a_i \in k$, $g_i \in G$ $(1 \leqq i \leqq n)$, we have

$$\begin{aligned}\langle f\sigma, x\rangle &= \langle \sigma, x \leftharpoonup f\rangle = \sum_{g\in G}\langle g, x\leftharpoonup f\rangle \\ &= \sum_{g\in G}\langle fg, x\rangle = \sum_{g\in G}\sum_{i=1}^{n}\langle a_i g_i g, x\rangle \\ &= \sum_{i=1}^{n} a_i \sum_{g\in G}\langle g_i g, x\rangle = \left(\sum_{i=1}^{n} a_i\right)\sigma(x).\end{aligned}$$

On the other hand, since

$$\langle f, 1\rangle = \sum_{i=1}^{n} a_i \langle g_i, 1\rangle = \sum_{i=1}^{n} a_i,$$

We get $f\sigma = \langle f, 1\rangle\sigma$. Now if we denote the order of G by $|G|$, then we have $\varepsilon(\sigma) = |G|$. Therefore $\varepsilon(\sigma) \neq 0$ if the characteristic of k is relatively prime to $|G|$.

EXAMPLE 3.14 Let G be a compact topological group and $H = \mathscr{R}(G)$ the $\mathbb{R}$-Hopf algebra of all real-valued continuous representative functions on G. Suppose ν is a Haar measure on G and set

$$\sigma(f) = \int_G f d\nu, \quad f \in H.$$

Then $\sigma \in H^*$ is an integral for H. Indeed, since a Haar measure is left invariant, for $x \in H^*$ and $f \in H$, we have

$$\begin{aligned}\langle x\sigma, f\rangle &= \langle \sigma, f \leftharpoonup x\rangle = \int_G (f\leftharpoonup x) d\nu = \int_G f d\nu \\ &= \langle \sigma, f\rangle.\end{aligned}$$

On the other hand, since $\langle x, 1\rangle = 1$, we have

$$x\sigma = \sigma = \langle x, 1\rangle\sigma.$$

EXAMPLE 3.15 Let $k[X]$ be the polynomial ring in one variable over a field k of characteristic $p > 0$. Then $k[X]$ is a k-Hopf algebra defined by $\Delta(X) = X\otimes 1 + 1\otimes X$, $\varepsilon(X) = 0$ and $S(X) = -X$. Let $\mathfrak{a}$ be the ideal of $k[X]$ generated by X^p, let $H = k[X]/\mathfrak{a}$, and write $H = k[x]$ where x is the image of X under the canonical projection $k[X] \to H$. Then H is a k-Hopf algebra where its structure is induced canonically from that of $k[X]$ since $\varepsilon(X^p) = 0$ and

$\Delta X^p = X^p \otimes 1 + 1 \otimes X^p$. (Note that $\mathfrak{a}$ is a Hopf ideal, which will be defined in Chapter 4, §1.) Now, $1, x, x^2, \ldots, x^{p-1}$ form a basis for H over k, and if we identify H with H^* via the bilinear form on H defined by $\langle x^i, x^j \rangle = \delta_{ij}$ $(0 \leqq i, j \leqq p-1)$, then x^{p-1} is an integral for H. Indeed we have

$$\langle x^i x^{p-1}, x^j \rangle = \langle x^i, 1 \rangle \langle x^{p-1}, x^j \rangle = \delta_{i0} \langle x^{p-1}, x^j \rangle.$$

Thus we get $x^i x^{p-1} = \langle x^i, 1 \rangle x^{p-1}$. In this situation, we have $\varepsilon(x^{p-1}) = 0$.

3.2 Integrals and complete reducibility

Let H be a k-Hopf algebra and M a left H-comodule. Suppose that for any left H-subcomodule N of M, there exists a left H-comodule morphism $p: M \to N$ such that $p \circ i = 1_N$ where $i: N \to M$ is the canonical embedding. Then we call M a **completely reducible left H-comodule**. This is equivalent to saying that M is a direct sum of simple left H-subcomodules.

THEOREM 3.3.2 For a k-Hopf algebra H, the following conditions are equivalent.

(i) Any left H-comodule is completely reducible.

(ii) Considering H as a left H-comodule via $\Delta: H \to H \otimes H$, H is completely reducible.

(iii) There exists a left integral σ for H satisfying $\sigma \circ \eta = 1_k$.

(iv) There exists a k-linear map $\rho: H \otimes H \to k$ satisfying

$$\rho \circ \Delta = \varepsilon, \quad (1 \otimes \rho)(\Delta \otimes 1) = (\rho \otimes 1)(1 \otimes \Delta).$$

Proof (i) $\Rightarrow$ (ii) is straightforward.

(ii) $\Rightarrow$ (iii) Now k and H are left H-comodules via η and Δ respectively. We have that $\eta: k \to H$ is a left H-comodule morphism and, by assumption (ii), H is completely reducible. Hence there exists a left H-comodule morphism $\sigma: H \to k$ such that $\sigma \circ \eta = 1_k$. The fact that σ is a left H-comodule morphism simply means that σ satisfies $(1 \otimes \sigma)\Delta = \eta \circ \sigma$.

(iii)⇒(iv) Let $\sigma : H \to k$ be a left integral for H such that $\sigma \circ \eta = 1_k$. Letting $\rho = \sigma \circ \mu \circ (1 \otimes S) : H \otimes H \to k$, we will show that ρ satisfies the condition of (iv). For $x \in H$, we have

$$\rho \circ \Delta(x) = \sigma\left(\sum_{(x)} x_{(1)} S(x_{(2)})\right) = \sigma \circ \varepsilon(x) = \varepsilon(x).$$

Thus we get $\rho \circ \Delta = \varepsilon$. Moreover, noting that

$$\sum_{(x)} x_{(1)} \sigma(x_{(2)}) = \eta\sigma(x) = \sigma(x)$$

since $(1 \otimes \sigma)\Delta = \eta \circ \sigma$, we have

$$\begin{aligned}(\rho \otimes 1)(1 \otimes \Delta)(x \otimes y) &= \sum_{(y)} \sigma(xS(y_{(1)}))y_{(2)} \\ &= \sum_{(x)(y)} \sigma(x_{(2)}S(y_{(1)}))x_{(1)}S(y_{(2)})y_{(3)} \\ &= \sum_{(x)(y)} \sigma(x_{(2)}S(y_{(1)}))x_{(1)}\varepsilon(y_{(2)}) \\ &= \sum_{(x)} \sigma(x_{(2)}S(y))x_{(1)} \\ &= (1 \otimes \rho)(\Delta \otimes 1)(x \otimes y).\end{aligned}$$

Therefore $(\rho \otimes 1)(1 \otimes \Delta) = (1 \otimes \rho)(\Delta \otimes 1)$ holds.

(iv)⇒(i) Suppose there exists a k-linear map $\rho : H \otimes H \to k$ satisfying the condition of (iv). Let N be a left H-subcomodule of a left H-comodule M and let $i : N \to M$ be the canonical embedding. Choose a k-linear map $p : M \to N$ such that $p \circ i = 1_N$, and let

$$q = (\rho \otimes 1_N)(1_H \otimes \psi_N)(1_H \otimes p)\psi_M.$$

Then we will show that $q : M \to N$ is a left H-comodule morphism satisfying $q \circ i = 1_N$. Now we have

$$\begin{aligned}q \circ i &= (\rho \otimes 1_N)(1_H \otimes \psi_N)(1_H \otimes p)\psi_M \circ i \\ &= (\rho \otimes 1_N)(1_H \otimes \psi_N)(1_H \otimes p)(1_H \otimes i)\psi_N \\ &= (\rho \otimes 1_N)(1_H \otimes \psi_N)\psi_N = (\rho \otimes 1_N)(\Delta \otimes 1_N)\psi_N \\ &= (\rho \circ \Delta \otimes 1_N)\psi_N = (\varepsilon \otimes 1_N)\psi_N = 1_N,\end{aligned}$$

$$\begin{aligned}
&(1_H\otimes q)\psi_M\\
&=(1_H\otimes\rho\otimes 1_N)(1_H\otimes 1_H\otimes\psi_N)(1_H\otimes 1_H\otimes p)(1_H\otimes\psi_M)\psi_M\\
&=(1_H\otimes\rho\otimes 1_N)(1_H\otimes 1_H\otimes\psi_N)(1_H\otimes 1_H\otimes p)(\Delta\otimes 1_M)\psi_M\\
&=(1_H\otimes\rho\otimes 1_N)(1_H\otimes 1_H\otimes\psi_N)(\Delta\otimes 1_N)(1_H\otimes p)\psi_M\\
&=(1_H\otimes\rho\otimes 1_N)(\Delta\otimes 1_H\otimes 1_N)(1_H\otimes\psi_N)(1_H\otimes p)\psi_M\\
&=(\rho\otimes 1_H\otimes 1_N)(1_H\otimes\Delta\otimes 1_N)(1_H\otimes\psi_N)(1_H\otimes p)\psi_M\\
&=(\rho\otimes 1_H\otimes 1_N)(1_H\otimes 1_H\otimes\psi_N)(1_H\otimes\psi_N)(1_H\otimes p)\psi_M\\
&=\psi_N(\rho\otimes 1_N)(1_H\otimes\psi_N)(1_H\otimes p)\psi_M=\psi_N\circ q.
\end{aligned}$$

Thus q is a left H-comodule morphism and satisfies $q\circ i=1_N$.

COROLLARY 3.3.3 With regard to a k-Hopf algebra H, the conditions in Theorem 3.3.2 and the following conditions are equivalent.

(v) Any rational right H^*-module is completely reducible.
(vi) The right H^*-module H is completely reducible.
(vii) H is a co-semi-simple k-coalgebra.

Proof (v)$\Leftrightarrow$(vii) is stated in Theorem 3.1.7. (i)$\Rightarrow$(v)$\Rightarrow$(vi)$\Rightarrow$(ii) is obvious.

Remark When H is a finite dimensional k-Hopf algebra, condition (vii) is equivalent to the semi-simplicity of H^*. As for Example 3.13, if $|G|$ is relatively prime to the characteristic of the field k, we see that the group k-algebra kG is semi-simple and that every kG-module is completely reducible (Maschke's theorem).

3.3 The uniqueness of integrals

We will show $\dim \mathscr{I}_H \leqq 1$ where $\mathscr{I}_H$ is the k-linear space consisting of integrals for a k-Hopf algebra H. This means that if a non-zero integral exists, it is unique up to scalar multiples. Let $\bar{k}$ be the algebraic closure of k and let $\bar{H}$ be the $\bar{k}$-Hopf algebra $H\otimes_k\bar{k}$. We can identify $\mathscr{I}_H$ with a k-linear subspace of $\mathscr{I}_{\bar{H}}$, so that $\dim_k \mathscr{I}_H \leqq \dim_{\bar{k}} \mathscr{I}_{\bar{H}}$. Thus, in order to show $\dim \mathscr{I}_H \leqq 1$, it does no harm to assume that the field k is algebraically closed. Throughout this section, k is assumed to be an algebraically closed field. Let corad $H = H_0$ and $H_n = \sqcap^{n+1} H_0$. Then we obtain a sequence $H_0 \subset H_1 \subset H_2 \subset \cdots$ of k-subcoalgebras of H. As seen in the proof of Theorem 2.3.9, the coradical is conilpotent and we have $H = \bigcup_{n=0}^{\infty} H_n$. Moreover, Theorem 2.3.11 ensures the existence of a projection $\pi : H \to H_0$. Let $K = \operatorname{Ker} \pi$ and let $\pi^* : H_0^* \to H^*$ be the dual k-algebra morphism of π. For a simple k-subcoalgebra N of H, its dual k-algebra N^* is a finite dimensional simple k-algebra and is isomorphic to the k-algebra $M_n(k)$ consisting of all $n \times n$ square matrices over k for some positive integer n. Hence N can be identified with the dual k-coalgebra $M_n(k)^*$ of $M_n(k)$. Let $e_{ij} \in M_n(k)$ be the matrix with (i, j)-entry $= 1$ and all other entries $= 0$, and let $e_{(N,i)}$ $(1 \leqq i \leqq n)$ be the element of N^* identified with $e_{ii} \in M_n(k)$, where we regard $e_{(N,i)}$ as an element of H_0^* identifying N^* with a simple k-subalgebra of H_0^*. Moreover, we denote the image of $e_{(N,i)}$ under $\pi^* : H_0^* \to H^*$ by the same notation, and, in particular, we denote $e_{(k,1)}$ by e_1. Then we have

$$e_{(N,i)} e_{(N',i')} = \delta_{NN'} \delta_{ii'} e_{(N,i)},$$

and thus $e_{(N,i)}$ is an idempotent of H^*.

LEMMA 3.3.4 With the same notation as above,

(i) $H \leftharpoonup e_{(N,i)}$ (resp. $e_{(N,i)} \rightharpoonup H$) is a right (resp. left) coideal of H.

(ii) $\Delta(H \leftharpoonup e_{(N,i)}) \subset N \otimes H + K \otimes H$,

(iii) $v \in H_n \leftharpoonup e_1 \Rightarrow \Delta v \in 1 \otimes v + K \otimes H_{n-1}$,

(iv) $v \in k \sqcap N \Rightarrow \Delta v \in 1 \otimes v + H \otimes N$,

(v) $v \in H \leftharpoonup e_1 \Rightarrow v \leftharpoonup e_1 = v$.

Proof (i) For $x \in H$, we observe that

$$\begin{aligned}
\Delta(x \leftharpoonup e_{(N,i)}) &= \Delta\Big(\sum_{(x)} \langle e_{(N,i)}, x_{(1)} \rangle x_{(2)}\Big) \\
&= \sum_{(x)} \langle e_{(N,i)}, x_{(1)} \rangle x_{(2)} \otimes x_{(3)} \\
&= \sum_{(x)} (x_{(1)} \leftharpoonup e_{(N,i)}) \otimes x_{(2)}.
\end{aligned}$$

Thus $H \leftharpoonup e_{(N,i)}$ is a right coideal. Likewise, $e_{(N,i)} \rightharpoonup H$ is a left coideal.

(ii) Taking note of the equality in (i), (ii) is clear.

(iii) For $x \in H_n$, let $v = x \leftharpoonup e_1$. Since $H_n = H_0 \sqcap H_{n-1}$, we have $\Delta x \in H_0 \otimes H + H \otimes H_{n-1}$. Hence

$$\begin{aligned}\Delta(x \leftharpoonup e_1) &= \sum_{(x)} (x_{(1)} \leftharpoonup e_1) \otimes x_{(2)} \\ &= \sum_{(x)} \langle e_1, x_{(1)} \rangle x_{(2)} \otimes x_{(3)} \in k1 \otimes H \\ &\quad + K \otimes H_{n-1}.\end{aligned}$$

On the other hand, since $(\varepsilon \otimes 1)\, \Delta v = v$, we have

$$\Delta v \in 1 \otimes v + K \otimes H_{n-1}.$$

(iv) For $v \in k \sqcap N$, we see that $\Delta v \in k \otimes H + H \otimes N$. Hence we get $\Delta v \in 1 \otimes v + H \otimes N$ in the same manner as in (iii).

(v) For $x \in H$, let $v = x \leftharpoonup e_1$. Then we have

$$v \leftharpoonup e_1 = (x \leftharpoonup e_1) \leftharpoonup e_1 = x \leftharpoonup e_1{}^2 = x \leftharpoonup e_1 = v.$$

LEMMA 3.3.5 $H = \coprod_{(N,i)} (e_{(N,i)} \rightharpoonup H)$.

Here the sum is taken over (N, i) where N is the component of H_0 expressed as a direct sum $H_0 = \coprod N$ of simple k-subcolagebras, and $1 \leqq i \leqq n$ where $n^2 = \dim N$.

Proof For $x \in H$, let C_x be the k-subcoalgebra of H generated by x. Then we get

$$\sum_{(N,i),\, N \subset C_x} e_{(N,i)} \rightharpoonup x = \varepsilon \rightharpoonup x = x \in \sum_{(N,i)} (e_{N,i)} \rightharpoonup H).$$

Meanwhile, since $\{e_{(N,i)}\}_{(N,i)}$ is a set of orthogonal idempotents, the sum on the right is clearly a direct sum.

THEOREM 3.3.6 If the antipode S of H is injective, then $\dim(e_1 \rightharpoonup H) = \dim(H \leftharpoonup e_1)$.

Proof Choose a basis $F = \{v_\lambda\}_{\lambda \in \Lambda}$ for $H \leftharpoonup e_1$ such that $F \cap (H_n \leftharpoonup e_1)$ is a basis for $H_n \leftharpoonup e_1$ for each n. If S is injective, we will show that

$\{e_1 \rightharpoonup S(v_\lambda)\}_{\lambda\in\Lambda}$ is linearly independent over k in $S(H) \subset H$. If such is the case, we get $\dim(e_1 \rightharpoonup H) \geq \dim(H \leftharpoonup e_1)$. By symmetry, we obtain the opposite inequality, and thus the equality holds.

From Lemma 3.3.4 (iii), it follows that for $v_\lambda \in H_n - H_{n-1}$, $e_1 \rightharpoonup S(v_\lambda) = S(v_\lambda) + q_\lambda$ where $q_\lambda \in S(H_{n-1})$. Hence, by Lemma 3.3.4 (v), we get

$$S^{-1}(e_1 \rightharpoonup S(v_\lambda)) \leftharpoonup e_1 = v_\lambda + t_\lambda, \quad t_\lambda \in H_{n-1} \leftharpoonup e_1.$$

Assume that $\{e_1 \rightharpoonup S(v_\lambda)\}_{\lambda\in\Lambda}$ is linearly dependent over k and let

$$\sum_{\beta\in\Lambda_1} \lambda_\beta (e_1 \rightharpoonup S(v_\beta)) = 0,$$

where Λ_1 is a finite set, $\lambda_\beta \neq 0$. Then we have

$$0 = \sum_{\beta\in\Lambda_1} \lambda_\beta S^{-1}(e_1 \rightharpoonup S(v_\beta)) \leftharpoonup e_1 = \sum_{\beta\in\Lambda_1} \lambda_\beta (v_\beta + t_\beta). \tag{3.18}$$

Let n_0 be the minimal positive integer for which $\{v_\beta; \beta\in\Lambda_1\} \subset H_{n_0}$ and let $\Lambda_2 = \{\gamma\in\Lambda_1; v_\gamma \in H_{n_0} - H_{n_0-1}\}$. Then $\Lambda_2 \neq \varnothing$ by the choice of n_0. Write the right side of (3.18) as

$$\sum_{\gamma\in\Lambda_2} \lambda_\gamma v_\gamma + \sum_{\gamma\in\Lambda_2} \lambda_\gamma t_\gamma + \sum_{\beta\in\Lambda_1-\Lambda_2} \lambda_\beta (v_\beta + t_\beta) = 0.$$

The sum of the second and third terms is an element of $H_{n_0-1} \leftharpoonup e_1$ and can be written as a linear combination of elements in $(H_{n_0-1} \leftharpoonup e_1) \cap F$ over k. This contradicts the choice of F. Therefore $\{e_1 \rightharpoonup S(v_\lambda)\}_{\lambda\in\Lambda}$ must be linearly independent over k.

LEMMA 3.3.7 Let σ be an integral for H. Then for any $a, b\in H$, we have

$$\sum_{(a)} \langle \sigma, a_{(2)} S(b) \rangle a_{(1)} = \sum_{(b)} \langle \sigma, a S(b_{(1)}) \rangle b_{(2)}. \tag{3.19}$$

Proof By Lemma 3.1.9, for any $f \in H^*$, we have

$$f(\sigma \leftharpoonup b) = \sum_{(b)} ((b_{(2)} \rightharpoonup f)\sigma) \leftharpoonup b_{(1)}.$$

On the other hand, since σ is an integral,

$$(b_{(2)} \rightharpoonup f)\sigma = \langle b_{(2)} \rightharpoonup f, 1 \rangle \sigma = \langle f, b_{(2)} \rangle \sigma.$$

Thus $f(\sigma \leftharpoonup b) = \sum_{(b)} \langle f, b_{(2)} \rangle \sigma \leftharpoonup b_{(1)}$.

Hence we obtain

$$\begin{aligned}\sum_{(b)}\langle\sigma, aS(b_{(1)})\rangle\langle f, b_{(2)}\rangle &= \sum_{(b)}\langle f, b_{(2)}\rangle\langle\sigma\leftharpoonup b_{(1)}, a\rangle\\ &= \langle f(\sigma\leftharpoonup b), a\rangle = \sum_{(a)}\langle f, a_{(1)}\rangle\langle\sigma\leftharpoonup b, a_{(2)}\rangle\\ &= \sum_{(a)}\langle\sigma, a_{(2)}S(b)\rangle\langle f, a_{(1)}\rangle,\end{aligned}$$

which gives the desired equality.

THEOREM 3.3.8 $\mathscr{I}_H \neq \{0\} \Leftrightarrow \dim(e_1 \rightharpoonup H) < \infty$.

Proof $\Leftarrow$Assume $\dim(e_1 \rightharpoonup H) < \infty$. Now $L = \coprod_{(N,i), N\neq k} (e_{(N,i)} \rightharpoonup H)$ is a left coideal of H and, by Lemma 3.3.5, $L \oplus (e_1 \rightharpoonup H) = H$. Hence L is of finite codimension and $\operatorname{codim} L = \dim(e_1 \rightharpoonup H)$. Since $1 \in e_1 \rightharpoonup H$, L is a proper left coideal of H and Theorem 3.3.1 implies $\mathscr{I}_H \neq \{0\}$.

$\Rightarrow$Suppose $\mathscr{I}_H \neq \{0\}$, and choose a non-zero integral σ. Then there exists $x \in H$ such that $\langle\sigma, x\rangle \neq 0$. We observe that

$$0 \neq \langle\sigma, x\rangle = \langle e_1, 1\rangle\langle\sigma, x\rangle = \langle e_1\sigma, x\rangle = \langle\sigma, x\leftharpoonup e_1\rangle.$$

Let W be the right H-subcomodule of the right H-comodule H generated by $w = x \leftharpoonup e_1$. Since $\mathscr{I}_H \neq \{0\}$, S is injective, and hence we get $\dim(H \leftharpoonup e_1) = \dim(e_1 \leftharpoonup H)$ by Theorem 3.3.6. Thus if we show $W = H \leftharpoonup e_1$, then we obtain $\dim(e_1 \rightharpoonup H) = \dim W < \infty$. By Lemma 3.3.4 (i), it follows that $W \subset H \leftharpoonup e_1$. If we suppose that $W \subsetneqq H \leftharpoonup e_1$, then there exists a positive integer n for which $H_n \leftharpoonup e_1 \not\subset W$. Let n be the smallest such n. Since $1 \in W$, we get $n \neq 1$. If we choose an element $v \in H_n \leftharpoonup e_1$ with $v \notin W$, then by Lemma 3.3.7, we have

$$\sum_{(w)}\langle\sigma, w_{(2)}S(v)\rangle w_{(1)} = \sum_{(v)}\langle\sigma, wS(v_{(1)})\rangle v_{(2)}.$$

Writing z for the left side, we get $z \in W$ by $\Delta w \in W \otimes H$. On the other hand, since $v \in H_n \leftharpoonup e_1$, using Lemma 3.3.4 (iii), we obtain $\Delta v \in 1 \otimes v + K \otimes H_{n-1}$. Hence we can write $v + t$ for $t \in H_{n-1}$ for the right side. By the choice of n, $t \leftharpoonup e_1 \in W$. Accordingly, $v = v \leftharpoonup e_1 = (z - t) \leftharpoonup e_1 = z - (t \leftharpoonup e_1) \in W$ and this contradicts the fact that $v \notin W$. This forces $W = H \leftharpoonup e_1$.

COROLLARY 3.3.9 Regardless of the choice of the projection π, $\dim(H \leftharpoonup e_1)$ is constant if it is finite.

Proof Let $\pi' : H \to H_0$ be another projection and construct e_1' as before. Assume that $\dim(H \leftharpoonup e_1') \leqq \dim(H \leftharpoonup e_1)$ and let λ be a non-zero integral for H. Now $(H \leftharpoonup e_1') \leftharpoonup e_1$ is a right H-subcomodule of H. If we pick $x \in H$ such that $\langle \lambda, x \rangle \neq 0$, then $\langle \lambda, x \leftharpoonup e_1' \rangle \neq 0$. Similarly we have $\langle \lambda, (x \leftharpoonup e_1') \leftharpoonup e_1 \rangle \neq 0$. Let W denote the right H-subcomodule of H generated by $(x \leftharpoonup e_1') \leftharpoonup e_1$. Then $W \subset (H \leftharpoonup e_1') \leftharpoonup e_1$, and we obtain $W = (H \leftharpoonup e_1') \leftharpoonup e_1$ in a way similar to the proof of the theorem. Thus we get $\dim(H \leftharpoonup e_1) \leqq \dim(H \leftharpoonup e_1')$. Therefore the equality holds.

THEOREM 3.3.10 (The uniqueness of integrals.) Given a k-Hopf algebra H,

$$\dim \mathscr{I}_H \leqq 1.$$

Proof Define a k-linear map $\varphi : \mathscr{I}_H \otimes (e_1 \rightharpoonup H) \to H^*$ by

$$\varphi(\sigma \otimes x) = \sigma \leftharpoonup x, \qquad \sigma \in \mathscr{I}_H,\ x \in (e_1 \rightharpoonup H).$$

Now let us show that

$$\operatorname{Im} \varphi \subset \left(\sum_{(N,i),\, N \neq k} H \leftharpoonup e_{(N,i)} \right)^{\perp}.$$

For $a \in (H \leftharpoonup e_{(N,i)})$ $(N \neq k)$, Lemma 3.3.4 (ii) implies $\Delta a \in N \otimes H + K \otimes H$, while for $b \in (e_1 \rightharpoonup H)$ we have $\Delta b \in H \otimes k + H \otimes K$. Thus, in the equality (3.19) of Lemma 3.3.7, the left side equals

$$\sum_{(a)} \langle \sigma, a_{(2)} S(b) \rangle a_{(1)} = v + y,\ v \in K,\ y \in N,$$

and the right side equals

$$\sum_{(b)} \langle \sigma, aS(b_{(1)}) \rangle b_{(2)} = w + z,\ w \in k,\ z \in K.$$

Therefore we have $y - w = z - v \in H_0 \cap K = \{0\}$, and hence $y = w \in k \cap N = \{0\}$. Thus

$$\langle \sigma \leftharpoonup b, a \rangle = \langle \sigma, aS(b) \rangle = \varepsilon(v + y) = \varepsilon(v) = 0.$$

Namely, $\sigma \leftharpoonup b \in \left(\sum_{(N,i),\, N \neq k} H \leftharpoonup e_{(N,i)} \right)$. Now φ is injective by Example 3.2. If $\mathscr{I}_H \neq \{0\}$, then S is injective, and applying Lemma 3.3.5 and Theorem 3.3.6, we see that $\dim(e_1 \rightharpoonup H) = \dim(H \leftharpoonup e_1)$ $= \operatorname{codim}\left(\sum_{(N,i),\, N \neq k} H \leftharpoonup e_{(N,i)} \right) \geqq \dim(\operatorname{Im}\varphi) = \dim(\mathscr{I}_H \otimes (e_1 \rightharpoonup H))$ $= \dim \mathscr{I}_H \cdot \dim(e_1 \rightharpoonup H)$. Therefore we obtain $\dim \mathscr{I}_H = 1$.

3.4 The existence of integrals

In Theorem 3.3.2 and Corollary 3.3.3, we observed the relationship between the existence of integrals and complete reducibility. Here we focus on the case of commutative k-Hopf algebras by adding more precision to the results thus far obtained, and study the conditions under which integrals exist.

THEOREM 3.3.11 Let k be a field of characteristic 0 and H a commutative k-Hopf algebra. Then we have

$$\mathscr{I}_H \neq \{0\} \Leftrightarrow H \text{ is co-semi-simple.}$$

Proof By Theorem 3.3.2 and Corollary 3.3.3, it follows that H is co-semi-simple $\Leftrightarrow$ there exists $\sigma \in \mathscr{I}_H$ such that $\sigma \circ \eta = 1_k$. Thus it suffices to show that $\langle \sigma, 1 \rangle \neq 0$ if $\sigma \in \mathscr{I}_H$, $\sigma \neq 0$. Assume that $\langle \sigma, 1 \rangle = 0$ for $\sigma \in \mathscr{I}_H$, $\sigma \neq 0$. If we let $\pi : H \to H_0$ be a projection and fix e_1, then we can choose a sequence $\{v_j\}_{1 \leqq j \leqq n}$ of elements in $H \leftharpoonup e_1$ and a sequence $\{n_i\}_{i \geqq 0}$ of non-negative integers in such a way that $v_1 = 1$, $n_0 = 0$, and furthermore so that $\{v_j\}_{1 \leqq j \leqq n_{i+1}}$ is a basis for $H_i \leftharpoonup e_1$ and $\varepsilon(v_j) = 0$ $(j \geqq 2)$. Pick any $v \in H$ such that $\langle \sigma, v \rangle = 1$. Since $\langle \sigma, 1 \rangle = 0$, we have $\langle \sigma, (v \leftharpoonup e_1) - \varepsilon(v \leftharpoonup e_1) \rangle = 1$. Meanwhile, since $w = v \leftharpoonup e_1 - \varepsilon(v \leftharpoonup e_1) \in H \leftharpoonup e_1$, we may write $\Delta w = \sum_{j=1}^{n} v_j \otimes u_j$, $u_j \in H$.

Now we proceed to show that $\langle \sigma, u_m S(v_m) \rangle = 1$ for each v_m $(1 \leqq m \leqq n)$. Suppose $n_i < m \leqq n_{i+1}$. Invoking Lemma 3.3.7 for $a = w$ and $b = v_m$ and taking note of the fact that $\Delta v_m \in 1 \otimes v_m + K \otimes H_{i-1}$, we get

$$\begin{aligned}\sum_{j=1}^{n} \langle \sigma, u_j S(v_m) \rangle v_j &= \sum_{(v_m)} \langle \sigma, wS((v_m)_{(1)}) \rangle (v_m)_{(2)} \\ &= y + v_m, \qquad y \in H_{i-1}.\end{aligned}$$

On the other hand, since $y \leftharpoonup e_1 = \sum_{j=1}^{n_i} \alpha_j v_j \in H_{i-1} \leftharpoonup e_1$, $v_m \leftharpoonup e_1 = v_m$, and $v_j \leftharpoonup e_1 = v_j$, we have

$$\sum_{j=1}^{n} \langle \sigma, u_j S(v_m) \rangle v_j = \sum_{j=1}^{n_i} \alpha_j v_j + v_m. \tag{3.20}$$

Since $\{v_j\}$ is linearly independent over k and $n_i < m$, a comparison of the coefficients of v_m in (3.20) shows that $\langle \sigma, u_m S(v_m) \rangle = 1$.

Now the commutativity of H implies that $\sum_{j=1}^{n} u_j S(v_j) = \sum_{j=1}^{n} S(v_j) u_j = \varepsilon(w) = 0$. Hence

$$n = \sum_{j=1}^{n} \langle \sigma, u_j S(v_j) \rangle = \langle \sigma, \sum_{j=1}^{n} u_j S(v_j) \rangle = 0.$$

This contradicts the fact that the characteristic of k is 0 and $n = \dim (H \leftharpoonup e_1) \neq 0$.

When k is of characteristic $p > 0$, if k is a perfect field and H is an integral domain, a similar result holds. We provide a lemma to verify this fact.

LEMMA 3.3.12 Let k be a perfect field of characteristic $p > 0$ and H a commutative k-Hopf algebra which is an integral domain. If $M \neq k$ is a simple k-subcoalgebra of H, then $k + \sum_{n=0}^{\infty} M^{p^n}$ is a direct sum.

Proof Since $k + \sum_{n=0}^{\infty} M^{p^n}$ is contained in the k-sub-Hopf algebra of H generated by M, we may assume that H is finitely generated. Now k is a perfect field and H has no non-zero nilpotent elements, so M^{p^n} is a simple k-coalgebra. If the sum is not direct, there exists j satisfying

$$M^{p^j} \cap \left(k + \sum_{n \neq j} M^{p^n} \right) \neq \{0\}.$$

The simplicity of M^{p^j} implies that $M^{p^j} = M^{p^n}$ for some $n \neq j$ or that

$M^{p^j} = k$. In either case, $M^{p^j} \subset \bigcap_{n=0}^{\infty} H^{p^n}$. On the other hand, if we set $\mathfrak{a} = H(\operatorname{Ker} \varepsilon)$, $\mathfrak{a}$ is a proper ideal of H, and we have $\bigcap_{n=0}^{\infty} \mathfrak{a}^{p^n} = \{0\}$ by Krull's intersection theorem (Theorem 1.5.9). Since $H^{p^n} = \mathfrak{a}^{p^n} + k$, we get $\bigcap_{n=0}^{\infty} H^{p^n} = k$. Hence $\dim M^{p^j} = 1$. Due to the absence of non-zero nilpotent elements in H, we also get $\dim M = 1$ and $M = ka$ for some $a \in G(H)$. Moreover, the fact that $M^{p^j} = k$ ensures the existence of an element α of k such that $a^{p^j} - \alpha = (a - \alpha^{p^{-j}})^{p^j} = 0$. This contradicts the assumption that $M \neq k$.

COROLLARY 3.3.13 Let k be a perfect field of characteristic $p > 0$ and let H be a commutative k-Hopf algebra as well as an integral domain. Let $M \neq k$ be a simple k-subcoalgebra of H and let $W \subset k \sqcap M$ be a k-subcoalgebra of H containing M and k. If we set corad $W = R_W$ and if

$$0 \to N_W \to W \to R_W \to 0 \tag{3.21}$$

is a split exact sequence of k-coalgebras obtained from Theorem 2.3.11, then

$$\sum_{n=0}^{\infty} W^{p^n} = k \oplus \left(\bigoplus_{n=0}^{\infty} M^{p^n} \right) \oplus \left(\bigoplus_{n=0}^{\infty} {N_W}^{p^n} \right) \quad \text{(direct sum)}. \tag{3.22}$$

Proof Let corad $H = R$ and let $0 \to N \to H \to R \to 0$ be a split exact sequence of k-coalgebras obtained from Theorem 2.3.11. Set $M^{p^j} = M_j$. For $x \in k \sqcap M_j$, we have $x = \left(\sum_i e_{(M_j, i)} + e_1 \right) \rightharpoonup x$, while for $y \in \sum_{n \neq j} k \sqcap M_n$, we get $\left(\sum_i e_{(M_j, i)} + e_1 \right) \rightharpoonup y \in k$ by Lemma 3.3.12. Hence, if $x = y$, then $x = y \in k$. In particular, if $x \in (k \oplus M_j \oplus {N_W}^{p^j}) \cap \left(k + \sum_{n \neq j} M_n + \sum_{n \neq j} {N_W}^{p^n} \right)$, then $x \in k$. Therefore the sum is direct.

THEOREM 3.3.14 Let k be a perfect field and let H be a commutative k-Hopf algebra which is an integral domain. Then

$$\mathscr{I}_H \neq \{0\} \Leftrightarrow H \text{ is co-semi-simple.}$$

Proof The case in which the characteristic of k is 0 has already been taken care of in Theorem 3.3.11, so suppose that the characteristic is $p>0$. It is enough to verify that $\langle \sigma, 1\rangle \neq 0$ if $\sigma \in \mathscr{I}_H$ and $\sigma \neq 0$ as in the proof of Theorem 3.3.11. Assume that $\langle \sigma, 1\rangle = 0$ and let corad $H = R_0$ and $R_i = \sqcap^{i+1} R_0$. Since $R_0 \leftharpoonup e_1 = k$, we have $\langle \sigma, R_0 \leftharpoonup e_1\rangle = 0$, while $\langle \sigma, H \leftharpoonup e_1\rangle \neq 0$. Thus we may take $x \in (R_1 \leftharpoonup e_1) - (R_0 \leftharpoonup e_1)$. By Lemma 3.3.5, there is a pair (M, i) consisting of a simple k-coalgebra M and i satisfying $e_{(M,i)} \rightharpoonup x \in (R_1 \leftharpoonup e_1) - (R_0 \leftharpoonup e_1)$. Now we easily get $v = e_{(M,i)} \rightharpoonup x \in k \sqcap M - (k+M)$.

(1) In the case $M = k$, we see that $k \sqcap k \subset R_1 \leftharpoonup e_1$. Thus $\{1, v, v^p, v^{p^2}, \ldots\} \subset k \sqcap k \subset R_1 \leftharpoonup e_1$. By Lemma 3.3.12, it is linearly independent over k. Hence $\dim (H \leftharpoonup e_1) = \dim (e_1 \rightharpoonup H) = \infty$. This contradicts Theorem 3.3.8.

(2) In the case $M \neq k$, let W be the k-subcoalgebra of H generated by v. By the decomposition in Corollary 3.3.13, (3.22), we can write $v = \alpha + m_v + n_v$, $\alpha \in k$, $m_v \in M$, $n_v \in N_W$, $n_v \neq 0$. By Lemma 3.3.4 (iv), it follows that $\Delta v^p \in 1 \otimes v^p + H \otimes M^p$, and hence $v^p \leftharpoonup e_1 = v^p + m_p$, $m_p \in M^p$. In general, we have

$$v^{p^n} \leftharpoonup e_1 = v^{p^n} + m_{(p^n)} = \alpha^{p^n} + m_v{}^{p^n} + n_v{}^{p^n} + m_{(p^n)}, \quad m_{(p^n)} \in M^{p^n}.$$

By Corollary 3.3.13, $\{v^{p^i} \leftharpoonup e_1\}_{i \in \mathbb{N}}$ is linearly independent over k. Hence $\dim (H \leftharpoonup e_1) = \dim (e_1 \rightharpoonup H) = \infty$. As in case (1), this is impossible by Theorem 3.3.8.

Remark If k is an algebraically closed field, the k-Hopf algebra which is the coordinate ring of $GL_n(k)$ is co-semi-simple if and only if the characteristic of k is zero or $n = 1$. (cf. Example 4.3.) This fact implies that there exists a non-zero integral for $GL_n(k)$ only in these cases.

4 The duality theorem

Assume that k is a field and let $\mathbf{H}_k$ be the category of commutative k-Hopf algebras. Given a group G, let $R_k(G)$ be the commutative k-Hopf algebra consisting of all representative functions on G with values in k. (cf. Chapter 2, §2.2.) It is easily seen that the correspondence

$$\Phi : \mathbf{Gr} \to \mathbf{H}_k$$

which carries G to $R_k(G)$ is a contravariant functor from the category **Gr** of groups to the category $\mathbf{H}_k$ of commutative k-Hopf algebras. Conversely, if H is a commutative k-Hopf algebra, the group $G(H^\circ)$ of all group-like elements in the dual k-Hopf algebra H° of H is simply $\mathbf{M}_k(H, k)$ regarded as a group with respect to convolution. The correspondence

$$\Psi : \mathbf{H}_k \to \mathbf{Gr}$$

which assigns $\mathbf{M}_k(H, k)$ to H is a contravariant functor from the category $\mathbf{H}_k$ to the category **Gr** and these two contravariant functors are adjoint to each other. Namely, we have

$$\mathbf{Gr}(G, \Psi(H)) \cong \mathbf{H}_k(H, \Phi(G)). \tag{3.23}$$

Given $\varphi \in \mathbf{Gr}(G, \Psi(H))$ and $\psi \in \mathbf{H}_k(H, \Phi(G))$, write $\varphi(g)(h) = \langle h, \varphi(g) \rangle$, $\psi(h)(g) = \langle g, \psi(h) \rangle$ for $g \in G$, $h \in H$. That φ corresponds to ψ by the correspondence of (3.23) is equivalent to the condition that

$$\langle h, \varphi(g) \rangle = \langle g, \psi(h) \rangle, \quad h \in H, \quad g \in G.$$

In particular, for an appropriate choice of k, the subcategory of $\mathbf{H}_k$ which corresponds to a subcategory of **Gr** consisting of, say, finite groups, compact topological groups, compact Lie groups, or affine algebraic groups can be characterized completely. Moreover, it can be shown that such corresponding subcategories are anti-equivalent to each other. The above fact applied to the case of compact topological groups is essentially Tannaka's duality theorem.

4.1 The duality theorem for finite groups

Let **Grf** be the category of finite groups and let k be an algebraically closed field of characteristic 0. Given a finite group G, the group k-Hopf algebra kG of G over k is a cocommutative co-semi-simple k-Hopf algebra of finite dimension (cf. Theorem 3.3.2, Corollary 3.3.3 and Example 3.13). Hence $\Phi(G) = (kG)^*$ is a finite dimensional commutative semi-simple k-Hopf algebra. We let $\mathbf{Hf}_k$ be the category consisting of such k-Hopf algebras. Conversely, if $H \in \mathbf{Hf}_k$, then $\Psi(H) = G(H^\circ)$ is a finite group. Now we go on to show that the two contravariant functors

$$\Phi : \mathbf{Grf} \to \mathbf{Hf}_k, \quad G \mapsto (kG)^* = \mathrm{Map}\,(G, k),$$
$$\Psi : \mathbf{Hf}_k \to \mathbf{Grf}, \quad H \mapsto G(H^\circ) = \mathbf{M}_k(H, k) \tag{3.24}$$

are inverse functors and that **Grf** and $\mathbf{Hf}_k$ are anti-equivalent.

LEMMA 3.4.1 Suppose that C is a pointed k-coalgebra. Then $kG(C) = \mathrm{corad}\ C$.

Proof That $kG(C) \subset \mathrm{corad}\ C$ is clear. Let $x \in \mathrm{corad}\ C$ and let D be the k-subcoalgebra of C generated by x. We can assume without loss of generality that D is simple. By assumption, we have $\dim D = 1$, and if $x_1 \in D$, $x_1 \neq 0$, then we get $\Delta(x_1) = c(x_1 \otimes x_1)$, $c \in k$. Since $x_1 = (\varepsilon \otimes 1)\Delta(x_1) = c\varepsilon(x_1)x_1$, $\varepsilon(x_1) \neq 0$. Choose $x_1 \in D$ so that $\varepsilon(x_1) = 1$. Then $(\varepsilon \otimes \varepsilon)\,\Delta(x_1) = c\varepsilon(x_1)\varepsilon(x_1) = c$, while $(\varepsilon \otimes \varepsilon)\Delta(x_1) = (\varepsilon \otimes 1)(1 \otimes \varepsilon)\Delta(x_1) = \varepsilon(x_1) = 1$. Thus $c = 1$. Therefore $x_1 \in G(C)$. Since $x \in D \subset kG(C)$, we obtain $kG(C) = \mathrm{corad}\ C$.

THEOREM 3.4.2 Let k be an algebraically closed field of characteristic 0. Then the two categories **Grf** and $\mathbf{Hf}_k$ are anti-equivalent by correspondences Φ and Ψ of (3.24).

Proof Given a finite group G, we have $\Psi\Phi(G) = G((kG)^{*\circ}) \cong G(kG)$. Clearly $G(kG) \supset G$. Since elements in G are linearly independent over k (cf. Theorem 2.1.2) and $\dim kG = |G|$, we get $G(kG) = G$. Conversely, if H is a commutative semi-simple k-Hopf algebra of finite dimension, H° is a finite dimensional cocommutative k-Hopf algebra as well as a pointed k-coalgebra by Theorem 2.3.3. Hence it follows from Lemma 3.4.1 that $kG(H^\circ) = \mathrm{corad}\ H^\circ$. On the other hand, since H° is co-semi-simple, we get $kG(H^\circ) = H^\circ$. Therefore $\Phi\Psi(H) = (kG(H^\circ))^\circ = H^{\circ\circ} \cong H$.

4.2 The duality theorem for compact topological groups

Let $\mathscr{G}$ be the category of compact topological groups and $\mathscr{H}$ the category of all commutative $\mathbb{R}$-Hopf algebras H over the real number field $\mathbb{R}$ with the following properties.

(1) There exists an integral σ for H satisfying the condition that, for $f \in H$ with $f \neq 0$, $\sigma(f^2) > 0$.

(2) For $f, g \in H$, if $f \neq g$, then there exists $x \in G(H^\circ)$ satisfying $\langle f, x \rangle \neq \langle g, x \rangle$, namely, $G(H^\circ)$ is dense in H^*.

Now the $\mathbb{R}$-Hopf algebra $\mathscr{R}(G) = R_{\mathbb{R}}(G) \cap C_{\mathbb{R}}(G)$ consisting of all real-valued continuous representative functions on a compact topological group G is a commutative $\mathbb{R}$-Hopf algebra satisfying conditions (1) and (2). Here we denote $\mathscr{R}(G)$ by $\Phi(G)$. Conversely, let H be a commutative $\mathbb{R}$-Hopf algebra satisfying (1) and (2). If we define a topology on $G(H^\circ) = \mathbf{Alg}_{\mathbb{R}}(H, \mathbb{R})$ as follows, $G(H^\circ)$ becomes a compact topological group, which we denote by $\Psi(H)$.

Given $f \in G(H^\circ)$, let a base of open neighborhoods of f be the family of all subsets of the form

$$\begin{aligned} &U(f; h_1, \ldots, h_n, \varepsilon) \\ &= \{g \in G(H^\circ); |g(h_i) - f(h_i)| < \varepsilon \ (1 \leqq i \leqq n)\}, \end{aligned}$$

where $h_1, \ldots, h_n$ $(n < \infty)$ are elements of H and $\varepsilon > 0$ is any positive real number. In this situation, we have the following theorem.

THEOREM 3.4.3 By correspondences Φ and Ψ, the category $\mathscr{G}$ of compact topological groups and the category $\mathscr{H}$ of commutative $\mathbb{R}$-Hopf algebras satisfying (1) and (2) correspond to each other anti-equivalently.

We omit the proof of this theorem. However, we note that the existence of Haar measures for compact topological groups and Peter–Weyl's theorem, which asserts that $\mathscr{R}(G)$ is dense in $C_{\mathbb{R}}(G)$, play key roles in the proof of the theorem. (cf. Hochschild [4], Theorem 3.5.)

We call a k-Hopf algebra which is finitely generated as a k-algebra a **finitely generated k-Hopf algebra**. In Theorem 3.4.3, if H is a finitely generated $\mathbb{R}$-Hopf algebra, $G(H^\circ)$ has a faithful representation and becomes a compact Lie group isomorphic to a closed subgroup of the orthogonal group $O_n(\mathbb{R})$ over $\mathbb{R}$. Conversely, if G is a compact Lie group, G has a faithful representation and $\Phi(G)$ is a finitely generated $\mathbb{R}$-Hopf algebra. Therefore we obtain the following corollary.

COROLLARY 3.4.4 By correspondences Φ and Ψ, the category of compact Lie groups and the category of finitely generated commutative $\mathbb{R}$-Hopf algebras satisfying (1) and (2) correspond to each other anti-equivalently.

LEMMA 3.4.5 Any k-Hopf algebra H is the union of finitely generated k-sub-Hopf algebras of H. Namely, considering the family $\{H_\lambda\}_{\lambda\in\Lambda}$ of all finitely generated k-sub-Hopf algebras of H as an inductive system via inclusion, we have

$$H = \varinjlim_{\lambda} H_\lambda.$$

Proof Take $h \in H$. The k-subcoalgebra C_λ generated by h is of finite dimension (cf. Corollary 2.2.14 (i)). Let H_λ be the k-subalgebra of H generated by C_λ and $S(C_\lambda)$. Then H_λ is a finitely generated k-sub-Hopf algebra of H. Thus H is the union of $\{H_\lambda\}_{\lambda\in\Lambda}$.

THEOREM 3.4.6 A compact topological group is a projective limit of compact Lie groups.

Proof Given $H \in \mathscr{H}$, we express H as the inductive limit $H = \varinjlim_{\lambda} H_\lambda$ of finitely generated $\mathbb{R}$-sub-Hopf algebras. Then we have $\Psi(H) = \varprojlim_{\lambda} \Psi(H_\lambda)$, and thus $G = \Psi(H)$ is a projective limit of compact Lie groups $G_\lambda = \Psi(H_\lambda)$ $(\lambda \in \Lambda)$.

Remark In Chapter 4, §2.1, we shall see that a similar duality exists for algebraic k-groups.

4

Applications to algebraic groups

We have shown in Chapter 3 that there exists a contravariant functor $\Psi : H \mapsto G(H^\circ)$ from the category of Hopf algebras over a field k to the category of groups. All of such groups $G(H^\circ)$ form a category whose morphisms are group morphisms $\Psi(f) : G(H'^\circ) \to G(H^\circ)$ where H and H' are k-Hopf algebras and $f : H \to H'$ is a k-Hopf algebra morphism. In particular, when k is an algebraically closed field and H is a finitely generated commutative k-Hopf algebra, $G(H^\circ)$ is called an affine algebraic k-group. If H is a commutative k-Hopf algebra, $G(H^\circ)$ is a projective limit of affine algebraic k-groups and is called an affine k-group or sometimes a pro-affine algebraic k-group. The category of affine algebraic k-groups is anti-equivalent to the category of finitely generated reduced commutative k-Hopf algebras, and the pair $(G(H^\circ), H)$ consisting of such $G(H^\circ)$ and H is an affine algebraic group in the sense of Hochschild [4]. Moreover, $G(H^\circ)$ has a faithful representation into the general linear group $GL_n(k)$, so we can look upon it as a linear algebraic group which was taken up by Chevalley [1], [2], Borel [3], etc. In general, a commutative k-Hopf algebra is a k-algebra representing an affine group scheme (cf. Appendix A.5) and we may reduce the properties of affine group schemes to those of commutative k-Hopf algebras.

In this chapter, among the various properties of affine algebraic k-groups $G(H^\circ)$, we will focus upon the basic ones obtained through the application of the theory of Hopf algebras. Throughout this chapter, k will denote an algebraically closed field.

1 Affine k-varieties

In this section, we briefly mention some properties of affine k-varieties that we will need in the sequel.

1.1 Affine k-varieties

If A is a commutative k-algebra, the set $X_A = \mathbf{M}_k(A, k)$ of all k-algebra morphisms from A to k is called an **affine k-variety**. In particular, when A is a finitely generated k-algebra, we say that X_A is an **affine algebraic k-variety**. For an element x of an affine algebraic k-variety X_A, Ker $x = \mathfrak{m}_x$ is a maximal ideal of A. Conversely, since $A/\mathfrak{m} \cong k$ (cf. Theorem 1.5.4) for an arbitrary maximal ideal $\mathfrak{m}$ of A, the canonical projection $A \to A/\mathfrak{m} \cong k$ is an element of X_A. Thus we obtain

$$X_A \cong \mathrm{Spm}\, A = \{\mathfrak{m}; \mathfrak{m} \text{ is a maximal ideal of } A\}.$$

When A is the polynomial ring $k[T_1, \ldots, T_n]$ in n variables over k, then

$$X_A \cong k^n = \{x = (x_1, \ldots, x_n); x_i \in k \ (1 \leqq i \leqq n)\},$$

and we call X_A the **affine space of dimension** n. In particular, when $n = 1$, it will be called the **affine line** and simply denoted by k. Moreover, if $\mathfrak{a}$ is an ideal of $k[T_1, \ldots, T_n]$ and $A = k[T_1, \ldots, T_n]/\mathfrak{a}$, then we can identify X_A with the subset of the affine space of dimension n as follows.

$$V(\mathfrak{a}) = \{x = (x_1, \ldots, x_n) \in k^n; f(x_1, \ldots, x_n) = 0 \quad \forall f \in \mathfrak{a}\}.$$

Let A, B be commutative k-algebras and let $u: A \to B$ be a k-algebra morphism. Then the map

$$^a u : X_B \to X_A, \quad ^a u(x) = x \circ u, \quad x \in X_B,$$

is called a **morphism of affine k-varieties**. Now $\mathbf{V}_k$ denotes the category of affine k-varieties, and the set of all morphisms from an affine k-variety X_B to X_A is written $\mathbf{V}_k(X_B, X_A)$. A morphism from an affine k-variety X_A to the affine line k is called a function on X_A. We see that $\mathbf{V}_k(X_A, k)$ turns out to be a k-algebra defined for f, $g \in \mathbf{V}_k(X_A, k)$ by

$$(f \pm g)(x) = f(x) \pm g(x), \quad (fg)(x) = f(x)g(x), \quad x \in X_A,$$
$$(\alpha f)(x) = \alpha f(x), \quad \alpha \in k.$$

A k-algebra morphism $u: k[T] \to A$ is uniquely determined by $u(T) = f \in A$, and the associated function $\hat{f} = {}^a u : X_A \to k$ on X_A is given by $\hat{f}(x) = x(f)$ for $x \in X_A$. Thus we obtain a surjective k-algebra morphism

$$\varphi : A \to \mathbf{V}_k(X_A, k)$$

defined by $f \mapsto \hat{f}$. In this situation, if X_A is an affine algebraic k-variety, from Theorem 1.5.5, we get

$$f \in \operatorname{Ker} \varphi \Leftrightarrow \bigcap_{\mathfrak{m} \in \operatorname{Spm} A} \mathfrak{m} = \operatorname{nil} A,$$

and hence $\mathbf{V}_k(X_A, k) \cong A/\operatorname{nil} A$. In particular, when A is reduced, namely when $\operatorname{nil} A = \{0\}$, then $\mathbf{V}_k(X_A, k) \cong A$. The two correspondences defined by

$$A \mapsto X_A = \mathbf{M}_k(A, k), \quad X_A \mapsto \mathbf{V}_k(X_A, k)$$

give an anti-equivalence between the category of finitely generated commutative reduced k-algebras and the category of affine algebraic k-varieties. A finitely generated commutative reduced k-algebra will be called an **affine k-algebra**.

If A and B are commutative k-algebras, from Chapter 1, §2.1, we get

$$\mathbf{M}_k(A, k) \times \mathbf{M}_k(B, k) \cong \mathbf{M}_k(A \otimes_k B, k).$$

Therefore the product $X_A \times X_B$ of X_A and X_B taken as sets admits a structure of an affine k-variety, and is called the product of X_A and X_B. If A and B are affine k-algebras, $A \otimes_k B$ is also an affine k-algebra.

1.2 The Zariski topology

When $\mathfrak{a}$ is an ideal of an affine k-algebra A, the morphism of affine algebraic k-varieties ${}^a u : X_{A'} \to X_A$ associated with the canonical k-algebra morphism $u : A \to A/\mathfrak{a} = A'$ is an injection, and we may consider $X_{A'} = V(\mathfrak{a})$ as a subset of X_A. Conversely, if E is a subset of X_A, then

$$\mathfrak{a}(E) = \{f \in A; x(f) = 0 \ \forall x \in E\}$$

is a radical ideal of A, and Theorems 1.5.4 and 1.5.5 can be rephrased as follows.

(i) $V(\mathfrak{a}) = \phi \Leftrightarrow \mathfrak{a} = A$.

(ii) $\mathfrak{a}(V(\mathfrak{a})) = \sqrt{\mathfrak{a}}$.

Indeed, if $\mathfrak{a} \subsetneqq A$, $\mathfrak{a}$ is contained in a maximal ideal $\mathfrak{m}$ of A and $x : A \to A/\mathfrak{m} \cong k$ is an element of $V(\mathfrak{a})$. Thus $V(\mathfrak{a}) \neq \phi$. This means that $V(\mathfrak{a}) = \phi$ implies $\mathfrak{a} = A$. The other direction is trivial. Moreover,

$\mathfrak{a}(V(\mathfrak{a})) \supset \sqrt{\mathfrak{a}}$ is obvious and the reverse inclusion is obtained from Theorem 1.5.5.

Given ideals $\mathfrak{a}$, $\mathfrak{b}$, $\mathfrak{a}_\lambda (\lambda \in \Lambda)$ of A, we have

(1) $V(\{0\}) = X_A$, $V(A) = \phi$,

(2) $\bigcap_{\lambda \in \Lambda} V(\mathfrak{a}_\lambda) = V\left(\sum_{\lambda \in \Lambda} \mathfrak{a}_\lambda\right)$,

(3) $V(\mathfrak{a}) \cup V(\mathfrak{b}) = V(\mathfrak{a} \cap \mathfrak{b})$.

Therefore we may define a topology on X_A by taking the closed sets to be the subsets $V(\mathfrak{a})$ for all ideals $\mathfrak{a}$ of A. This is called the **Zariski topology** of X_A. By the correspondence $\mathfrak{a} \mapsto V(\mathfrak{a})$, the family of all radical ideals of A is in one-to-one correspondence with the family of all closed subsets of X_A. In the Zariski topology, a set with a single element is closed and corresponds to a maximal ideal of A. If A and B are affine k-algebras and $u: A \to B$ is a k-algebra morphism, the morphism ${}^a u: X_B \to X_A$ of affine algebraic k-varieties is a continuous map. The Zariski topology on $X_A \times X_B \cong X_{A \otimes_k B}$ is stronger than the product topology of topological spaces X_A and X_B. For $f \in A$, an open set $U_f = X_A - V(f)$ of X_A will be called a **principal open set** of X_A. It is easy to see that $U_f \cong \mathbf{M}_k(A_f, k)$, where A_f is the quotient ring of A by $S = \{f^n; n = 0, 1, 2, \ldots\}$. Therefore U_f may be regarded as an affine algebraic k-variety.

EXERCISE 4.1 Show that $\{U_f; f \in A\}$ forms a base of open sets for the topological space X_A.

1.3 Connected sets and irreducible sets

A subset F of a topological space X is said to be **connected** if F cannot be written as a disjoint union of two proper, non-empty, closed subsets. Further, F is said to be **irreducible** if it is not a union of two proper, non-empty, closed subsets. Evidently, if F is irreducible, then it is connected.

EXERCISE 4.2 (i) The following conditions are equivalent to saying that a topological space X is irreducible.

(1) Any open subset in X is connected.

(2) Any non-empty open subset is dense in X.

(ii) If a subset F of X is irreducible, then so is the closure $\bar{F}$ of F.

(iii) The image of an irreducible set under a continuous map is irreducible.

THEOREM 4.1.1 Any irreducible (resp. connected) subset of X is contained in a maximal irreducible (resp. maximal connected) subset of X and X is the union of the maximal irreducible (resp. maximal connected) subsets, which are called the **irreducible** (resp. **connected**) **components** of X.

Proof We give the proof only for the case of irreducible subsets. (The connected case may be proven similarly.) Let S be an irreducible subset of X, and let Σ be the family of all irreducible subsets of X containing S. Then Σ is an ordered set relative to inclusion, and we have $S \in \Sigma$ since S is irreducible. If $\{F_\lambda\}_{\lambda \in \Lambda}$ is a totally ordered subset of Σ, $F = \bigcup_{\lambda \in \Lambda} F_\lambda$ is also irreducible. Indeed, for non-empty open subsets U and V of F, there exist F_λ, F_μ satisfying $U \cap F_\lambda \neq \phi$ and $V \cap F_\mu \neq \phi$. Provided $F_\lambda \subseteq F_\mu$, $U \cap F_\mu$ and $V \cap F_\mu$ are non-empty open subsets of F_μ, so that $U \cap V \cap F_\mu \neq \phi$. Therefore $U \cap V \neq \phi$. Namely, F is irreducible. Using Zorn's lemma, we can conclude that Σ has a maximal element. Moreover, for any point $x \in X$, there exists a maximal irreducible subset which contains x because $\{x\}$ is irreducible, and hence X is the union of the maximal irreducible subsets.

THEOREM 4.1.2 Let A be an affine k-algebra. Then for the affine k-variety $X_A = \mathbf{M}_k(A, k)$, the following conditions are equivalent.

(1) There exist r closed (and also open) subsets F_i $(1 \leqq i \leqq r)$ such that any two are disjoint and $X_A = \bigcup_{i=1}^{r} F_i$.

(2) There are r elements $e_1, \ldots, e_r$ of A satisfying

$$e_i e_j = \delta_{ij} e_i \quad (1 \leqq i, j \leqq r), \quad e_1 + \cdots + e_r = 1.$$

Proof (2)$\Rightarrow$(1). Let $\mathfrak{a}_i := A(1 - e_i)$. Then $F_i = V(\mathfrak{a}_i)$ $(1 \leqq i \leqq r)$ satisfy the condition of (1). In fact, since $\prod_{i=1}^{r} (1 - e_i) = 1 - \sum_{i=1}^{r} e_i = 0$, we have

$$F_1 \cup F_2 \cup \cdots \cup F_r = V(\mathfrak{a}_1 \mathfrak{a}_2 \cdots \mathfrak{a}_r) = V(\{0\}) = X_A.$$

Further, since $i \neq j$ implies $\mathfrak{a}_i + \mathfrak{a}_j = A$, it follows that $F_i \cap F_j = V(\mathfrak{a}_i + \mathfrak{a}_j) = V(A) = \phi$.

(1) $\Rightarrow$ (2) Let $F_i = V(\mathfrak{a}_i)$, and, for $V_i = \bigcup_{j \neq i} F_j$, set $V_i = V(\mathfrak{b}_i)$ (where $\mathfrak{a}_i$ and $\mathfrak{b}_i$ are ideals of A). Then

$$V(\mathfrak{a}_i \mathfrak{b}_i) = V(\mathfrak{a}_i \cap \mathfrak{b}_i) = V(\mathfrak{a}_i) \cup V(\mathfrak{b}_i) = F_i \cup V_i = X_A.$$

Hence $\mathfrak{a}_i \mathfrak{b}_i = \mathfrak{a}_i \cap \mathfrak{b}_i = \{0\}$. Meanwhile, $\mathfrak{a}_i + \mathfrak{b}_i = A$ for $F_i \cap V_i = \phi$ (cf. Theorem 1.5.5). Therefore $A = \mathfrak{a}_i \oplus \mathfrak{b}_i$ (direct sum). Write $1 = e_i + b_i$ for $e_i \in \mathfrak{a}_i, b_i \in \mathfrak{b}_i$. Then $e_i b_i = 0$ implies $e_i^2 = e_i$, and $\{e_1, \ldots, e_r\}$ satisfies $e_i e_j = \delta_{ij} e_i (1 \leq i, j \leq r)$. Setting $e = e_1 + \cdots + e_r$, we get $V(Ae) \subset F_i (1 \leq i \leq r)$, so $V(Ae) = \phi$ and $Ae = A$ (cf. Theorem 1.5.5). Therefore, there exists $a \in A$ such that $ae = 1$. Moreover, $e^2 = e$ implies $e = ae^2 = ae = 1$.

THEOREM 4.1.3 Let A be an affine k-algebra. The affine k-variety X_A is irreducible $\Leftrightarrow A$ is an integral domain.

Proof Assume A is not an integral domain. Then we have $f, g \in A$ such that $f \neq 0$, $g \neq 0$, and $fg = 0$. Set $F_1 = V(Af)$ and $F_2 = V(Ag)$. Then F_1 and F_2 are proper closed subsets of X_A because f and g are not nilpotent. Since $F_1 \cap F_2 = V(Afg) = V(\{0\}) = X_A$, we conclude that X_A is not irreducible. Conversely, if X_A is not irreducible, then there exist proper closed subsets F_1, F_2 of X_A such that $X_A = F_1 \cup F_2$. If we pick the radical ideals $\mathfrak{a}_i$ $(i = 1, 2)$ so that $F_i = V(\mathfrak{a}_i)$, then we have $\mathfrak{a}_1 \not\supset \mathfrak{a}_2$ and $\mathfrak{a}_2 \not\supset \mathfrak{a}_1$. If we take elements f, g of A such that $f \in \mathfrak{a}_1$, $f \notin \mathfrak{a}_2$, and $g \in \mathfrak{a}_2$, $g \notin \mathfrak{a}_1$, then we get $f \neq 0$, $g \neq 0$, and $fg \in \mathfrak{a}_1 \mathfrak{a}_2$. On the other hand, $V(\mathfrak{a}_1 \mathfrak{a}_2) = V(\mathfrak{a}_1) \cup V(\mathfrak{a}_2) = X_A$ implies that $\mathfrak{a}_1 \mathfrak{a}_2 = \{0\}$, which shows that $fg = 0$. Thus A is not an integral domain.

EXERCISE 4.3 (i) X_A is connected $\Leftrightarrow A$ has no idempotents other than 0 and 1.

(ii) A closed subset $V(\mathfrak{a})$ of X_A is irreducible $\Leftrightarrow \mathfrak{a}$ is a prime ideal of A.

EXERCISE 4.4 Let X be a topological space. We say that X is a

Noetherian space if any strictly descending chain of closed subsets of X

$$F_1 \supsetneqq F_2 \supsetneqq F_3 \supsetneqq \cdots$$

is finite. Now prove the following statements.

(i) An affine algebraic k-variety X_A is a Noetherian space.

(ii) The number of irreducible components of a Noetherian space is finite. (cf. references for Chapter 1, Bourbaki [4], II–4.2.)

Let A be an affine k-algebra and assume X_A is irreducible. Then A is an integral domain and the transcendental degree of the quotient field $Q(A)$ of A over k, abbreviated trans. $\deg_k Q(A)$ is called the **dimension** of X_A, and written $\dim X_A$. When X_A has irreducible components $X_1, \ldots, X_r$, we will define $\dim X_A = \max\limits_{1 \leqq i \leqq r} (\dim X_i)$. The dimension of X_A is precisely the Krull dimension of A.

THEOREM 4.1.4 Let A, B be affine k-algebrs. If both X_A and X_B are irreducible, $X_A \times X_B$ is also irreducible.

Proof If $x \in X_A$, $y \in X_B$, then $\{x\}$, $\{y\}$ are closed sets, and we have $\{x\} \times X_B \cong X_B$, and $X_A \times \{y\} \cong X_A$ (isomorphisms of topological spaces). Suppose that $X_A \times X_B$ is the union of two closed subsets Z_1 and Z_2. Since $X_A \times \{y\} = (Z_1 \cap (X_A \times \{y\})) \cup (Z_2 \cap (X_A \times \{y\}))$ is irreducible, $X_A \times \{y\} \subset Z_1 \cap (X_A \times \{y\})$ or $X_A \times \{y\} \subset Z_2 \cap (X_A \times \{y\})$. So we have $X_B = W_1 \cup W_2$ where $W_i = \{y \in X_B;\ X_A \times \{y\} \subset Z_i\}$ $(i = 1, 2)$. On the other hand, the W_i are closed since the isomorphism $X_B \to \{x\} \times X_B$ sends W_i to the closed subset $Z_i \cap (\{x\} \times X_B)$ of $\{x\} \times X_B$. The fact that X_B is irreducible implies that either $X_B = W_1$ or $X_B = W_2$, namely, either $X_A \times X_B = Z_1$ or $X_A \times X_B = Z_2$. Hence $X_A \times X_B$ is irreducible.

EXERCISE 4.5 Show that $X_A \times X_B$ is connected if X_A and X_B are connected.

THEOREM 4.1.5 (Chevalley) Let A and B be affine k-algebras, and let $u : A \to B$ be a k-algebra morphism. Then the image of X_B under the associated morphism ${}^a u : X_B \to X_A$ of affine algebraic k-varieties contains a non-empty open subset of $\overline{{}^a u(X_B)}$.

Proof We may assume X_B and X_A are irreducible and that $\overline{{}^a u(X_B)} = X_A$. In this situation, A and B are integral domains and $u : A \to B$ is injective. Since B can be viewed as a finitely generated A-algebra, we may write $B = A[x_1, \ldots, x_n]$ where, say, $x_1, \ldots, x_r$ are algebraically independent over A and $x_{r+1}, \ldots, x_n$ are algebraic over $A' = A[x_1, \ldots, x_r]$. For $r + 1 \leqq j \leqq n$, let

$$g_{j0}(x)x_j^{d_j} + g_{j1}(x)x_j^{d_j - 1} + \cdots = 0,$$
$$g_{ji}(x) \in A'(0 \leqq i \leqq d_j), \quad g_{j0}(x) \neq 0,$$

and let $g = \prod_{j=r+1}^{n} g_{j0}(x_1, \ldots, x_r)$. If we choose a non-zero coefficient $f \in A$ of g, the open set $U_f = \{x \in X_A; x(f) \neq 0\}$ of X_A satisfies the required condition of the theorem. Indeed, for $x \in U_f$, if $\mathfrak{m}$ is the maximal ideal associated with x, then $f \notin \mathfrak{m}$. Setting $\mathfrak{m}' = \mathfrak{m}A'$, we see that $g \notin \mathfrak{m}'$ and $B_{\mathfrak{m}'}$ is integral over $A'_{\mathfrak{m}'}$. Hence, by Theorem 1.5.6, there exists a maximal ideal $\mathfrak{n}'$ of $B_{\mathfrak{m}'}$ such that $\mathfrak{n}' \cap A'_{\mathfrak{m}'} = \mathfrak{m}'A'_{\mathfrak{m}'}$. Now, $\mathfrak{n}' \cap A = \mathfrak{n}' \cap A' \cap A = \mathfrak{m}$. Since $\mathfrak{m} = \mathfrak{n} \cap A$, for a point y of X_B which is associated with $\mathfrak{n} = \mathfrak{n}' \cap B$, we have $x = {}^a u(y)$. Therefore $U_f \subseteq {}^a u(X_B)$.

2 Affine k-groups

In this section, we define affine k-groups and its subgroups, and present some examples.

2.1 Affine k-groups

Given an affine k-variety $G = \mathbf{M}_k(H, k)$ which admits a group structure, if the two maps

$$m : G \times G \to G, \quad (x, y) \mapsto xy,$$
$$s : G \to G, \qquad x \mapsto x^{-1},$$

are morphisms of affine k-varieties, then we call G an **affine k-group**. In particular, when G is an affine algebraic k-variety, G is called an **affine algebraic k-group**. The maps m, s, and the embedding $e \to G$ of the identity element e into G, correspond to the k-algebra morphisms

$$\Delta : H \to H \otimes H, \quad S : H \to H, \quad \varepsilon : H \to k.$$

In this situation, to say that the multiplication map $m : G \times G \to G$ satisfies the group axioms is equivalent to saying that Δ and ε satisfy the axioms of k-coalgebras and that S is the antipode, in which case, H turns out to be a k-Hopf algebra. Therefore an affine k-group is precisely $G = \mathbf{M}_k(H, k)$, where H is a commutative reduced k-Hopf algebra, and G is regarded as a group where the multiplication is given via the convolution defined by

$$(x * y)(f) = \sum_{(f)} x(f_{(1)})y(f_{(2)}), \quad f \in H, \quad x, y \in G.$$

In other words, G is the group $G(H^\circ)$ of all group-like elements of the dual k-Hopf algebra H° of H.

Let $G = \mathbf{M}_k(H, k)$ and $E = \mathbf{M}_k(K, k)$ be affine k-groups. The map $\varphi : G \to E$ is called a **morphism of affine k-groups** if φ is a morphism of affine k-varieties as well as a group morphism. A morphism of affine k-groups is given by $\varphi = {}^a u$, which corresponds to a k-Hopf algebra morphism $u : K \to H$. We denote the category of affine k-groups by $\mathbf{AG}_k$, and the set of all morphisms of affine k-groups from G to E by $\mathbf{AG}_k(G, E)$. The two correspondences

$$H \mapsto \mathbf{M}_k(H, k), \quad G \mapsto \mathbf{V}_k(G, k)$$

give an anti-equivalence between the category of affine algebraic k-groups and the category of finitely generated commutative reduced k-Hopf algebras. A k-Hopf algebra whose underlying k-algebra is an affine k-algebra will be called an **affine k-Hopf algebra**. By Lemma 3.4.5, any reduced commutative k-Hopf algebra H may be regarded as an inductive limit $\varinjlim_{\lambda} H_\lambda$ of affine k-Hopf algebras H_λ $(\lambda \in \Lambda)$. Thus the affine k-group $G = \mathbf{M}_k(H, k)$ can be regarded as a projective limit of affine algebraic k-groups $G_\lambda = \mathbf{M}_k(H_\lambda, k)$ $(\lambda \in \Lambda)$, and hence we can write $G = \varprojlim_{\lambda} G_\lambda$. Sometimes an affine k-group is called a pro-affine algebraic k-group.

Remark (1) In general, given a commutative k-Hopf algebra H,

$$G : R \mapsto \mathbf{M}_k(H, R), \quad R \in \mathbf{M}_k,$$

is a representable covariant functor from the category of commutative k-algebras to the category of groups, and such functors are called affine k-group schemes. We may consider an affine k-group as the group $G(k)$ of k-rational points of an affine k-group scheme G.

(2) For a group G, the k-Hopf algebra $R_k(G)$ of all representative functions on G with values in k gives rise to an affine k-group $G^* = \mathbf{M}_k(R_k(G), k)$, which is said to be the affine k-group associated with G. We will see that G^* is useful in studying the representations of G.

EXAMPLE 4.1 Let $H = k[T]$ be the polynomial ring in one variable over k. If we define Δ, S, and ε by

$$\Delta(T) = T \otimes 1 + 1 \otimes T, \quad S(T) = -T, \quad \varepsilon(T) = 0,$$

then H becomes a k-Hopf algebra. Then $G_a = \mathbf{M}_k(H, k)$ is the **additive group** of the affine line k.

EXAMPLE 4.2 Let $H = k[T, T^{-1}]$ be the quotient ring of the polynomial ring $k[T]$ in one variable over k by the multiplicative set $S = \{T^n; n = 0, 1, 2, \dots\}$. Define Δ, S and ε by

$$\Delta(T) = T \otimes T, \quad S(T) = T^{-1}, \quad \varepsilon(T) = 1.$$

Then H becomes a k-Hopf algebra. Here $G_m = \mathbf{M}_k(H, k)$ is the **multiplicative group** $k^* = k - \{0\}$ of the non-zero elements of k.

EXAMPLE 4.3 Let $H = k[T_{ij}, \det(T_{ij})^{-1}]_{1 \leq i, j \leq n}$ be the quotient ring of the polynomial ring $k[T_{ij}]_{1 \leq i, j \leq n}$ in n^2 variables over k by the multiplicative set $S = \{\det(T_{ij})^n; n = 0, 1, 2, \dots\}$. Define Δ, S and ε by

$$\Delta(T_{ij}) = \sum_{k=1}^{n} T_{ik} \otimes T_{kj}, \quad S(T_{ij}) = \det(T_{ij})^{-1} A_{ji}, \quad \varepsilon(T_{ij}) = \delta_{ij}$$

(where A_{ji} is the cofactor of T_{ji} in the matrix (T_{ij})). Then H is a k-Hopf algebra. Now $GL_n(k) = \mathbf{M}_k(H, k)$ consists of all non-singular $n \times n$ matrices with entries in k, regarded as a group with respect to matrix multiplication. This is called the **general linear group**.

EXAMPLE 4.4 The symmetric k-algebra $S(V)$ over a k-linear space

V admits a k-Hopf algebra structure (cf. Example 2.10). In this case,

$$D_a(V) = \mathbf{M}_k(S(V), k) \cong \mathbf{Mod}_k(V, k) = V^*$$

is the group relative to the addition of the dual k-linear space V^* of V. If V is finite dimensional, the additive group V_a of V is the affine algebraic k-group represented by $S(V^*)$, namely

$$V_a \cong D_a(V^*) = \mathbf{M}_k(S(V^*), k).$$

Moreover, let $\mathscr{L}(V)$ be the additive group of all k-linear endomorphisms of a finite dimensional k-linear space V. Then

$$\mathscr{L}(V) = \mathbf{Mod}_k(V, V) \cong \mathbf{Mod}_k(V^* \otimes V, k) \cong \mathbf{M}_k(S(V^* \otimes V), k).$$

Therefore $\mathscr{L}(V)$ is an affine algebraic k-group.

EXAMPLE 4.5 (The character group.) Given an affine k-group $G = \mathbf{M}_k(H, k)$, $D(G) = \mathbf{AG}_k(G, G_m)$ has a structure of an abelian group, whose multiplication is defined by

$$(\chi_1\chi_2)(x) = \chi_1(x)\chi_2(x), \quad \chi_1, \chi_2 \in D(G), \quad x \in G,$$

and is called the **character group** of G. Since

$$D(G) \cong \mathbf{Hopf}_k(k[T, T^{-1}], H) \cong G(H),$$

we may consider the character group of G as the group $G(H)$ of all group-like elements of the k-Hopf algebra H.

EXAMPLE 4.6 (The diagonalizable group.) Let H be a commutative k-Hopf algebra. If $H = kG(H)$, in other words, if H is cocommutative pointed co-semi-simple, then $G = \mathbf{M}_k(H, k)$ is called a **diagonalizable group**. In this case, H is the group k-Hopf algebra $k\Gamma$ of Γ, where Γ is the character group of G isomorphic to $G(H)$. Conversely, assume that Γ is an abelian group and that $H = k\Gamma$ is the group k-Hopf algebra of Γ. Then H is a commutative cocommutative pointed co-semi-simple k-Hopf algebra and $G = \mathbf{M}_k(H, k)$ is an affine k-group with character group Γ.

2.2 Subgroups

Let $G = \mathbf{M}_k(H, k)$ be an affine k-group. If an affine k-subvariety

$F = \mathbf{M}_k(H/\mathfrak{a}, k)$ of G, where $\mathfrak{a}$ is an ideal of H, happens to be a subgroup of G, then we call F an **affine *k*-subgroup** of G, or simply a **closed subgroup** of G. Since the structure maps of the k-Hopf algebra $H/\mathfrak{a}$ are induced by the structure maps Δ, ε and S of the k-Hopf algebra H, the ideal $\mathfrak{a}$ of H satisfies the following conditions.

(1) $\Delta(\mathfrak{a}) \subset H \otimes \mathfrak{a} + \mathfrak{a} \otimes H$.

(2) $\varepsilon(\mathfrak{a}) = 0$.

(3) $S(\mathfrak{a}) \subset \mathfrak{a}$.

In general, a two-sided ideal $\mathfrak{a}$ of a k-Hopf algebra H with properties (1) and (2) is called a **bi-ideal**, and further is called a **Hopf ideal** when it also satisfies property (3). Now we have the following theorem.

THEOREM 4.2.1 Let H and L be k-bialgebras (resp. k-Hopf algebras) and $\varphi : H \to L$ be a morphism of k-bialgebras (resp. k-Hopf algebras). Assume that $\mathfrak{a}$ is a bi-ideal (resp. Hopf ideal) of H and let $\pi : H \to H/\mathfrak{a}$ be the canonical projection. Then

(i) $H/\mathfrak{a}$ has a unique k-bialgebra (resp. k-Hopf algebra) structure which makes π a k-bialgebra (resp. k-Hopf algebra) morphism.

(ii) $\operatorname{Ker} \varphi = \{x \in H; \varphi(x) = 0\}$ is a bi-ideal (resp. Hopf ideal) of H.

(iii) For a bi-ideal (resp. Hopf ideal) $\mathfrak{a}$ contained in $\operatorname{Ker} \varphi$, there is a unique k-bialgebra (resp. k-Hopf algebra) morphism $\bar{\varphi} : H/\mathfrak{a} \to L$ such that $\bar{\varphi} \circ \pi = \varphi$.

The proof is straightforward. The k-bialgebra (resp. k-Hopf algebra) $H/\mathfrak{a}$ defined in (i) is said to be a **factor *k*-bialgebra** (resp. **factor *k*-Hopf algebra**) of H. The family of all closed subgroups of an affine algebraic k-group $G = \mathbf{M}_k(H, k)$ corresponds one-to-one to the family of all factor k-Hopf algebras of H by its radical Hopf ideals.

THEOREM 4.2.2 Let $G = \mathbf{M}_k(H, k)$ be an affine k-group and let X be a subgroup of G. Then the closure $\bar{X}$ of X with respect to the Zariski topology is a closed subgroup of G. In particular, if X contains a dense open subset of $\bar{X}$, then $X = \bar{X}$ and X is a closed subgroup.

Proof The multiplication map $m : (x, y) \mapsto xy$ of G is a continuous map, so $\bar{X}\bar{X} \subset \overline{XX} = \bar{X}$. Similarly, $s : x \mapsto x^{-1}$ is a homeomorphism from G onto G, so $(\bar{X})^{-1} \subset \bar{X}$. Thus $\bar{X}$ is a closed subgroup of G. In turn, assume that $U \subset X$ is a dense open subset of $\bar{X}$. Then for any $g \in \bar{X}$, $U \cap gU^{-1}$ is a dense open subset of $\bar{X}$. Hence we may write $x = gy^{-1}$ $(x, y \in U)$. This forces $\bar{X} = UU = X$.

THEOREM 4.2.3 A connected affine algebraic k-group G is irreducible.

Proof Denote by G_1 the irreducible component of G containing the identity element e of G. Since $\bar{G}_1$ is also irreducible, G_1 is a closed subset. For each $x \in G_1$, $x^{-1}G_1$ is also an irreducible subset containing e, so $x^{-1}G_1 \subset G_1$. Moreover, if $x \in G$, then xG_1x^{-1} is also an irreducible subset containing e, and hence $xG_1x^{-1} \subset G_1$. Therefore G_1 is a closed normal subgroup and $[G : G_1] < \infty$ (cf. Exercise 4.4). Hence the irreducible components of G are the cosets of G_1 and G can be decomposed into the disjoint union of the irreducible components. Therefore, if $G_1 \subsetneqq G$, G is not connected, namely, if G is connected, then G is irreducible.

Henceforth, we call an irreducible affine algebraic k-group a **connected affine algebraic k-group**.

THEOREM 4.2.4 Let G be a connected affine algebraic k-group and let $\{X_\lambda\}_{\lambda \in \Lambda}$ be a family of irreducible subsets of G such that each X_λ contains the identity element e as well as a dense open subset of $\bar{X}_\lambda$. Then the subgroup X of G generated by $\{X_\lambda\}_{\lambda \in \Lambda}$ is a connected closed subgroup of G. Moreover, we can choose a sequence $\{Y_1, \ldots, Y_n\}$ consisting of X_λs or X_λ^{-1}s, where n is at most 2dim G, so that X can be expressed in the form $X = Y_1 Y_2 \ldots Y_n$.

Proof We may assume that X_λ^{-1} is contained in $\{X_\lambda\}_{\lambda \in \Lambda}$. Let $(i_1, \ldots, i_p)$ be a finite subset of Λ. Then by Theorem 4.1.4, the closed subsets of type $\overline{X_{i_1} \cdots X_{i_p}}$ of G are irreducible. If $G = \mathbf{M}_k(H, k)$, then H is an integral domain satisfying the descending chain condition on prime ideals (cf. Theorem 1.5.4), and hence there is a maximal element among such irreducible closed subsets, say $Z = \overline{X_{j_1} \cdots X_{j_q}}$. We can pick X_{j_i} $(1 \leqq i \leqq q)$ such that

$$\overline{X_{j_1}} \subsetneqq \overline{X_{j_1} X_{j_2}} \subsetneqq \cdots \subsetneqq \overline{X_{j_1} \cdots X_{j_q}}.$$

Then, since each closed set is irreducible, we have $q \leqq \dim G = n$. For any $(i_1, \ldots, i_p)$, we obtain

$$\overline{X_{j_1} \cdots X_{j_q}}\, \overline{X_{i_1} \cdots X_{i_p}} \subset \overline{X_{j_1} \cdots X_{j_q} X_{i_1} \cdots X_{i_p}} = \overline{X_{j_1} \cdots X_{j_q}},$$

so that Z is a closed subgroup of G containing X and $\bar{X} \subset Z$. In turn, suppose that $U_{j_1}, \ldots, U_{j_q}$ are respectively dense open subsets of $\overline{X_{j_1}}, \ldots, \overline{X_{j_q}}$ contained in $X_{j_1}, \ldots, X_{j_q}$. Then $U = U_{j_1} \cdots U_{j_q}$ is a dense open set of Z contained in X, and, by Theorem 4.2.2, $\bar{X} = UU = X$. Therefore X is a connected closed subgroup of G and can be expressed as a product of at most 2dim G X_λs.

Let X and Y be subgroups of a group G. The subgroup of G generated by the commutators $[x, y] = xyx^{-1}y^{-1}$ where $x \in X$ and $y \in Y$ is called the commutator subgroup generated by X and Y, and is denoted by $[X, Y]$. In particular, we call $[G, G]$ the **commutator subgroup** of G.

COROLLARY 4.2.5 Let X, Y be closed subgroups of a connected affine algebraic k-group G, and assume Y is connected. Then $[X, Y]$ is a connected closed subgroup of G and any element of $[X, Y]$ can be written as a product of at most 2dim G commutators. In particular, $[G, G]$ is a connected closed subgroup of G.

Proof Associate to each $a \in X$, the morphism $\varphi : Y \to G$ of affine k-varieties defined by

$$\varphi : y \mapsto [a, y] = aya^{-1}y^{-1}$$

and let $\varphi(Y) = Y_a$. Then Y_a is an irreducible subset containing the identity element e of G, and since φ is a morphism of affine algebraic k-varieties, by Theorem 4.1.5, Y_a contains a dense open subset of $\overline{Y_a}$. Applying Theorem 4.2.4 to $\{Y_a\}_{a \in X}$, we have that $[X, Y]$ is a connected closed subgroup and each element can be written as a product of at most 2dim G commutators.

We will define the notion of solvable groups and nilpotent groups for affine k-groups in much the same way as they are defined for abstract groups. Let G be an affine k-group and let

$$D_0(G)=G,\quad D_1(G)=[G,G],\ D_n(G)=[D_{n-1}(G),D_{n-1}(G)]\,(n\geqq 1),$$
$$C_0(G)=G,\quad C_1(G)=[G,G],\quad C_n(G)=[G,C_{n-1}(G)]\,(n\geqq 1).$$

For the following sequences

$$\overline{D_0(G)}\supset\overline{D_1(G)}\supset\cdots,\quad \overline{C_0(G)}\supset\overline{C_1(G)}\supset\cdots$$

of closed subgroups of G, if there exists a positive integer n such that $\overline{C_n(G)}=\{e\}$ (resp. $\overline{D_n(G)}=\{e\}$), then we call G a **nilpotent group** (resp. a **solvable group**). Corollary 4.2.5 guarantees that a connected affine algebraic k-group is nilpotent (resp. solvable) if and only if it is nilpotent (resp. solvable) as an abstract group.

2.3 Actions of affine k-groups on affine k-varieties

Given an affine k-group $G=\mathbf{M}_k(H,k)$ and an affine k-variety $X=\mathbf{M}_k(A,k)$, we say that G acts on X on the left if there is a morphism of affine k-varieties $\sigma:G\times X\to X$ which makes the following diagrams commute, and then G is called a **left transformation group** of an affine k-variety X.

$$\begin{array}{ccc} G\times G\times X & \xrightarrow{m\times 1} & G\times X \\ {\scriptstyle 1\times\sigma}\downarrow & & \downarrow{\scriptstyle \sigma} \\ G\times X & \xrightarrow{\sigma} & X \end{array} \qquad\qquad \begin{array}{ccc} \{e\}\times X & \longrightarrow & G\times X \\ & \searrow^{\wr} & \downarrow{\scriptstyle \sigma} \\ & & X \end{array}$$

This is equivalent to saying that A is a left H-comodule k-algebra via the k-algebra morphism $\psi:A\to H\otimes A$ associated with σ. Similarly, we may define the action of G on X on the right. For instance, let $X=G$ and $\sigma=m$. Then G acts on itself on the left (resp. right), and this action is called the **left** (resp. **right**) **translation** of G. This action is tantamount to regarding the k-Hopf algebra H as a left (resp. right) H-comodule k-algebra via Δ.

Let $E=\mathbf{M}_k(K,k)$ and $F=\mathbf{M}_k(L,k)$ be affine k-groups, $m:F\times F\to F$ the multiplication map of F, and $\delta:E\to E\times E$ the diagonal map of E defined by $x\mapsto(x,x)$. Suppose that a morphism of affine

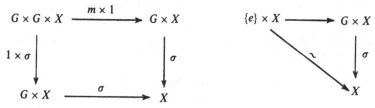

k-varieties

$$\sigma : E \times F \rightarrow F$$

defines an action of E on F, denoted by $\sigma(x, y) = x \rightharpoonup y$. If

$$x \rightharpoonup (y_1 y_2) = (x \rightharpoonup y_1)(x \rightharpoonup y_2), \quad x \in E, \quad y_1, y_2 \in F,$$

holds, namely if

$$\sigma(1 \times m) = m(\sigma \times \sigma)(1 \times \tau \times 1)(\delta \times 1 \times 1),$$

E is said to act on F by automorphisms of F, and we call E a **left transformation group** of an affine k-group F. (Here τ is the map $E \times F \rightarrow F \times E$ defined by $(x, y) \mapsto (y, x)$. Henceforth, we will let τ denote such a map as well as the map $M \otimes N \rightarrow N \otimes M$ defined by $x \otimes y \mapsto y \otimes x$ etc.) In this situation, the k-algebra morphism $\psi : L \rightarrow K \otimes L$ associated with σ satisfies

$$(1 \otimes \Delta_L)\psi = (\mu_K \otimes 1 \otimes 1)(1 \otimes \tau \otimes 1)(\psi \otimes \psi)\Delta_L.$$

That is to say, the comultiplication map Δ_L of L is a left K-comodule morphism and L is a left K-comodule k-Hopf algebra. Conversely, if L has such a structure, E turns out to be a left transformation group of F.

EXAMPLE 4.7 Let $G = \mathbf{M}_k(H, k)$ be an affine k-group and set $E = F = G$. Define a map $\sigma_0 : G \times G \rightarrow G$ by

$$\sigma_0(g, h) = g \rightharpoonup h = ghg^{-1}, \quad g, h \in G,$$

namely,

$$\sigma_0 = m(m \times 1)(1 \times 1 \times s)(1 \times \tau)(\delta \times 1).$$

Then G is a left transformation group of G via σ_0. We call this action of G the **left adjoint action**. Corresponding to this action, H has a structure of a left H-comodule k-Hopf algebra whose structure map $\psi_0 : H \rightarrow H \otimes H$ is given by

$$\psi_0 = (\mu \otimes 1)(1 \otimes \tau)(1 \otimes 1 \otimes S)(\Delta \otimes 1)\Delta,$$

in other words, $\psi_0(x) = \sum_{(x)} x_{(1)} S(x_{(3)}) \otimes x_{(2)}$ for $x \in H$ (cf. Example 3.8).

Similarly, the action of G defined by

$$\sigma_0' : G \times G \to G, \quad \sigma_0'(h, g) = h \leftharpoonup g = g^{-1}hg$$

is called the **right adjoint action** of G. Via this action, G becomes a right transformation group of G, and the structure map $\psi_0' : H \to H \otimes H$ of the right H-comodule k-Hopf algebra H associated with σ_0' is given by $\psi_0'(x) = \sum_{(x)} x_{(2)} \otimes S(x_{(1)})x_{(3)} \quad (x \in H)$.

Normal subgroups Let $G = \mathbf{M}_k(H, k)$ be an affine k-group. The closed subgroup $F = \mathbf{M}_k(H/\mathfrak{a}, k)$ of G is called a **closed normal subgroup** if F is a normal subgroup as an abstract group. In this situation, F is stable under the left adjoint action of G, so the structure map $\psi_0 : H \to H \otimes H$ of the left H-comodule k-algebra H induces the structure map of $H/\mathfrak{a}$ as a left H-comodule k-algebra. Hence $\mathfrak{a}$ is a left H-subcomodule of H and satisfies

$$\psi_0(\mathfrak{a}) \subset H \otimes \mathfrak{a}.$$

In general, a Hopf ideal $\mathfrak{a}$ of a commutative k-Hopf algebra H is called a **normal Hopf ideal** if $\mathfrak{a}$ satisfies this condition. The family of all closed normal subgroups of an affine algebraic k-group $G = \mathbf{M}_k(H, k)$ and the family of all normal radical Hopf ideals of the affine k-Hopf algebra H are in one-to-one correspondence.

LEMMA 4.2.6 Let H be a commutative k-Hopf algebra and K a k-sub-Hopf algebra and set $K^+ = \operatorname{Ker} \varepsilon_K$. Then $\mathfrak{a} = K^+H$ is a normal Hopf ideal of H.

Proof Since Δ is a k-algebra morphism, we see that

$$\Delta(\mathfrak{a}) = \Delta(K^+)\Delta(H) \subset (K \otimes K^+ + K^+ \otimes K)(H \otimes H) \subset \mathfrak{a} \otimes H + H \otimes \mathfrak{a},$$

and we clearly have $S(\mathfrak{a}) \subset \mathfrak{a}$, $\varepsilon(\mathfrak{a}) = 0$. Therefore $\mathfrak{a}$ is a Hopf ideal of H. In turn, for $x \in K^+$,

$$\begin{aligned}(1 \otimes \varepsilon_k)\psi_0(x) &= \sum_{(x)} x_{(1)} S(x_{(3)}) \varepsilon(x_{(2)}) = \sum_{(x)} x_{(1)} S(x_{(2)}) \\ &= \eta \circ \varepsilon(x) = 0\end{aligned}$$

implies $\psi_0(K^+) \subset \text{Ker}(1 \otimes \varepsilon_K) = K \otimes K^+$. Hence

$$\psi_0(\mathfrak{a}) = \psi_0(K^+H) \subset (K \otimes K^+)(H \otimes H) = H \otimes \mathfrak{a}.$$

Let H be an affine k-Hopf algebra and K a k-sub-Hopf algebra. Then we have the morphism of affine algebraic k-groups

$$\rho : G = \mathbf{M}_k(H, k) \to E = \mathbf{M}_k(K, k)$$

associated with the canonical embedding $K \to H$. Here the image $\rho(G)$ of G is dense in E, so Theorem 4.1.5 shows that $\rho(G)$ contains a dense open subset of E. Hence, by Theorem 4.2.2, $\rho(G) = E$, that is to say, ρ is surjective. Moreover, $\text{Ker}\,\rho = \mathbf{M}_k(H/\mathfrak{a}, k)$ where $\mathfrak{a} = K^+H$, and hence the factor group $G/\text{Ker}\,\rho$ is isomorphic to $E = \mathbf{M}_k(K, k)$ as an abstract group. Thus identifying $G/\text{Ker}\,\rho$ with E, we can regard it as an affine k-group. Thanks to the next theorem, $G/\text{Ker}\,\rho$ is an affine algebraic k-group.

THEOREM 4.2.7 Let H be a k-Hopf algebra and an integral domain. If H is a finitely generated k-algebra, then an arbitrary k-sub-Hopf algebra K of H is also a finitely generated k-algebra.

The proof of the theorem rests on two lemmas.

LEMMA 4.2.8 Let H be a k-Hopf algebra as well as an integral domain, and let K be a k-sub-Hopf algebra of H.

(i) Denote by $Q(K)$ the quotient field of K. Then $Q(K) \cap H = K$.

(ii) If $Q(K) = Q(H)$, then $K = H$.

Proof Pick $f = a/b \in Q(K) \cap H$ where $a, b \in K$. It is harmless to assume that H is generated by a, b and f as a k-Hopf algebra. Set $G = \mathbf{M}_k(H, k)$, and let $M = kGf$ be the kG-module generated by f. Then M is of finite dimension and lies in $Q(K) \cap H$. Now, $\mathfrak{a} = \{c \in K; cM \subset K\}$ is an ideal of K and $\mathfrak{a} \neq \{0\}$ since M is finite dimensional. To show $\mathfrak{a} = K$, assume $\mathfrak{a} \subsetneqq K$. For $x \in G$, since $xM = M$ and $xK = K$, we have $x\mathfrak{a} = \mathfrak{a}$. This shows that $\mathfrak{a}$ is a kG-module. In turn, since the canonical morphism of affine k-groups $\varphi : G \to E = \mathbf{M}_k(K, k)$ is surjective, we have $E\mathfrak{a} = \mathfrak{a}$. Since $\mathfrak{a} \subsetneqq K$, by Theorem 1.5.5 (Hilbert's *Nullstellensatz*), we can find $x \in G$ such that $x(\mathfrak{a}) = 0$.

But $x(\mathfrak{a}) = xE(\mathfrak{a}) = E(\mathfrak{a}) = 0$. So again we invoke Theorem 1.5.5, which shows $\mathfrak{a} = \{0\}$. This contradicts the fact that $\mathfrak{a} \neq \{0\}$. Therefore $\mathfrak{a} = K$ and $f \in K$. Namely, (i) holds. If $Q(K) = Q(H)$, then from (i), we get $K = Q(K) \cap H = Q(H) \cap H = H$. Thus we have (ii).

LEMMA 4.2.9 Let L be an extension field of k and let $K \supset k$ be a subfield of L. If L is finitely generated over k, K is also finitely generated over k.

Proof First assume that K/k is an algebraic extension. Let $\{x_1, \ldots, x_t\}$ be a transcendental basis for L over K. Then $\{x_1, \ldots, x_t\}$ is algebraically independent over k. Now if a subset $\{y_1, \ldots, y_n\}$ of K is linearly independent over k, then it is also linearly independent over $K(x_1, \ldots, x_t)$. Indeed, the existence of non-trivial rational functions $f_i(x_1, \ldots, x_t) \in K(x_1, \ldots, x_t)$ $(1 \leqq i \leqq n)$ such that $\sum_{i=1}^{n} f_i(x_1, \ldots, x_t) y_i = 0$ contradicts the transcendence of $\{x_1, \ldots, x_t\}$ over K. Hence $[L : K(x_1, \ldots, x_t)] \geqq [K : k]$. Therefore K/k is a finite extension and K is finitely generated over k. In the general case, given a transcendental basis $\{x_1, \ldots, x_r\}$ for K over k, pick $x_{r+1}, \ldots, x_n$ so that $\{x_1, \ldots, x_r, x_{r+1}, \ldots, x_n\}$ becomes a transcendental basis for L over k. Set $K_1 = k(x_1, \ldots, x_r)$. Then, since K/K_1 is algebraic, K is finitely generated over K_1, and hence over k.

***Proof of Theorem* 4.2.7** Denote by $Q(H)$, $Q(K)$, the quotient fields of H, K, respectively. Now $Q(H)$ is a finitely generated extension field of k, so that, by Lemma 4.2.9, $Q(K)$ is finitely generated over k. Let $\{x_1, \ldots, x_n\}$ be a set of generators for $Q(K)$ consisting of elements of K, and let K_1 be the k-sub-Hopf algebra of K generated by $\{x_1, \ldots, x_n\}$. Then $K_1 \subset K$ and $Q(K_1) = Q(K)$ imply $K_1 = K$ by Lemma 4.2.8. Thus K is a finitely generated k-Hopf algebra.

The connected component of the identity element Let A be an affine k-algebra. If $X_1, \ldots, X_r$ are the connected components of $X_A = \mathbf{M}_k(A, k)$, by Theorem 4.1.2, there exists a set $\{e_1, \ldots, e_r\}$ of orthogonal idempotents of A such that $A = Ae_1 \oplus \cdots \oplus Ae_r$. Now $\pi_0(A) = ke_1 \oplus \cdots \oplus ke_r$ is a k-subalgebra of A and we have that $\pi_0(A)$

$=k \Leftrightarrow X_A$ is connected. Moreover, for affine k-algebras A and B, $\pi_0(A \otimes B) \cong \pi_0(A) \otimes \pi_0(B)$. In fact, it is easy to see that the canonical k-linear map $\pi_0(A) \otimes \pi_0(B) \to \pi_0(A \otimes B)$ is an injective k-algebra morphism, and the surjectivity is obtained from the fact that $X_A \times X_B$ is connected when X_A and X_B are connected.

Let H be an affine k-Hopf algebra. Since $\Delta : H \to H \otimes H$ is a k-algebra morphism, $\Delta(\pi_0(H)) \subset \pi_0(H \otimes H) \cong \pi_0(H) \otimes \pi_0(H)$. Similarly, $S(\pi_0(H)) \subseteq \pi_0(H)$. Therefore $\pi_0(H)$ is a k-sub-Hopf algebra of H. By Lemma 4.2.6, $\mathfrak{a} = \pi_0(H)^+ H$ is a normal Hopf ideal of H and $\pi_0(H/\mathfrak{a}) = k$. Let ${}^a u = \varphi : G \to E = \mathbf{M}_k(\pi_0(H), k)$ be the morphism of affine algebraic k-groups associated with the canonical embedding $u : \pi_0(H) \to H$. Then it follows that E is a finite group and that Ker $\varphi = G_0 = \mathbf{M}_k(H/\mathfrak{a}, k)$ is a connected closed normal subgroup of G such that $G/G_0 \cong E$. Therefore we conclude that G_0 is the **connected component** containing the identity element of G.

2.4 The representation of affine algebraic k-groups

Let $G = \mathbf{M}_k(H, k)$ be an affine algebraic k-group and V a finite dimensional k-linear space. We will call a morphism of affine k-groups

$$\rho : G \to GL(V)$$

a **representation** of G on V and call V the **representation space** of the representation ρ. Now let $\sigma : V^* \otimes V \to H$ be the k-coalgebra morphism associated with ρ via the bijection

$$\mathbf{AG}_k(G, GL(V)) \cong \mathbf{Big}_k(S(V^* \otimes V), H) \cong \mathbf{Cog}_k(V^* \otimes V, H)$$

(where $S(V^* \otimes V)$ is the k-bialgebra which represents the underlying semigroup relative to multiplication of $\mathrm{End}_k(V)$). If we take a basis $\{v_1, \ldots, v_n\}$ for V and its dual basis $\{f_1, \ldots, f_n\}$ for V^*, and for $x \in G$ the matrix $(\rho_{ij}(x))$ of $\rho(x)$ relative to $\{v_1, \ldots, v_n\}$, then to say that ρ is a morphism of affine k-group is equivalent to saying that $\sigma(f_i \otimes v_j) = \rho_{ij} \in H$. Moreover, the k-linear map $\psi : V \to V \otimes H$ corresponding to σ by the bijection $\mathbf{Mod}_k(V^* \otimes V, H) \cong \mathbf{Mod}_k(V, V \otimes H)$ is given by

$$\psi : v_j \mapsto \sum_{i=1}^{n} v_i \otimes \rho_{ij},$$

and V becomes a right H-comodule via ψ. By (3.9), V has a left rational H^*-module structure induced by ψ. Therefore the representations ρ of an affine algebraic k-group G on V correspond one-to-one to the structures of rational left H^*-modules on V.

LEMMA 4.2.10 For an affine k-Hopf algebra H, $G = \mathbf{M}_k(H, k)$ is dense in H^*.

Proof For any $f \in H$, if $x(f) = 0$ for all $x \in G$, then f lies in every maximal ideal of H. Thus, thanks to Theorem 1.5.5, $f \in \text{rad } H = \text{nil } H = \{0\}$. This proves the density of G in H^*.

The lemma above shows that if H is an affine k-Hopf algebra, a rational left H^*-module structure on V is uniquely determined by the corresponding rational left kG-module structure on V.

THEOREM 4.2.11 An affine algebraic k-group $G = \mathbf{M}_k(H, k)$ has a faithful representation ρ on a finite dimensional k-linear space V, and G is isomorphic to a closed subgroup of $GL(V)$.

Proof We will consider H as a right H-comodule with the structure map $\Delta : H \to H \otimes H$. Since H is a finitely generated k-algebra, there is a finite dimensional k-linear subspace V which is an H-subcomodule of H and generates H as k-algebra. Let $\rho : G \to GL(V)$ be the representation of G on V corresponding to the H-comodule structure on V. Then the k-Hopf algebra morphism $\psi_\rho : H(GL(V)) \to H$, where $H(GL(V))$ is the k-Hopf algebra which is associated with $GL(V)$, is surjective due to the choice of V. Hence ρ is injective. Moreover, $\text{Im } \rho$ is a closed subgroup of $GL(V)$ (cf. Theorem 4.1.5, Theorem 4.2.2) and $\text{Im } \rho \cong \mathbf{M}_k(H(GL(V))/\mathfrak{a}, k)$, where $\mathfrak{a} = \text{Ker } \psi_\rho$. Therefore G is isomorphic to a closed subgroup of $GL(V)$.

We obtain the next theorem from Corollary 3.1.5 and Theorem 3.1.7 since, for an affine k-Hopf algebra H, the representations on V of the affine algebraic k-group $G = \mathbf{M}_k(H, k)$ correspond one-to-one to the right H-comodule structures on V.

THEOREM 4.2.12 Let $G = \mathbf{M}_k(H, k)$ be an affine algebraic k-group. Then

(1) Any representation of G is completely reducible $\Leftrightarrow H$ is co-semi-simple.

(2) Any irreducible representation of G is 1-dimensional $\Leftrightarrow H$ is pointed.

(3) Any irreducible representation of G is trivial $\Leftrightarrow H$ is irreducible.

(4) Any representation of G is diagonalizable $\Leftrightarrow H$ is pointed co-semi-simple.

Moreover, in the connected case, we have the following.

THEOREM 4.2.13 Let H be an affine k-Hopf algebra, and assume $G = \mathbf{M}_k(H, k)$ is a connected affine algebraic k-group. Then

(1) G is solvable $\Leftrightarrow H$ is pointed.

(2) G is unipotent $\Leftrightarrow H$ is irreducible.

(3) G is a torus $\Leftrightarrow H$ is pointed co-semi-simple.

Here we call G a **unipotent group** if G consists of **unipotent elements** x (elements such that $x - 1$ is nilpotent), and call G a **torus** if G is isomorphic to a direct product of a finite number of G_ms.

EXAMPLE 4.8 (Representation of the diagonalizable group.) Let Γ be a finite abelian group, and $k\Gamma$ the group k-Hopf algebra of Γ. Let $\rho : G \to GL(V)$ be a representation of the diagonalizable group $G = \mathbf{M}_k(k\Gamma, k)$ on a k-linear space V and $\psi : V \to V \otimes k\Gamma$ the $k\Gamma$-comodule structure map of V corresponding to ρ. For any $v \in V$, set

$$\psi(v) = \sum_{\gamma \in \Gamma} p_\gamma(v) \otimes \gamma.$$

Then $p_\gamma \in \mathrm{End}_k(V)$, and in this situation, we have

$$(\psi \otimes 1)\psi(v) = \sum_{\gamma, \gamma' \in \Gamma} p_\gamma(p_{\gamma'}(v)) \otimes \gamma \otimes \gamma',$$

$$(1 \otimes \Delta)\psi(v) = \sum_{\gamma \in \Gamma} p_\gamma(v) \otimes \gamma \otimes \gamma, \quad (1 \otimes \varepsilon)\psi(v) = \sum_{\gamma \in \Gamma} p_\gamma(v) = v.$$

Hence

$$p_\gamma \circ p_{\gamma'} = 0 \quad (\gamma \neq \gamma'), \quad p_\gamma \circ p_\gamma = p_\gamma, \quad \sum_{\gamma \in \Gamma} p_\gamma = 1,$$

and, if we set $p_\gamma(V) = V_\gamma$, it is clear that $V = \coprod_{\gamma \in \Gamma} V_\gamma$ and

$$V_\gamma = \{v \in V;\ \psi(v) = v \otimes \gamma\}.$$

Hence $v \in V_\gamma$ implies $\rho(g)v = g(\gamma)v$ for $g \in G$ and we may identify Γ with the character group of G.

EXAMPLE 4.9 (Representation of the additive group.) Let $\rho : G_a \to GL(V)$ be a representation of $G_a = \mathbf{M}_k(k[T], k)$ on V and $\psi : V \to V \otimes k[T]$ the $k[T]$-comodule structure map of V corresponding to ρ. For $v \in V$, set

$$\psi(v) = \sum_i \rho_i(v) \otimes T^i.$$

Then $\rho_i \in \mathrm{End}_k(V)$, and, for each $v \in V$, $\rho_i(v) = 0$ except for a finite number of i, while $(\psi \otimes 1)\psi(v) = \sum_{i;j} \rho_j(\rho_i(v)) \otimes T^j \otimes T^i$, $(1 \otimes \Delta)\psi(v) = \sum_i \rho_i(v) \otimes (T \otimes 1 + 1 \otimes T)^i$, $(1 \otimes \varepsilon)\psi(v) = \rho_0(v) = v$. Hence we have

$$\rho_j \circ \rho_i = \binom{i+j}{i} \rho_{i+j}, \quad \rho_0 = 1_V.$$

Conversely, a sequence $\{\rho_i\}_{i \in I}$ of elements of $\mathrm{End}_k(V)$ with such properties determines a $k[T]$-comodule structure on V, and, if such is the case, for $t \in k$, we see that

$$\rho(t) = \sum_{i=0}^{\infty} t^i \rho_i.$$

Now suppose the characteristic of k is 0. Set $\rho_1 = X$. Then we have $\rho_n = X^n/n!$, and we may choose for each $v \in V$ a positive integer n satisfying $X^n(v) = 0$. That is to say, X is nilpotent and we obtain $\rho(t) = \exp tX$. Therefore the representations of G_a on V correspond one-to-one to the nilpotent endomorphisms of V. If the characteristic of k is $p > 0$, then we have $s_i = \rho_{p^i} \in \mathrm{End}_k(V)$ and

(a) for each $v \in V$, except for a finite number of i, $s_i(v) = 0$ and $s_j \circ s_i = s_i \circ s_j$, $s_i{}^p = 0$.

Let $n = n_0 + n_1 p + \cdots + n_r p^r$ $(0 \leqq n_i < p)$ be the p-adic expansion of a positive integer n. Then it is easy to see that

$$\text{(b)} \quad \rho_n = \frac{s_0^{\,n_0} s_1^{\,n_1} \cdots s_r^{\,n_r}}{n_0!\, n_1! \cdots n_r!}.$$

Conversely, if a sequence $\{s_i\}_{i \in I}$ of elements of $\mathrm{End}_k(V)$ satisfies (a), we may define a $k[T]$-comodule structure on V with ρ_n as in (b). If we set

$$\exp(s_i X) = 1 + s_i X + \cdots + \frac{s_i^{\,p-1}}{(p-1)!} X^{p-1},$$

then

$$\psi(v) = \sum_{n=0}^{\infty} \rho_n(v) \otimes T^n = \prod_{i=0}^{\infty} \exp(s_i \otimes T^{p^i})(v)$$

holds, and for $t \in k$, we may write $\rho(t) = \prod_{i=0}^{\infty} \exp(s_i t^{p^i})$.

3 Lie algebras of affine algebraic k-groups

In this section, we will define the Lie algebra of an affine k-group and study the relationship between affine k-groups and their Lie algebras.

3.1 Lie algebras of affine k-groups

Given a commutative reduced k-Hopf algebra H, the k-Lie algebra $P(H^\circ)$ is called the **k-Lie algebra** of the affine k-group $G = \mathbf{M}_k(H, k)$, and is denoted by Lie G. If $\rho: G = \mathbf{M}_k(H, k) \to E = \mathbf{M}_k(K, k)$ is a morphism of affine k-groups, the k-Hopf algebra morphism $u^\circ: H^\circ \to K^\circ$ which is the dual of the k-Hopf algebra morphism $u: K \to H$ corresponding to ρ induces a k-Lie algebra morphism $d\rho: \mathrm{Lie}\, G \to \mathrm{Lie}\, E$, which is called the **differential** of ρ.

EXERCISE 4.6 The correspondence $G \mapsto \mathrm{Lie}\ G$ is a covariant functor from the category of affine k-groups to the category of k-Lie algebras.

We see that a commutative k-Hopf algebra H has a left (resp. right)

H-comodule structure with Δ as the structure map. Now let ${}^HE$ (resp. E^H) be the k-linear subspace of $E = \mathrm{End}_k(H)$ consisting of all left (resp. right) H-comodule morphisms from H to H, namely,

$$
\begin{aligned}
{}^HE &= \{\sigma \in \mathrm{End}_k(H);\ \Delta \circ \sigma = (1 \otimes \sigma) \circ \Delta\},\\
E^H &= \{\sigma \in \mathrm{End}_k(H);\ \Delta \circ \sigma = (\sigma \otimes 1) \circ \Delta\}.
\end{aligned}
$$

Recall that a left (resp. right) H-comodule structure on H yields a rational right (resp. left) H^*-module structure on H. If we denote the action by $\leftharpoonup$ (resp. $\rightharpoonup$), then we have

$$
\begin{aligned}
{}^HE &= \{\sigma \in \mathrm{End}_k(H); \sigma(x \leftharpoonup f) = \sigma(x) \leftharpoonup f \quad \forall x \in H \quad \forall f \in H^*\},\\
E^H &= \{\sigma \in \mathrm{End}_k(H); \sigma(f \rightharpoonup x) = f \rightharpoonup \sigma(x) \quad \forall x \in H \quad \forall f \in H^*\}.
\end{aligned}
$$

Here ${}^HE$ and E^H are k-subalgebras of E.

THEOREM 4.3.1 Let H be a commutative k-bialgebra and H^* the dual k-algebra of the underlying k-coalgebra of H. Then $H^* \cong {}^H\mathrm{End}_k(H)$.

Proof Define a k-linear map $\varphi : {}^H\mathrm{End}_k(H) \to H^*$ by $\sigma \mapsto \varepsilon \circ \sigma$. Then, for σ, $\rho \in {}^H\mathrm{End}_k(H)$,

$$
\begin{aligned}
\varphi(\sigma) * \varphi(\rho) &= (\varphi(\sigma) \otimes \varphi(\rho))\Delta = (\varepsilon \otimes \varepsilon)(\sigma \otimes 1)(1 \otimes \rho)\Delta\\
&= (\varepsilon \otimes \varepsilon)(\sigma \otimes 1)\Delta\rho\\
&= (\varepsilon \otimes 1)(\sigma \otimes 1)(\rho \otimes 1) = \varphi(\sigma\rho).
\end{aligned}
$$

Thus φ is a k-algebra morphism. Conversely, define a k-linear map $\psi : H^* \to \mathrm{End}_k(H)$ by $f \mapsto (1 \otimes f)\Delta$. Then $\psi(H^*) \subset {}^H\mathrm{End}_k(H)$ since

$$
\begin{aligned}
(1 \otimes \psi(f))\Delta &= (1 \otimes (1 \otimes f)\Delta)\Delta = (1 \otimes 1 \otimes f)(1 \otimes \Delta)\Delta\\
&= (1 \otimes 1 \otimes f)(\Delta \otimes 1)\Delta = \Delta(1 \otimes f)\Delta = \Delta\psi(f).
\end{aligned}
$$

Moreover, ψ is the inverse of φ by the calculation below.

$$
\begin{aligned}
(\psi \circ \varphi)(\sigma) &= (1 \otimes \varepsilon \circ \sigma)\Delta = (1 \otimes \varepsilon)(1 \otimes \sigma)\Delta\\
&= (1 \otimes \varepsilon)\Delta \circ \sigma = \sigma\\
(\varphi \circ \psi)(f) &= \varepsilon(1 \otimes f)\Delta = (1 \otimes f)(\varepsilon \otimes 1)\Delta = f.
\end{aligned}
$$

Hence φ is an isomorphism.

COROLLARY 4.3.2 Let H be a commutative k-bialgebra, set

$E = \mathrm{End}_k(H)$, and let $\mathrm{Aut}_k(H)$ be the group of all k-algebra automorphisms of H. Then

$$G(H^\circ) \cong {}^H E \cap \mathrm{Aut}_k(H), \quad P(H^\circ) \cong {}^H E \cap \mathrm{Der}_k(H).$$

Proof We proceed to show that the desired isomorphisms are obtained by taking the restrictions of the isomorphism $H^* \cong {}^H E$ in Theorem 4.3.1 to $G(H^\circ)$ and $P(H^\circ)$ respectively. For $f \in H^*$, set $\sigma = \psi(f)$. If $f \in G(H^\circ)$, then we have $\sigma(1) = 1$ and

$$\begin{aligned}\sigma(xy) &= \sum_{(x)(y)} x_{(1)} y_{(1)} \langle f, x_{(2)} y_{(2)} \rangle \\ &= \sum_{(x)(y)} x_{(1)} y_{(1)} \langle f \otimes f, x_{(2)} \otimes y_{(2)} \rangle \\ &= \sum_{(x)(y)} x_{(1)} y_{(1)} \langle f, x_{(2)} \rangle \langle f, y_{(2)} \rangle = \sigma(x)\sigma(y).\end{aligned}$$

Hence $\sigma \in \mathrm{Aut}_k(H)$. If $f \in P(H^\circ)$, then we obtain

$$\begin{aligned}\sigma(xy) &= \sum_{(x)(y)} x_{(1)} y_{(1)} \langle f \otimes 1 + 1 \otimes f, x_{(2)} \otimes y_{(2)} \rangle \\ &= \sum_{(x)(y)} x_{(1)} y_{(1)} \varepsilon(x_{(2)}) \langle f, y_{(2)} \rangle + x_{(1)} y_{(1)} \varepsilon(y_{(2)}) \langle f, x_{(2)} \rangle \\ &= x\sigma(y) + \sigma(x) y.\end{aligned}$$

Thus $\sigma \in \mathrm{Der}_k(H)$. Conversely, for $\sigma \in {}^H E$, set $\varphi(\sigma) = f$. If $\sigma \in \mathrm{Aut}_k(H)$, for any $x, y \in H$, we have

$$\langle \Delta(f) - f \otimes f, x \otimes y \rangle = \langle f, xy \rangle - \langle f, x \rangle \langle f, y \rangle = 0.$$

Hence $f \in G(H^\circ)$. If $\sigma \in \mathrm{Der}_k(H)$, we have

$$\begin{aligned}&\langle \Delta(f) - f \otimes 1 - 1 \otimes f, x \otimes y \rangle \\ &= \langle f, xy \rangle - \varepsilon(x) \langle f, y \rangle - \varepsilon(y) \langle f, x \rangle \\ &= 0.\end{aligned}$$

Hence $f \in P(H^\circ)$. Therefore $G(H^\circ)$, $P(H^\circ)$ correspond bijectively to ${}^H E \cap \mathrm{Aut}_k(H)$, ${}^H E \cap \mathrm{Der}_k(H)$, respectively. It is easy to see that these correspondences give isomorphisms of groups and k-Lie algebras respectively.

A k-derivation $D: H \to H$ from H to H is said to be **left invariant** if D satisfies $\Delta \circ D = (1 \otimes D) \circ \Delta$. By Corollary 4.3.2, Lie G is isomorphic to the k-Lie algebra ${}^H\mathrm{Der}_k(H) = {}^H E \cap \mathrm{Der}_k(H)$ consisting of all left invariant k-derivations of H. Moreover, Lie G coincides with the k-linear subspace

$$\mathrm{Der}_k(H, \varepsilon_* k) = \{\delta \in H^*;\ \delta(fg) = \delta(f)\varepsilon(g) + \varepsilon(f)\delta(g)\ \forall f, g \in H\}$$

of H^* as k-linear spaces.

If $F = \mathbf{M}_k(H/\mathfrak{a}, k)$ is a closed subgroup of an affine k-group $G = \mathbf{M}_k(H, k)$, the k-Lie algebra morphism di: Lie $F \to$ Lie G associated with the canonical embedding $i: F \to G$ is an injection and we can consider Lie F as a k-sub-Lie algebra of Lie G.

EXAMPLE 4.10 (The k-Lie algebra of the additive group G_a.) A k-derivation D from $k[T]$ to $k[T]$ is uniquely determined by $D(T) = f \in k[T]$ since we have $D(T^n) = nT^{n-1}f$. If D is left invariant, then

$$\Delta(f) = \Delta D(T) = (1 \otimes D)\Delta(T) = 1 \otimes D(T) + T \otimes D(1) = 1 \otimes f,$$

and hence $f = (1 \otimes \varepsilon)\Delta f = \varepsilon(f) \in k$. Therefore it follows that $D \in$ Lie $G_a \Leftrightarrow D(T) = \alpha \in k$, and we obtain Lie $G_a \cong k$.

EXAMPLE 4.11 (The k-Lie algebra of the multiplicative group G_m.) A k-derivation D from the k-Hopf algebra $k[T, T^{-1}]$ to $k[T, T^{-1}]$ is uniquely determined by $D(T) = f \in k[T, T^{-1}]$. If D is left invariant, then

$$\Delta(f) = \Delta D(T) = (1 \otimes D)\,\Delta\,(T) = T \otimes D(T) = T \otimes f,$$

and hence $f = (1 \otimes \varepsilon)\Delta(f) = \varepsilon(f)T$. Therefore we obtain $D \in$ Lie $G_m \Leftrightarrow D(T) = \alpha T$, $\alpha \in k$, which gives Lie $G_m \cong k$.

EXAMPLE 4.12 (The k-Lie algebra of the general linear group $GL_n(k)$.) A k-derivation D from the k-Hopf algebra $H = k[T_{ij}, \det(T_{ij})^{-1}]_{1 \leq i, j \leq n}$ to H is uniquely determined by $D(T_{ij}) = t_{ij} \in H$ $(1 \leq i, j \leq n)$. If D is left invariant, then we have

$$\Delta(t_{ij}) = \Delta D(T_{ij}) = (1 \otimes D)\Delta(T_{ij}) = \sum_{l=1}^{n} T_{il} \otimes t_{lj},$$

so $t_{ij}=(1\otimes\varepsilon)\Delta(t_{ij})=\sum_l T_{il}\varepsilon(t_{lj})$. Conversely, for an arbitrary $a=(\alpha_{ij})\in M_n(k)$, define $D\in \mathrm{Der}_k(H)$ by $D(T_{ij})=\sum_l T_{il}\alpha_{lj}$. Then $\Delta D(T_{ij})=\sum_{l,m} T_{il}\otimes T_{lm}\alpha_{mj}=(1\otimes D)\Delta(T_{ij})$, namely, D is left invariant. We will denote this derivation by D_a. The underlying k-linear space of Lie $GL_n(k)$ is isomorphic to the k-linear space $M_n(k)$ via the correspondence $a\mapsto D_a$. Further, for $a=(\alpha_{ij})$, $b=(\beta_{ij})\in M_n(k)$, if we set $ab=c=(\gamma_{ij})$, then

$$\begin{aligned} D_a\circ D_b(T_{ij}) &= D_a\left(\sum_l T_{il}\beta_{lj}\right)=\sum_{l,m} T_{il}\alpha_{lm}\beta_{mj} \\ &= \sum_l T_{il}\gamma_{lj}=D_c(T_{ij}), \end{aligned}$$

so $[D_a, D_b]=D_{ab-ba}$, and we get Lie $GL_n(k)\cong M_n(k)_L$ as k-Lie algebras.

EXERCISE 4.7 With the same notation as in Example 4.2, show that $D_a(\det(T_{ij}))=(\mathrm{Tr}\,a)\det(T_{ij})$ for $a\in M_n(k)$ and verify that Lie $SL_n(k)$ $\cong\{a\in M_n(k);\,\mathrm{Tr}\,a=0\}\subset M_n(k)_L$.

Let $u:K\to H$ be the k-Hopf algebra morphism associated with a morphism of affine k-groups $\varphi:G=\mathbf{M}_k(H,k)\to E=\mathbf{M}_k(K,k)$. The differential $d\varphi:\text{Lie } G\to \text{Lie } E$ of φ is given by

$$d\varphi(\delta)(f)=\delta(f\circ u),\qquad \delta\in\text{Lie } G,\qquad f\in K.$$

For $a\in G$, let $\hat{a}$ be the inner automorphism of G defined by $x\mapsto axa^{-1}$. Then its differential $d\hat{a}:\text{Lie } G\to\text{Lie } G$ is given by

$$d\hat{a}(\delta)(f)=\sum_{(f)} a(f_{(1)})a(S(f_{(3)}))\delta(f_{(2)}),\quad \delta\in\text{Lie } G,\quad f\in H.$$

We call the representation $Ad:G\to GL_k(\text{Lie } G)$ of G defined by $a\mapsto d\hat{a}$ the **adjoint representation** of G.

EXERCISE 4.8 Show that the adjoint representation of $G=GL_n(k)$ is given by

$$d\hat{a}:D\mapsto aDa^{-1}\quad (D\in M_n(k)_L),$$

where Lie G is identified with $M_n(k)_L$.

3.2 The criterion of separability

First, we study the relationship between field extensions and derivations. We note that from Theorem 4.3.3 up to Lemma 4.3.9, the field k is not necessarily assumed to be algebraically closed.

THEOREM 4.3.3 Let $k \subset k' \subset K$ be fields and K/k' a finite separable extension. Then every k-derivation from k' to K extends uniquely to a k-derivation from K to K.

Proof We may assume $K = k'(s)$ and $f'(s) \neq 0$ where $f(X) = \sum_{i=0}^{n} a_i X^i$ ($a_i \in k'$) is the minimal polynomial of $s \in K$ over k'. Take $D \in \mathrm{Der}_k(K)$ and let D' be the restriction of D to k'. Then

$$\begin{aligned} 0 = D(f(s)) = \sum_{i=0}^{n} D(a_i s^i) &= \sum_{i=0}^{n} D(a_i) s^i + \sum_{i=1}^{n} a_i i s^{i-1} D(s) \\ &= f^{D'}(s) + f'(s) D(s), \end{aligned}$$

where $f^{D'}(X) = \sum_{i=0}^{n} D'(a_i) X^i$ and $f'(X) = \sum_{i=1}^{n} a_i i X^{i-1} \in k'[X]$. This forces $D(s) = -f^{D'}(s)/f'(s)$, and hence D is uniquely determined by D'. Conversely, if $D' \in \mathrm{Der}_k(k', K)$, let $D(s) = -f^{D'}(s)/f'(s)$ and define D by

$$D\left(\sum_{i=0}^{n-1} b_i s^i\right) = \sum_{i=0}^{n-1} D'(b_i) s^i + \sum_{i=1}^{n-1} b_i i s^{i-1} D(s),\ b_0, \ldots, b_{n-1} \in k'.$$

Then D is a k-linear map from K to K whose restriction to k' is D'. Moreover, the choice of $D(s)$ implies that, for any positive integer m,

$$D\left(\sum_{i=0}^{m} b_i s^i\right) = \sum_{i=0}^{m} D'(b_i) s^i + \sum_{i=1}^{m} b_i i s^{i-1} D(s), \quad b_i \in k\ (1 \leqq i \leqq m),$$

so we have that $D \in \mathrm{Der}_k(K)$.

COROLLARY 4.3.4 Let K/k' be a finite field extension. Then K/k' is separable $\Leftrightarrow \mathrm{Der}_{k'}(K) = \{0\}$.

Proof Suppose K/k' is separable. Note that the zero derivation of k' extends to the zero derivation of K, and it is unique by Theorem 4.3.3.

Hence, if $D \in \mathrm{Der}_{k'}(K)$, $D = 0$, and we have $\mathrm{Der}_{k'}(K) = \{0\}$. Conversely, assume K/k' is not separable. Let k'_s be the subfield of K consisting of all elements of K separable over k' (k'_s is called the separable closure of k' in K). Take a maximal subfield of K containing k'_s, say k''. Then K/k'' is purely inseparable. If we pick $t \in K - k''$, then $K \supset k''(t) \supsetneqq k''(t^p) \supset k''$, and the maximality of k'' implies $K = k''(t)$ and that $\{1, t, \ldots, t^{p-1}\}$ is a basis for K over k''. Now we may define $D \in \mathrm{End}_{k'}(K)$ by $D(t^i) = it^{i-1} (0 \leqq i \leqq p-1)$. Then $D \in \mathrm{Der}_{k'}(K)$ and $D \neq 0$. Hence $\mathrm{Der}_{k'}(K) \neq \{0\}$.

LEMMA 4.3.5 If K/k' is a purely transcendental field extension, then

$$\dim_K \mathrm{Der}_{k'}(K) = \text{trans. deg}_{k'} K.$$

Proof Let $K = k'(x_1, \ldots, x_r)$ where $x_1, \ldots, x_r$ are algebraically independent over k'. The partial derivation $\partial/\partial x_i$ with regard to an element x_i of $k'[x_1, \ldots, x_r]$ extends uniquely to a k'-derivation D_i of $K = k'(x_1, \ldots, x_r)$. Now we will show that $\{D_1, \ldots, D_r\}$ is a basis for $\mathrm{Der}_{k'}(K)$ over K. Indeed, for any $D \in \mathrm{Der}_{k'}(K)$, $D(f) = \sum_{i=1}^{r} D_i(f)D(x_i)$ for $f \in K$ implies $D = \sum_{i=1}^{r} D(x_i)D_i$. In turn, if $\sum_{i=1}^{r} f_iD_i = 0 \, (f_i \in K, 1 \leqq i \leqq r)$, we get $\sum_{i=1}^{r} f_iD_i(x_j) = f_j = 0 \; (1 \leqq j \leqq r)$ since $D_i(x_j) = \delta_{ij} \; (1 \leqq i, j \leqq r)$. We thus conclude that $\dim_K \mathrm{Der}_{k'}(K) = r = \text{trans. deg}_{k'} K$.

THEOREM 4.3.6 If K/k' is a finitely generated field extension, then $\mathrm{Der}_{k'}(K) = \{0\} \Leftrightarrow K/k'$ is a finite separable extension.

Proof If K/k' is a finite separable extension, then Corollary 4.3.4 forces $\mathrm{Der}_{k'}(K) = \{0\}$. Conversely, suppose $\mathrm{Der}_{k'}(K) = \{0\}$ and assume that K/k' is not algebraic. Pick a transcendental basis $\{x_1, \ldots, x_d\}$ $(d \geqq 1)$ for K over k' and set $k'' = k'(x_1, \ldots, x_d)$. If K/k'' is not separable, by Corollary 4.3.4, we have $\mathrm{Der}_{k''}(K) \neq \{0\}$, so that $\mathrm{Der}_{k'}(K) \neq \{0\}$. This contradiction forces K/k'' to be separable. On the other hand, $\mathrm{Der}_{k'}(k'')$ is spanned by $D_1, \ldots, D_d$ over k'' by Lemma 4.3.5, and $D_i \neq 0 \, (1 \leqq i \leqq d)$ extends uniquely to an element of $\mathrm{Der}_{k'}(K)$

by Theorem 4.3.3. Hence $\mathrm{Der}_{k'}(K) \neq \{0\}$ and again we get a contradiction. Therefore K/k' must be algebraic as well as separable, and we may conclude that K/k' is a finite extension since K is finitely generated over k'.

In general, given a finitely generated field extension K/k, we say that K/k is **separably generated** when we can choose an appropriate transcendental basis $\{x_1, \ldots, x_d\}$ for K over k which makes $K/k(x_1, \ldots, x_d)$ a separable extension. When the characteristic of k is $p > 0$, a set $\{s_1, \ldots, s_d\}$ of elements of K satisfying $[K^p k(s_1 \ldots, s_d) : K^p k] = p^d$ is said to be p-**independent** over k. Furthermore, a subset S of K is called a p-**basis** for K over k if S satisfies the following conditions.

(1) Any finite subset of S is p-independent over k.

(2) $K^p k(S) = K$.

(The existence of a p-basis follows from Zorn's lemma.) In general, we have $|S| \geqq \mathrm{trans.deg}_k K$.

LEMMA 4.3.7 Let K/k be a finitely generated field extension. If S is a p-basis for K over k, then $|S| = \dim_K \mathrm{Der}_k(K)$. Moreover, K/k is separably generated $\Leftrightarrow |S| = \mathrm{trans.deg}_k K$.

Proof Let $S = \{s_1, \ldots, s_d\}$ and $k_i = k(K^p)(S - \{s_i\})$ $(1 \leqq i \leqq d)$. Then $k_i \subsetneqq K$ and $k_i(s_i) = K$. If we define $D_i \in \mathrm{Der}_{k_i}(K)$ by $D_i(s_i^n) = ns_i^{n-1}$, then $\{D_1, \ldots, D_d\}$ forms a basis for $\mathrm{Der}_k(K)$ over K. In fact, $\{D_1, \ldots, D_d\}$ is linearly independent over K and, moreover, any $D \in \mathrm{Der}_k(K)$ can be written in the form $D = \sum_{i=1}^{n} D(s_i)D_i$. Hence $|S| = \dim_K \mathrm{Der}_k(K)$. Now, if $D \in \mathrm{Der}_{k(S)}(K)$, then $D(S) = 0$ and $D(K^p k) = 0$, so that $D = 0$. Thus, by Theorem 4.3.6, it follows that $K/k(S)$ is a finite separable extension. Thus $|S| = \mathrm{trans.deg}_k K$ implies that K is separably generated over k. Conversely, assume K/k is separably generated. Since $\mathrm{trans.deg}_k K = \dim_K \mathrm{Der}_k(K)$ by Theorem 4.3.3 and Lemma 4.3.5, we have that S is algebraically independent over k and that $|S| = \mathrm{trans.deg}_k K$.

LEMMA 4.3.8 Let K/k be a finitely generated field extension.

Assume that for any subset $\{x_1, \ldots, x_d\}$ of K which is linearly independent over k, $\{x_1{}^p, \ldots x_d{}^p\}$ is also linearly independent over k. Then K/k is separably generated. In particular, when k is perfect, K/k is always separably generated.

Proof Let S be a p-basis for K/k. By Lemma 4.3.7, $K/k(S)$ is a finite separable field extension. Therefore it suffices to show that S is algebraically independent over k. Suppose the contrary and let m be the minimal integer such that there exists a subset $\{s_1, \ldots, s_m\}$ of S which is not algebraically independent over k. Now we take $\Gamma = \{e = (e_1, \ldots, e_m); e_i$ is zero or a positive integer$\}$ as follows.

(1) $\{s^e = s_1{}^{e_1} \cdots s_m{}^{e_m}; e \in \Gamma\}$ is linearly dependent over k.

(2) The number of elements in the set Γ satisfying (1) is the least possible.

(3) For the set Γ satisfying (1), $\sum_{e \in \Gamma} (e_1 + \cdots + e_m)$ is the least possible.

In this situation, if $\sum_{e \in \Gamma} c_e s^e = 0$ ($c_e \in k$, not all the c_e are zero), condition (2) says that for any $e \in \Gamma$, $c_e \neq 0$. Here taking $D_i \in \mathrm{Der}_k(K)$ as in the proof of Lemma 4.3.7, we get

$$D_i\left(\sum_{e \in \Gamma} c_e s^e\right) = \frac{e_i}{s_i} \sum_{e \in \Gamma} c_e s^e = 0.$$

Conditions (2), (3) imply $e_i c_e = 0$ ($e \in \Gamma$). In turn, e_i is divisible by p because $c_e \neq 0$. Set $t_e = s^{e/p} = s_1{}^{e_1/p} \cdots s_m{}^{e_m/p}$. Then $\{t_e{}^p; e \in \Gamma\}$ is linearly dependent over k, and, from the assumption of the lemma, it is clear that $\{t_e; e \in \Gamma\}$ is also linearly dependent over k. This contradicts condition (3). Thus S is algebraically independent.

LEMMA 4.3.9 Let A be a finitely generated k-algebra as well as an integral domain, and K the quotient field of A. Then $\mathrm{rank}_A \mathrm{Der}_k(A) = \dim_K \mathrm{Der}_k(K)$ if $\mathrm{Der}_k(A)$ is a free A-module.

Proof Clearly any element D of $\mathrm{Der}_k(A)$ extends uniquely to an element $\bar{D}$ of $\mathrm{Der}_k(K)$. Moreover, if $D_1, \ldots, D_d \in \mathrm{Der}_k(A)$ are linearly independent over A, then $\bar{D}_1, \ldots, \bar{D}_d$ are also linearly independent over K. Indeed, suppose $\sum_{i=1}^{d} \frac{b_i}{a_i} \bar{D}_i = 0$ where $a_i, b_i \in A, a_i \neq 0\ (1 \leq i \leq d)$.

Then we have $\sum_{i=1}^{d} ab_i\bar{D}_i = 0$, where $a = a_1 a_2 \cdots a_d$. Therefore $\sum_{i=1}^{d} ab_i D_i = 0$ where $ab_i \in A$ $(1 \leqq i \leqq d)$. Thus $b_i = 0$ $(1 \leqq i \leqq d)$ for $ab_i = 0$ and $a \neq 0$. Conversely, suppose $D_1, \ldots, D_d \in \mathrm{Der}_k(K)$ are linearly independent over K. Let $\{a_1, \ldots, a_n\}$ be generators of A as a k-algebra and let $D_i(a_j) = c_{ij}/b_{ij} \in K$ where $c_{ij}, b_{ij} \in A$, $b_{ij} \neq 0$ $(1 \leqq i \leqq d,\ 1 \leqq j \leqq n)$. Set $b = \prod b_{ij}$. Then the restrictions of $bD_1, \ldots, bD_d$ to A are elements of $\mathrm{Der}_k(A)$ which are linearly independent over A. This proves $\mathrm{rank}_A\, \mathrm{Der}_k(A) = \dim_K \mathrm{Der}_k(K)$.

We now return to the subject of k-Lie algebras of affine algebraic k-groups and in what follows we assume that k is algebraically closed.

LEMMA 4.3.10 Let H be an affine k-Hopf algebra which is an integral domain. Then $\mathrm{Der}_k(H)$ is free over H and

$$\mathrm{rank}_H\, \mathrm{Der}_k(H) = \dim_k {}^H\mathrm{Der}_k(H).$$

Proof Denote by K the quotient field of H. Then K/k is separably generated thanks to Lemma 4.3.8. Choose a transcendental basis $\{x_1, \ldots, x_d\}$ for K over k lying in $\mathfrak{m} = \mathrm{Ker}\, \varepsilon$. If we take $D_1, \ldots, D_d \in \mathrm{Der}_k(K)$ satisfying $D_i(x_j) = \delta_{ij}$ $(1 \leqq i,\ j \leqq d)$, then the restrictions of $D_1, \ldots, D_d$ to H form a basis for $\mathrm{Der}_k(H)$ over H (cf. Lemma 4.3.7 and Lemma 4.3.9). Now the map

$$\varphi : \mathrm{Der}_k(H) \to \varepsilon_*({}^H\mathrm{Der}_k(H)), \quad D \mapsto (1 \otimes \varepsilon D)\Delta = \bar{D}$$

is an H-module morphism and is surjective since $\varphi(D) = D$ for $D \in {}^H\mathrm{Der}_k(H)$. Furthermore, if $\sum_{i=1}^{d} c_i \bar{D}_i = 0$ $(c_i \in k,\ 1 \leqq i \leqq d)$, then $0 = \sum_{i=1}^{d} c_i \bar{D}_i(x_j) \equiv c_j \pmod{\mathfrak{m}}$ because $\Delta(x_j) \equiv x_j \otimes 1 + 1 \otimes x_j \pmod{\mathfrak{m} \otimes \mathfrak{m}}$. Therefore $c_j = 0$ $(1 \leqq j \leqq d)$, which is to say that $\bar{D}_1, \ldots, \bar{D}_d$ are linearly independent over k. Thus we have $\mathrm{rank}_H\, \mathrm{Der}_k(H) = \dim_k {}^H\mathrm{Der}_k(H)$.

THEOREM 4.3.11 Let G be an affine algebraic k-group. Then $\dim G = \dim_k \mathrm{Lie}\, G$.

Proof It is harmless to assume that G is connected. Take an affine k-Hopf algebra H such that $G=\mathbf{M}_k(H, k)$. Then H is an integral domain and we denote by K the quotient field of H. Since K/k is separably generated, according to Lemmas 4.3.7, 4.3.9 and 4.3.10, we have

$$\begin{aligned}\dim G &= \text{trans.deg}_k K = \dim_K \mathrm{Der}_k(K) = \mathrm{rank}_H \mathrm{Der}_k(H)\\ &= \dim_k {}^H\mathrm{Der}_k(H) = \dim_k \mathrm{Lie}\, G.\end{aligned}$$

Let $G=\mathbf{M}_k(H, k)$, $E=\mathbf{M}_k(K, k)$ be connected affine algebraic k-groups such that H, K are integral domains. Also let $\varphi: G\to E$ be a morphism of affine k-groups and $u: K\to H$ the k-Hopf algebra morphism associated with φ. We say that the morphism of affine k-groups φ is **separable** if the quotient field $Q(H)$ of H is separably generated over the quotient field $Q(u(K))$ of $u(K)$. Now we have the following theorem.

THEOREM 4.3.12 Let G, E be connected affine algebraic k-groups and let $\varphi: G\to E$ be a morphism of affine k-groups. Then φ is separable $\Leftrightarrow d\varphi: \mathrm{Lie}\, G\to \mathrm{Lie}\, E$ is surjective.

Proof Let H, K be integral domains such that $G=\mathbf{M}_k(H, k)$, $E=\mathbf{M}_k(K, k)$ respectively. We may assume that the k-Hopf algebra morphism $u: K\to H$ associated with $\varphi: G\to E$ is injective.

Now $d\varphi$ naturally induces an H-module morphism ${}^H\mathrm{Der}_k(H)\otimes_k H \to {}^K\mathrm{Der}_k(K)\otimes_k H$, and we obtain an H-module morphism

$$\psi: \mathrm{Der}_k(H)\to \mathrm{Der}_k(K, H), \quad D\mapsto D|_K.$$

Then

$$\mathrm{Ker}\,\psi = \mathrm{Der}_K(H) = \{D\in \mathrm{Der}_k(H);\ D(K)=0\},$$

and, since we have $Q(K)\cap H = K$ by Lemma 4.2.8 (ii), we get

$$\mathrm{rank}_H \mathrm{Der}_K(H) = \dim_{Q(H)} \mathrm{Der}_{Q(K)}(Q(H)).$$

If $d\varphi$ is surjective, then so is ψ, and $Q(H)$ and $Q(K)$ are separably generated over k by Lemma 4.3.8. Therefore, thanks to Lemma 4.3.7, we have

$$\dim_{Q(H)} \mathrm{Der}_{Q(K)}(Q(H)) = \mathrm{trans.deg}_k Q(H) - \mathrm{trans.deg}_k Q(K) = \mathrm{trans.deg}_{Q(K)} Q(H).$$

Hence, again by Lemma 4.3.7, we can conclude that $Q(H)/Q(K)$ is separably generated. The converse is verified by tracing along the arguments above in the reverse order.

3.3 The hyperalgebras of affine algebraic k-groups

Given an element x of an affine algebraic k-group $G = \mathbf{M}_k(H, k)$, set $\mathrm{Ker}\, x = \mathfrak{m}_x$ and

$$D_x{}^n(H, k) = \{\alpha \in H^*;\ \alpha(\mathfrak{m}_x{}^{n+1}) = 0\} = (\mathfrak{m}_x{}^{n+1})^\perp = (H/\mathfrak{m}_x{}^{n+1})^*.$$

Applying Corollary 2.3.21 to the affine k-Hopf algebra H and $\mathfrak{m}_x$, we see that $D_x{}^n(H, k)$ is a k-subcoalgebra of the dual k-coalgebra H° of H since $\mathfrak{m}_x{}^{n+1}$ is a finite codimensional ideal. In particular, we have

$$\mathrm{Der}_k(H, x_* k) = \{\delta \in D_x{}^1(H, k);\ \delta(1) = 0\}.$$

Indeed, $\mathfrak{m}_x$ is generated by $\{f - x(f);\ f \in H\}$ and since

$$\delta((f - x(f))(g - x(g))) = \delta(fg) - x(f)\delta(g) - \delta(f)x(g) + \delta(x(fg)),$$

we obtain

$$\delta \in \mathrm{Der}_k(H,\ x_* k) \Leftrightarrow \delta(\mathfrak{m}^2) = 0, \quad \delta(1) = 0.$$

Thus, if we take x as ε (the identity element of G), we have

$$\mathrm{Lie}\, G = \mathrm{Der}_k(H, \varepsilon_* k) \cong (\mathfrak{m}_\varepsilon/\mathfrak{m}_\varepsilon{}^2)^*.$$

Now define

$$D_x(H, k) = \bigcup_{n=0}^{\infty} D_x{}^n(H, k).$$

Then $D_x(H, k)$ is a k-subcoalgebra of H°. Moreover, for $x, y \in G$, $\alpha \in D_x{}^n(H, k)$, $\beta \in D_y{}^m(H, k)$, we can verify that the product $\alpha\beta$ in H^* of α and β lies in $D_{xy}{}^{n+m}(H, k)$. Hence $D_\varepsilon(H, k)$ is a k-sub-bialgebra of H°.

THEOREM 4.3.13 Let H be an affine k-Hopf algebra and let

$x \in G(H^\circ)$. Then $D_x(H, k)$ is the irreducible component $(H^\circ)_x$ of H° containing x and $H^\circ = \coprod_{x \in G} D_x(H, k)$ (k-coalgebra direct sum). Furthermore, $D_\varepsilon(H, k)$ is a kG-module k-bialgebra and

$$H^\circ = D_\varepsilon(H, k) \# kG.$$

Proof Clearly, $x \in D_x(H, k)$. First we will prove that $D_x(H, k)$ is irreducible. It suffices to show that $x \in C$ for any simple k-subcoalgebra C of $D_x(H, k)$. Now since there exists a positive integer n satisfying $C \subseteq (\mathfrak{m}_x^{n+1})^\perp$, we have $C^{\perp(H)} \supset (\mathfrak{m}_x^{n+1})^{\perp\perp(H)} = \mathfrak{m}_x^{n+1}$. Further, the maximality of the ideal $C^{\perp(H)}$ of H forces $C^{\perp(H)} = \mathfrak{m}_x$. Thus $C^{\perp(H)\perp} = \mathfrak{m}_x^\perp = kx$, and we have $x \in C$. Now in order to prove $D_x(H, k) = (H^\circ)_x$, let C be a finite dimensional k-subcoalgebra of $(H^\circ)_x$. Then $\mathfrak{a} = C^{\perp(H)}$ is a finite codimensional ideal of H and $\mathfrak{m}_x = (kx)^{\perp(H)} \supset C^{\perp(H)} = \mathfrak{a}$. Hence $H/\mathfrak{a}$ is a local k-algebra of finite dimension and, taking sufficiently large n, we have $\mathfrak{a} \supset \mathfrak{m}_x^{n+1}$. Since C is finite dimensional,

$$C = \mathfrak{a}^\perp \subset (\mathfrak{m}_x^{n+1})^\perp = D_x^n(H, k) \subset D_x(H, k).$$

Hence $(H^\circ)_x \subset D_x(H, k)$, and, from the irreducibility of $D_x(H, k)$, we get $(H^\circ)_x = D_x(H, k)$. Since H° is a pointed cocommutative k-Hopf algebra, Theorem 2.4.7 allows us to conclude that

$$H^\circ = \coprod_{x \in G} D_x(H, k) = D_\varepsilon(H, k) \# kG.$$

The irreducible cocommutative k-Hopf algebra $D_\varepsilon(H, k) = (H^\circ)_1$ is called the **hyperalgebra** of the affine algebraic k-group $G = \mathbf{M}_k(H, k)$. When k has characteristic zero, we have $(H^\circ)_1 = U(P(H^\circ)) \cong U(\text{Lie } G)$ by Theorem 2.5.3, and $(H^\circ)_1$ plays essentially the same role as Lie G.

THEOREM 4.3.14 Let k be a perfect field of characteristic $p > 0$, H a finitely generated commutative k-Hopf algebra, and let $\mathfrak{m} = \text{Ker } \varepsilon$, $\hat{H} = \varprojlim_n H/\mathfrak{m}^{n+1}$. Then the following conditions are equivalent.

(i) $\hat{H} \cong k[[x_1, \ldots, x_n]]$ (the power series ring in n variables over k).

(ii) $(H^\circ)_1 \cong B(V)$ as k-coalgebras for some k-linear space V.
(iii) $\hat{H}$ is an integral domain.
(iv) $\hat{H}$ is reduced.
(v) H is reduced.

Proof Since $(H^\circ)_1 = \varinjlim_n (H/\mathfrak{m}^{n+1})^* = \varinjlim_n (\mathfrak{m}^{n+1})^\perp$ and $H/\mathfrak{m}^{n+1}$ is a k-algebra of finite dimension, we have $(H^\circ)_1{}^* = \varprojlim_n H/\mathfrak{m}^{n+1} = \hat{H}$. Applying Corollary 2.5.16 to the cocommutative irreducible k-Hopf algebra $(H^\circ)_1$, we obtain the equivalence of (i), (ii), (iii) and (iv). Now we will show (ii)$\Leftrightarrow$(v). If $(H^\circ)_1 \cong B(V)$, $\hat{H} = (H^\circ)_1{}^*$ is reduced since (ii)$\Leftrightarrow$(iv). In turn, since we may canonically embed H in $\hat{H}$, H is also reduced. To prove the converse, we provide the following lemmas.

LEMMA 4.3.15 Let $A \subset B$ be commutative k-algebras, where A is a finitely generated k-algebra and B is a finitely generated A-module. Then the dual k-coalgebra morphism $i^\circ : B^\circ \to A^\circ$, induced by the canonical embedding $i : A \to B$, is surjective.

Proof For a finite codimensional ideal $\mathfrak{a}$ of A, we can choose a positive integer l satisfying $\mathfrak{a}^{l+1}B \cap A = \mathfrak{a}(\mathfrak{a}^l B \cap A)$ by Artin–Rees' Lemma 1.5.8. According to Lemma 2.3.20, the ideal $\mathfrak{b} = \mathfrak{a}^{l+1}B$ of B is of finite codimension and $\mathfrak{b} \cap A \subset \mathfrak{a}$. Hence there is a k-linear map $\varphi : B/\mathfrak{b} \to A/\mathfrak{a}$. If $f \in A^\circ$ and Ker $f \supset \mathfrak{a}$, then f induces a k-linear map $\bar{f} : A/\mathfrak{a} \to k$. Now let $\pi : B \to B/\mathfrak{b}$ be the canonical projection and set

$$g : B \xrightarrow{\pi} B/\mathfrak{b} \xrightarrow{\varphi} A/\mathfrak{a} \xrightarrow{\bar{f}} k.$$

Then we have Ker $g \supset \mathfrak{b}$, so $g \in B^\circ$ and $i^\circ(g) = f$. Therefore i° is surjective.

LEMMA 4.3.16 Let k be a perfect field of characteristic $p > 0$ and A a finitely generated commutative reduced k-algebra. Then $\mathscr{V} : p_* A^\circ \to A^\circ$ is surjective.

Proof Since k is a perfect field and A is reduced, the Frobenius map $\mathscr{F} : A \to p_* A$, $a \mapsto a^p$ is injective. Let $A^{(p)}$ be the k-subalgebra of A

spanned by $\{a^p; a \in A\}$ over k. Then $\mathscr{F}$ yields a bijection $\varphi : A \to p_* A^{(p)}$. For $f \in A^\circ$, define

$$g : p_* A^{(p)} \xrightarrow{\varphi^{-1}} A \xrightarrow{f} k \xrightarrow{\mathscr{F}} p_* k.$$

Then g is a k-linear map. If we take a finite codimensional ideal $\mathfrak{a}$ of A which is contained in Ker f, $\mathscr{F}(\mathfrak{a})$ is a finite codimensional ideal of $A^{(p)}$ and Ker $g \supset \mathscr{F}(\mathfrak{a})$. Therefore $g \in (A^{(p)})^\circ$. Now A is a finitely generated k-algebra and, since each $a \in A$ satisfies $X^p - a^p = 0$, A is integral over $A^{(p)}$. Hence A is a finitely generated $A^{(p)}$-module. Thanks to Lemma 4.3.15, the dual k-coalgebra morphism $i^\circ : A^\circ \to (A^{(p)})^\circ$ of the canonical embedding $i : A^{(p)} \to A$ is surjective. Here, if we choose $h \in A^\circ$ such that $i^\circ(h) = g$, we get $\mathscr{V}(h) = f$. In fact, let $\lambda_A : A \to A^{\circ *}$ be the canonical embedding. Then we have

$$\lambda_A(a)^p(h) = h(a^p) = g(a^p) = \mathscr{F}(f(a)) = \mathscr{F}(\lambda_A(a)(f)).$$

Since $\lambda_A(A)$ is dense in $A^{\circ *}$, Lemma 2.5.7 forces $\mathscr{V}(h) = f$. Therefore $\mathscr{V}$ is surjective.

***Proof of* (v) ⇒ (ii) *in Theorem* 4.3.14** If H is reduced, $\mathscr{V}_{H^\circ} : p_* H^\circ \to H^\circ$ is surjective. Moreover, the cocommutativity of H° gives $H^\circ = \coprod_{x \in G(H^\circ)} (H^\circ)_x = (H^\circ)_1 \oplus C$, where $C = \coprod_{x \neq 1} (H^\circ)_x$. In this situation, we can verify that $\mathscr{V}_{H^\circ} = \mathscr{V}_{(H^\circ)_1} \oplus \mathscr{V}_C$. Hence $\mathscr{V}_{(H^\circ)_1}$ is also surjective and Corollary 2.5.15 implies $(H^\circ)_1 \cong B(V)$.

Given an affine algebraic k-group G, a k-sub-Lie algebra of Lie G is said to be an **algebraic k-sub-Lie algebra** if it is the k-Lie algebra of some closed subgroup F of G. Not every k-sub-Lie algebra of Lie G is necessarily algebraic. When k is of characteristic 0, we can characterize algebraic k-sub-Lie algebras by the structure of k-Lie algebras (cf. Chevalley [1]). Moreover, there is a lattice isomorphism between the lattice of connected closed subgroups of a connected affine algebraic k-group G and the lattice of algebraic k-sub-Lie algebras of Lie G. However, when the characteristic of k is $p > 0$, such an isomorphism does not exist, and sometimes k-Lie algebras of two non-isomorphic connected closed subgroups of G coincide in Lie G. Yet, even then, we can construct a correspondence analogous to that for the case where the characteristic is 0 by substituting the hyperalgebra $(H^\circ)_1$ of G for Lie G (cf. Takeuchi [15]).

4 Factor groups

Given an affine k-group $G = \mathbf{M}_k(H, k)$ and its normal closed subgroup $N = \mathbf{M}_k(H/\mathfrak{a}, k)$, we will endow the factor group G/N with a structure of an affine k-group. In §2, we have shown that for a k-sub-Hopf algebra K of H, $\mathfrak{a} = K^+H$ is a normal Hopf ideal of H and that the morphism of affine k-groups $\varphi = {}^a u : G = \mathbf{M}_k(H, k) \to E = \mathbf{M}_k(K, k)$ associated with the canonical embedding $u : K \to H$ is surjective and Ker $\varphi = N = \mathbf{M}_k(H/\mathfrak{a}, k)$. In this section, we will conversely show that for a normal Hopf ideal $\mathfrak{a}$ of an affine k-Hopf algebra H, there exists a k-sub-Hopf algebra $K = K(\mathfrak{a})$ of H such that $K^+H = \mathfrak{a}$, where the two assignments $K \mapsto K^+H, \mathfrak{a} \mapsto K(\mathfrak{a})$ give a one-to-one correspondence between the family of all normal Hopf ideals of H and that of k-sub-Hopf algebras of H (cf. Theorem 4.4.7). Furthermore, due to the construction of K, $K = K(\mathfrak{a})$ is the largest k-sub-Hopf algebra among the k-sub-Hopf algebras M of H satisfying $M^+H \subset \mathfrak{a}$. Therefore, given a normal closed subgroup $N = \mathbf{M}_k(H/\mathfrak{a}, k)$ (where $\mathfrak{a}$ is a normal radical ideal of H) of an affine algebraic k-group $G = \mathbf{M}_k(H, k)$, if we set $K = K(\mathfrak{a})$, then the morphism of affine k-groups $\varphi = {}^a u : G = \mathbf{M}_k(H, k) \to E = \mathbf{M}_k(K, k)$ associated with the canonical embedding $u : K \to H$ is surjective and Ker $\varphi = N$. We will consider G/N as an affine k-group, identifying it with E which is isomorphic to G/N, and call it the **factor group** of G by N. Then G/N has the following properties.

(1) The canonical projection $\pi : G \to G/N$ is a morphism of affine k-groups.

(2) Given an affine k-group G' and a morphism of affine k-groups $\varphi : G \to G'$ satisfying Ker $\varphi \supset N$, there exists a unique morphism of affine k-groups $\psi : G/N \to G'$ such that $\varphi = \psi \circ \pi$.

Indeed, part (1) is trivial and, in order to show part (2), let $G' = \mathbf{M}_k(H', k)$ and let $u : H' \to H$ be the k-Hopf algebra morphism associated with φ. Then $M = u(H')$ is a k-sub-Hopf algebra of H and Ker $\varphi = \mathbf{M}_k(H/\mathfrak{b}, k)$ where $\mathfrak{b} = M^+H$ is a normal Hopf ideal of H. Since Ker $\varphi \supset N$, we have $\mathfrak{b} \subset \mathfrak{a}$. So $M^+ \subset \mathfrak{a}$. Since K is the largest among such k-sub-Hopf algebras of H, we have $M \subset K$. Therefore the k-Hopf algebra morphism $u : H' \to H$ can be decomposed into the composition of $H' \to K$ and the canonical embedding $K \to H$. This completes the verification of (2).

As shown above, the structure of an affine k-group for G/N satisfying (1) and (2) is uniquely determined. Now Theorem 4.2.7 guarantees that G/N is also an affine algebraic k-group.

Remark If G is connected, the morphism of affine algebraic k-groups $\pi : G \to G/N$ is separable.

In what follows, we will demonstrate the existence of a correspondence between k-sub-Hopf algebras of an affine k-Hopf algebra H and normal Hopf ideals (cf. Theorem 4.4.7). The field k is not necessarily assumed to be algebraically closed unless indicated otherwise. We start by constructing $K(\mathfrak{a})$.

Let H be a commutative k-Hopf algebra and $\mathfrak{a}$ a normal Hopf ideal of H. For the canonical k-Hopf algebra morphism $p : H \to H/\mathfrak{a} = L$,

$$(1 \otimes p)\Delta : H \to H \otimes L$$
$$(\text{resp. } (p \otimes 1)\Delta : H \to L \otimes H)$$

defines on H a right (resp. left) L-comodule structure. Now

$$H^L = \{x \in H;\, (1 \otimes p)\Delta(x) = x \otimes 1\}$$
$$(\text{resp. } {}^L H = \{x \in H;\, (p \otimes 1)\Delta(x) = 1 \otimes x\})$$

is a left (resp. right) coideal of H. In fact,

$$x \in H^L \Leftrightarrow \Delta x - x \otimes 1 \in H \otimes \mathfrak{a}.$$

Therefore, for $x \in H^L$, we get

$$(\Delta \otimes 1)(\Delta x - x \otimes 1)$$
$$= \sum x_{(1)} \otimes x_{(2)} \otimes x_{(3)} - \sum x_{(1)} \otimes x_{(2)} \otimes 1 \in H \otimes H \otimes \mathfrak{a}.$$

Hence $\sum x_{(1)} \otimes x_{(2)} \in H \otimes H^L$, namely, $\Delta(H^L) \subset H \otimes H^L$. Similarly we get $\Delta({}^L H) \subset {}^L H \otimes H$. Now we will show that $H^L = {}^L H$. Since $\mathfrak{a}$ is a normal Hopf ideal, for $x \in \mathfrak{a}$, we have $\psi_0(x) = \sum x_{(1)} S(x_{(3)}) \otimes x_{(2)} \in H \otimes \mathfrak{a}$. Thus, for $x \in {}^L H$,

$$(\psi_0 \otimes 1)(\Delta x - 1 \otimes x)$$
$$= \sum x_{(1)} S(x_{(3)}) \otimes x_{(2)} \otimes x_{(4)} - 1 \otimes 1 \otimes x \in H \otimes \mathfrak{a} \otimes H.$$

If we define the map $\omega : H \otimes \mathfrak{a} \otimes H \to H \otimes \mathfrak{a}$ by $x \otimes y \otimes z \mapsto xz \otimes y$, then

$$\omega(\psi_0 \otimes 1)(\Delta x - 1 \otimes x) = \sum x_{(1)} \otimes x_{(2)} - x \otimes 1 \in H \otimes \mathfrak{a}.$$

Thus $x \in H^L$, which implies ${}^L H \subset H^L$. Similarly we get $H^L \subset {}^L H$, and hence ${}^L H = H^L$. Here if we set $K(\mathfrak{a}) = H^L$, $K(\mathfrak{a})$ is a right and left coideal of H, and hence a k-sub-Hopf algebra of H. Moreover, $K = K(\mathfrak{a})$ is the largest k-sub-Hopf algebra of H such that $K^+ = \operatorname{Ker} \varepsilon \subset \mathfrak{a}$. In fact, for $x \in K^+$, we have $\Delta x - 1 \otimes x \in H \otimes \mathfrak{a}$, which shows that $(\varepsilon \otimes 1)(\Delta x - x \otimes 1) = x \in \mathfrak{a}$. Hence $K^+ \subset \mathfrak{a}$. On the other hand, if M is a k-sub-Hopf algebra of H satisfying $M^+ \subset \mathfrak{a}$, then, for $x \in M$, we have $\Delta x - x \otimes 1 \in M \otimes M^+ \subset H \otimes \mathfrak{a}$. Hence $x \in H^L = K$. Namely, $M \subset K$.

Next, we show that $K(\mathfrak{a})^+ H = \mathfrak{a}$; but first we need some lemmas.

LEMMA 4.4.1 Let C be a k-coalgebra and V a right C-comodule. For any k-linear subspace W of V, there exists a unique smallest coideal $\mathfrak{a}$ of C such that W is a $C/\mathfrak{a}$-subcomodule of V.

Proof Let $\psi : V \to V \otimes C$ be the right C-comodule structure map of V and define the map $\bar{\psi} : V^* \otimes V \to C$ by $\bar{\psi}(f \otimes x) = \sum \langle f, x_{(0)} \rangle x_{(1)}$. Set $W^\perp = (V/W)^* = \mathbf{Mod}_k(V/W, k)$ and $\mathfrak{a} = \bar{\psi}(W^\perp \otimes W)$. Then, for any $f \in C^*$, we have

$$f \rightharpoonup W \subset W \Leftrightarrow \langle W^\perp, f \rightharpoonup W \rangle = \langle f, \bar{\psi}(W^\perp \otimes W) \rangle = 0.$$

Therefore $\mathfrak{a}^\perp$ is the largest k-subalgebra of C^* such that W becomes an $\mathfrak{a}^\perp$-submodule.This means that $\mathfrak{a}$ is the smallest coideal of C such that W is a $C/\mathfrak{a}$-subcomodule.

With the same notation as in the lemma, we set $N(V, W) = H\mathfrak{a} + HS(\mathfrak{a})$. Then $N(V, W)$ is the smallest Hopf ideal of H such that W is an $H/N(V, W)$-subcomodule of V. If V is a right C-comodule, $\otimes^n V$, $\wedge^n V$ are right C-comodules and $\otimes^n V \to \wedge^n V \to 0$ is an exact sequence of C-comodules.

LEMMA 4.4.2 Let H be a commutative k-Hopf algebra, and let V, V' be right H-comodules.

(i) If W is an n-dimensional k-linear subspace of V, then $N(V, W) = N(\wedge^n V, \wedge^n W)$.

(ii) If $W \subset V$ and $W' \subset V'$ are k-linear subspaces and $\dim W = \dim W' = 1$, then $N(V \otimes V', W \otimes W') = N(V, W) + N(V', W')$.

Proof For a Hopf ideal $\mathfrak{a}$ of H, if we define k-linear maps

$$\rho, \sigma : V \otimes H/\mathfrak{a} \to V \otimes H/\mathfrak{a}$$

$$\text{by } \rho(v \otimes h) = \sum v_{(0)} \otimes v_{(1)} h, \quad \sigma(v \otimes h) = \sum v_{(0)} \otimes S(v_{(1)}) h,$$

then ρ and σ are inverses of each other and are k-linear isomorphisms.

(i) Trivially $N(V, W) \supset N(\wedge^n V, \wedge^n W) = \mathfrak{a}$. Take $\wedge$ as the exterior product with respect to the $H/\mathfrak{a}$-module structure, and consider the following commutative diagram.

$$\begin{array}{ccc} \wedge^n(V \otimes H/\mathfrak{a}) & \xrightarrow{\wedge^n \rho} & \wedge^n(V \otimes H/\mathfrak{a}) \\ \uparrow & & \uparrow \\ \wedge^n(W \otimes H/\mathfrak{a}) & \longrightarrow & \wedge^n(W \otimes H/\mathfrak{a}) \end{array}$$

Applying the next lemma 4.4.3 (i) to $\rho(W \otimes H/\mathfrak{a})$ and $W \otimes H/\mathfrak{a}$, we get $\rho(W \otimes H/\mathfrak{a}) = W \otimes H/\mathfrak{a}$. Therefore $\mathfrak{a} \supset N(V, W)$.

(ii) That $N(V, W) + N(V', W') \supset N(V \otimes V', W \otimes W') = \mathfrak{a}$ is straightforward. Consider the following commutative diagram.

$$\begin{array}{ccc} (V \otimes H/\mathfrak{a}) \otimes_{H/\mathfrak{a}} (V' \otimes H/\mathfrak{a}) & \xrightarrow{\rho \otimes_{H/\mathfrak{a}} \rho} & (V \otimes H/\mathfrak{a}) \otimes_{H/\mathfrak{a}} (V' \otimes H/\mathfrak{a}) \\ \uparrow & & \uparrow \\ (W \otimes H/\mathfrak{a}) \otimes_{H/\mathfrak{a}} (W' \otimes H/\mathfrak{a}) & \longrightarrow & (W \otimes H/\mathfrak{a}) \otimes_{H/\mathfrak{a}} (W' \otimes H/\mathfrak{a}) \end{array}$$

If we apply the next Lemma 4.4.3 (ii) to $\rho(W \otimes H/\mathfrak{a})$, $W \otimes H/\mathfrak{a}$, $V \otimes H/\mathfrak{a}$, and $\rho(W' \otimes H/\mathfrak{a})$, $W' \otimes H/\mathfrak{a}$, $V' \otimes H/\mathfrak{a}$, we obtain $\rho(W \otimes H/\mathfrak{a}) = W \otimes H/\mathfrak{a}$ and $\rho(W' \otimes H/\mathfrak{a}) = W' \otimes H/\mathfrak{a}$. Therefore $\mathfrak{a} \supset N(V, W) + N(V', W')$.

LEMMA 4.4.3 Let R be a commutative ring, V, V' R-modules, and let W_1, W_2 (resp. W_1', W_2') be direct summands of V (resp. V').

(i) If W_1 and W_2 are free R-modules of rank n, then $\wedge^n W_1 = \wedge^n W_2 \Rightarrow W_1 = W_2$.

(ii) If W_1, W_2, W_1' and W_2' are isomorphic to R, then $W_1 \otimes_R W_1' = W_2 \otimes_R W_2'$ in $V \otimes_R V' \Leftrightarrow W_1 = W_2$ and $W_1' = W_2'$.

Proof (i) Let $\{e_1, \ldots, e_n\}$ be a basis for W_1 over R and let $V = W_2 \oplus W_2''$. Set $e_i = f_i + f_i''$ where $f_i \in W_2$, $f_i'' \in W_2''$. Since $\wedge^n V = \bigoplus_{i=0}^{n} (\wedge^i W_2) \otimes_R (\wedge^{n-i} W_2'')$, we have $e_1 \wedge \cdots \wedge e_n = f_1 \wedge \cdots \wedge f_n$. So $\{f_1, \ldots, f_n\}$ is a basis for W_2 over R. Taking the component of $e_1 \wedge \cdots \wedge e_n$ at $(\wedge^{n-1} W_2) \otimes_R W_2''$, we easily get $f_i'' = 0$. Hence $W_1 = W_2$.

(ii) Let $V = W_2 \oplus W_2''$ and $V' = W_2' \oplus W_2'''$. Moreover, let e, e' be bases for W_1, W_1' over R respectively, and write $e = f + f''$ where $f \in W_2, f'' \in W_2''$ and $e' = f' + f'''$ where $f' \in W_2'$, $f''' \in W_2'''$. Then f and f' are bases for W_2 and W_2' respectively. Thus $f \otimes_R f''' = f'' \otimes_R f' = 0$ forces $f'' = 0 = f'''$, which implies $W_1 = W_2$ and $W_1' = W_2'$.

LEMMA 4.4.4 Let H be a finitely generated commutative k-Hopf algebra and $\mathfrak{a}$ a Hopf ideal of H. Then there exist a finite dimensional right H-comodule V and a one-dimensional k-linear subspace W of V such that $\mathfrak{a} = N(V, W)$.

Proof Now H is a right H-comodule with Δ as the structure map. If $p: H \to H/\mathfrak{a}$ is the canonical projection, then H is a right $H/\mathfrak{a}$-comodule with $(1 \otimes p)\Delta$ as the structure map and $\mathfrak{a}$ is its $H/\mathfrak{a}$-subcomodule. It is possible to write $\mathfrak{a} = HW$ for some finite dimensional $H/\mathfrak{a}$-subcomodule W of $\mathfrak{a}$. Let V be the H-subcomodule of H generated by W. If $\mathfrak{b}$ is a Hopf ideal of H satisfying $\Delta W \subset W \otimes H + H \otimes \mathfrak{b}$, then $(\varepsilon \otimes 1)\Delta W = W \subset \mathfrak{b}$. Thus $N(V, W) = \mathfrak{a}$. From Lemma 4.4.2, we get $\mathfrak{a} = N(\wedge^n V, \wedge^n W)$ where $n = \dim W$. Thus $\wedge^n V$, $\wedge^n W$, taken to be V, W respectively, give the desired result.

LEMMA 4.4.5 Let H be a commutative k-Hopf algebra and V a right H-comodule. Moreover, let $W = kv$ be a one-dimensional k-linear subspace of V and set $\mathfrak{a} = N(V, W)$. Then there exists an element $g \in H$ satisfying

$$\varepsilon(g) = 1, \quad \Delta(g) \equiv g \otimes g \pmod{H \otimes \mathfrak{a}},$$

and

$$\psi(v) \equiv v \otimes g \pmod{V \otimes \mathfrak{a}}.$$

Proof Since W is a one-dimensional $H/\mathfrak{a}$-comodule, there exists a group-like element $g'(\operatorname{mod} \mathfrak{a})$ of $H/\mathfrak{a}$ such that $\psi(v) \equiv v \otimes g'$ $(\operatorname{mod} V \otimes \mathfrak{a})$. If we take $f \in V^*$ such that $\langle f, v \rangle = 1$, then

$$g = \sum_{(v)} \langle f, v_{(0)} \rangle v_{(1)} \equiv g' \ (\operatorname{mod} \mathfrak{a}),$$

$$\Delta g = \sum_{(v)} \langle f, v_{(0)} \rangle v_{(1)} \otimes v_{(2)} \equiv \sum_{(v)} \langle f, v_{(0)} \rangle v_{(1)} \otimes g'$$
$$\equiv g \otimes g \ (\operatorname{mod} H \otimes \mathfrak{a}).$$

Hence g satisfies the required conditions.

LEMMA 4.4.6 Assume that k is an algebraically closed field. Let H be a finitely generated commutative reduced k-Hopf algebra and $\mathfrak{a}$ its normal Hopf ideal. Then there exist a right H-comodule V and a one-dimensional k-linear subspace $W = kv$ of V satisfying $\mathfrak{a} = N(V, W)$ and $\psi(v) \equiv v \otimes 1 \ (\operatorname{mod} V \otimes \mathfrak{a})$.

Proof Let V and W be as in Lemma 4.4.4 and $g \in H$ as in Lemma 4.4.5. From Lemma 4.2.10, we get

$$H^* \rightharpoonup g = kG(H^\circ) \rightharpoonup g.$$

In turn, for $x \in G(H^\circ)$ and $y \in (H/\mathfrak{a})^*$, we have $xyx^{-1} \in (H/\mathfrak{a})^*$. Hence, by the condition in Lemma 4.4.5, kg is an $H/\mathfrak{a}$-subcomodule of H and, for any $x \in G(H^\circ)$,

$$(H/\mathfrak{a})^* \rightharpoonup (x \rightharpoonup g) = x \rightharpoonup ((H/\mathfrak{a})^* \rightharpoonup g) \subset k(x \rightharpoonup g).$$

Thus $k(x \rightharpoonup g)$ becomes an $H/\mathfrak{a}$-subcomodule of H. Now $U = H^* \rightharpoonup g = kG(H^\circ) \rightharpoonup g$ can be expressed as a sum of one-dimensional $H/\mathfrak{a}$-subcomodules. Given $\bar{h} \in G(H/\mathfrak{a})$, let

$$U_{\bar{h}} = \{x \in U;\ \psi(x) = x \otimes \bar{h} \in U \otimes H/\mathfrak{a}\}.$$

Then as $H/\mathfrak{a}$-comodules, U is the direct sum of the $U_{\bar{h}}(\bar{h} \in G(H/\mathfrak{a}))$ and $U_{\bar{g}}$ is a direct summand of the $H/\mathfrak{a}$-comodule U where $\bar{g}$ is the image of g in $H/\mathfrak{a}$. If we regard U^* as a right H-comodule using the antipode S, then $(U_{\bar{g}})^* = (U^*)_{\bar{g}^{-1}}$. We see that $U_{\bar{g}} \neq \{0\}$ since $g \in U_{\bar{g}}$, and, if v' is a non-zero element of $(U^*)_{\bar{g}^{-1}}$, then $N(U^*, kv') \subset \mathfrak{a}$ implies

$$N(V \otimes U^*, W \otimes kv') = N(V, W) + N(U^*, kv') = \mathfrak{a}.$$

Therefore $V \otimes U^*$ and $v \otimes v'$ satisfy the conditions of the lemma.

THEOREM 4.4.7 Assume that k is an algebraically closed field. Then normal Hopf ideals of an affine k-Hopf algebra H are in one-to-one correspondence with k-sub-Hopf algebras of H in the following manner.

$$\mathfrak{a} \mapsto K(\mathfrak{a}), \qquad K \mapsto K^+H.$$

Proof Given a normal Hopf ideal $\mathfrak{a}$ of H, let a right H-comodule V and its one-dimensional k-linear subspace $kv = W$ be as in Lemma 4.4.6. Viewing V^* trivially as an H-comodule, the map

$$\bar{\psi} : V^* \otimes V \to H, \quad f \otimes v \mapsto \sum_{(v)} \langle f, v_{(0)} \rangle v_{(1)}$$

is a right H-comodule morphism, and we have

$$\bar{\psi}(W^\perp \otimes W) \subset \{x \in \mathfrak{a};\ \Delta(x) \equiv x \otimes 1 \bmod H \otimes \mathfrak{a}\} \subset K(\mathfrak{a})^+.$$

From the definition of $N(V, W)$, we obtain $\mathfrak{a} = N(V, W) \subset K(\mathfrak{a})^+H \subset \mathfrak{a}$. Therefore $K(\mathfrak{a})^+H = \mathfrak{a}$. In turn, let K be a k-sub-Hopf algebra of H and $\mathfrak{a} = K^+H$. Since K is reduced, we can identify K with the k-Hopf algebra $\mathbf{V}_k(L, k)$ where L is the affine k-group $\mathbf{M}_k(K, k)$, and $\mathbf{V}_k(L, k)$ is determined uniquely by $\mathfrak{a}$. Thus we see that $K = K(\mathfrak{a})$ since $K^+H = K(\mathfrak{a})^+H$.

Extensions of affine algebraic k-groups Given a normal Hopf ideal $\mathfrak{a}$ of a commutative k-Hopf algebra H, set $L = H/\mathfrak{a}$. We call $K = K(\mathfrak{a})$ the **Hopf kernel** of the canonical k-Hopf algebra morphism $p : H \to L$, and denote it by Hker p. Conversely, if K is a k-sub-Hopf algebra of H and $i : K \to H$ is the canonical embedding, $\mathfrak{a} = K^+H$ is a normal Hopf ideal of H and K turns out to be the Hopf kernel of the canonical k-Hopf algebra morphism $p : H \to H/\mathfrak{a}$. In these circumstances, (H, i, p) is called an **extension** of K by L and the sequence

$$(0 \to) K \xrightarrow{i} H \xrightarrow{p} L (\to 0)$$

is said to be an **exact sequence** of k-Hopf algebras. When k is an algebraically closed field, if H is an affine k-Hopf algebra, K is also an affine k-Hopf algebra (cf. Theorem 4.2.7), and, moreover, if $\mathfrak{a}$ is a radical ideal, then L is an affine k-Hopf algebra as well. Now we get the exact sequence of affine algebraic k-groups

$$1 \to G(L^\circ) \to G(H^\circ) \to G(K^\circ) \to 1.$$

Given an extension (H, i, p) of K by L, if there exists a k-Hopf algebra (resp. k-algebra) morphism $q: H \to K$ such that $q \circ i = 1_K$, we say the extension (H, i, p) is **split** (resp. **semi-split**) and call q its **section**. If an extension of affine k-Hopf algebras is split, then its corresponding exact sequence of affine algebraic k-groups is also split. We proceed to show conversely that an extension of affine algebraic k-groups is obtained from an extension of affine k-Hopf algebras.

In general, let $F = \mathbf{M}_k(L, k)$ and $E = \mathbf{M}_k(K, k)$ be affine k-groups. Assume that E acts on F as automorphisms and denote the action by $\sigma: E \times F \to F$. Then L becomes a left K-comodule k-algebra where the k-algebra morphism $\psi: L \to K \otimes L$ associated with σ is taken as the structure map. Now define multiplication on the affine k-variety $F \times E = \mathbf{M}_k(L \otimes K, k)$ by

$$m_{F \times E} = (m_F \times m_E)(1 \times \sigma \times 1 \times 1)(1 \times 1 \times \tau \times 1)(1 \times \delta \times 1 \times 1),$$

namely, by

$$(y, x)(y', x') = (y\sigma(x, y'), xx'), \quad x, x' \in E, \quad y, y' \in F.$$

Then $F \times E$ becomes an affine k-group and the inverse of any element can be obtained by the map

$$s_{F \times E} = (s_F \times s_E)(\sigma \times 1)(s_F \times 1 \times 1)(\tau \times 1)(1 \times \delta),$$

namely,

$$(y, x)^{-1} = (\sigma(x^{-1}, y)^{-1}, x^{-1}), \quad x \in E, \quad y \in F.$$

This affine k-group is called the **semi-direct product** of F and E with respect to σ, and denoted by $F \times_\sigma E$. Then F and E are closed subgroups of $F \times_\sigma E$ and F is furthermore a normal closed subgroup of $F \times_\sigma E$. Now letting $H = L \flat K$ be the co-semi-direct product of k-Hopf algebras L and K with respect to ψ (cf. Chapter 3, §2.5), we see that $F \times_\sigma E = \mathbf{M}_k(H, k)$ and that (H, i, p) is an extension of K by L, where $p: H \to L$ and $i: K \to H$ are the canonical projection and embedding respectively. The canonical projection $q: H \to K$ is a section of this extension, so that the extension (H, i, p) is split.

Let $G = \mathbf{M}_k(H, k)$ be an affine k-group and $F = \mathbf{M}_k(H/\mathfrak{a}, k)$ a closed normal subgroup of G. Then $\mathfrak{a}$ is a normal Hopf ideal of H. Moreover, if we let $K = K(\mathfrak{a})$, $L = H/\mathfrak{a}$ and $p: H \to L$, $i: K \to H$ be the canonical projection and embedding respectively, then (H, i, p) is an extension of K by L and we get $G/F \cong E = \mathbf{M}_k(K, k)$. Assume that this extension is split and let $q: H \to K$ be its section. Then H becomes a right K-bimodule with structure maps $\varphi = \mu(1 \otimes i)$ and $\psi = (1 \otimes g)\psi_0'$ (where the map ψ_0' is the one defined in Example 4.7, cf. Exercise 3.3), and the K-module morphism $H^K \otimes K \to H$ defined by $a \otimes b \mapsto ai(b)$ is an isomorphism of k-algebras (cf. Example 3.3). Now we define a k-Hopf algebra structure on H^K so that the canonical projection $p: H \to H^K$ turns out to be a k-Hopf algebra morphism, and denote it by L. In other words, L is a k-Hopf algebra whose underlying k-algebra is that of H^K and whose k-coalgebra structure maps and antipode are defined by

$$\Delta_L = (p \otimes p)\Delta_H, \qquad \varepsilon_L = \varepsilon_H|_L, \qquad S_L = p \circ S_H.$$

In this situation, L admits a structure of a right K-bimodule k-Hopf algebra with the structure map $\psi = (1 \otimes q)\psi_0'$, and we have $H \cong L \flat K$. This corresponds to the fact that G can be expressed as a semi-direct product of F and E.

When an extension (H, i, p) of K by L is semi-split, its section $q: H \to K$ is a k-algebra morphism and is not necessarily a k-Hopf algebra morphism. We see that H is a right K-bimodule k-algebra with structure maps $\varphi = \mu(1 \otimes i)$ and $\psi = (1 \otimes q)\psi_0$ and that $H \cong L \otimes K$ as k-algebras. Let (H, i, p), (H', i', p') be extensions of K by L which are semi-split and q, q' be their respective sections. If there exists a k-Hopf algebra isomorphism $u: H \to H'$ such that $u \circ i = i'$, $p' \circ u = p$, and $q' \circ u = q$ as k-algebra morphisms, we say that the two extensions above are equivalent. (This relation defines an equivalence relation.) Let $\otimes^n K = K \otimes \cdots \otimes K$ (the n-fold tensor product of K), $\otimes^0 K = k$, and assume that L is cocommutative. Then $\mathbf{M}_k(L, \otimes^n K)$ is an abelian group with respect to convolution. If we define k-module morphisms

$$\delta^{n-1}: \mathbf{M}_k(L, \otimes^{n-1} K) \to \mathbf{M}_k(L, \otimes^n K) \qquad (n \geqq 1)$$

by $\delta^{n-1}(f) = ((1 \otimes \cdots \otimes 1 \otimes f)\psi) * ((\Delta \otimes 1 \otimes \cdots \otimes 1) f^{-1})$
$$* ((1 \otimes \Delta \otimes 1 \otimes \cdots \otimes 1) f) * \cdots * ((1 \otimes 1 \otimes \cdots \otimes \Delta) f^{\pm})$$
$$* (f^{\mp} \otimes \eta_k),$$

then $\delta^n \delta^{n-1} = 0$, and $Z^n(L, K) = \operatorname{Ker} \delta^n$ as well as $B^n(L, K) = \operatorname{Im} \delta^{n-1}$ are both k-linear subspaces of $\mathbf{M}_k(L, \otimes^n K)$. We call

$$H^n(L, K) = Z^n(L, K)/B^n(L, K)$$

the ***n*th cohomology group** of the K-bimodule L. Now we see that $H^2(L, K)$ corresponds one-to-one with the equivalence classes of extensions of K by L which are semi-split.

5 Unipotent groups and solvable groups

It is a standard result in the theory of connected linear algebraic groups that a unipotent algebraic k-group has a composition series with factor groups isomorphic to the additive groups G_a. It is also known that a connected solvable group admits a decomposition into a semi-direct product of a torus and a unipotent group. We devote this section to proving the above properties by applying the theory of Hopf algebras. Here k is assumed to be an algebraically closed field.

5.1 Unipotent groups

A commutative pointed irreducible k-Hopf algebra H is called a **unipotent k-Hopf algebra**, and the affine k-group $G = \mathbf{M}_k(H, k)$ is said to be a **unipotent k-group**. Any irreducible representation of a unipotent k-group is trivial.

THEOREM 4.5.1 Let H be a unipotent k-Hopf algebra whose underlying k-algebra is a finitely generated integral domain. If $\operatorname{Kdim} H = 1$, then $H \cong k[x]$ for some $x \in P(H)$. If $\operatorname{Kdim} H \geqq 1$, then there exist a k-sub-Hopf algebra K of H and an element x of H satisfying the following properties.

(i) $\operatorname{Kdim} H = \operatorname{Kdim} K + 1$.

(ii) x is transcendental over K and H is integral over $K[x]$. Moreover, $q(x) = \Delta(x) - x \otimes 1 - 1 \otimes x \in K^+ \otimes K^+$.

(iii) $H/K^+ H$ is an integral domain whose Krull dimension is one.

As a result of this theorem, we see that a connected unipotent algebraic k-group has a composition series whose factor groups are isomorphic to G_a. To prove this theorem, we will provide some lemmas. Recall that if we set $R_0 = k1$ (the coradical of H) and $R_n = R_0 \sqcap R_{n-1} (n \geqq 1)$, then $\{R_n\}_{n \geqq 0}$ is the coradical filtration on H and satisfies

$$R_n \subset R_{n+1}, \quad H = \bigcup_{n=0}^{\infty} R_n \quad \text{and} \quad \Delta R_n \subset \sum_{i=0}^{n} R_i \otimes R_{n-i}.$$

LEMMA 4.5.2 If a unipotent k-Hopf algebra H is algebraic over its k-sub-Hopf algebra K, then H is integral over K.

Proof Let $\{R_n\}_{n \geqq 0}$ be the coradical filtration on H and A_n the k-subalgebra of H generated by R_n. Then $L_n = A_n K$ is a k-sub-Hopf algebra of H and we have $A_0 = k$, $L_0 = K$ and $\bigcup_{n=0}^{\infty} L_n = H$. Therefore it is sufficient to prove that L_n is integral over L_{n-1}. Moreover, since A_n is generated by R_n, it is enough to show that any element of R_n is integral over L_{n-1}. Let $y \in R_n - L_{n-1}$ and suppose that

$$a_m y^m + \cdots + a_1 y + a_0 = 0, \quad a_i \in L_{n-1} \; (0 \leqq i \leqq m), \quad a_m \neq 0,$$

is the minimal polynomial of y over L_{n-1}. If $a_m \in k$, then y is integral over L_{n-1}. Assume $a_m \notin k$. Let $\{C_n\}_{n \geqq 0}$ be the coradical filtration on L_{n-1} and suppose $a_m \in C_{j_0} - C_{j_0 - 1}$. Since $C_0 = R_0 = k1$, it follows that L_{n-1} is pointed irreducible, and, by Lemma 2.4.13, $\Delta a_m = a_m \otimes 1 + 1 \otimes a_m + s$, where $s \in C_{j_0 - 1} \otimes C_{j_0 - 1}$. Moreover, since $\Delta y = y \otimes 1 + 1 \otimes y + t$, $t \in R_{n-1} \otimes R_{n-1} \subset L_{n-1} \otimes L_{n-1}$, we have

$$\Delta y \in (L_{n-1} y \otimes L_{n-1}) + (L_{n-1} \otimes L_{n-1} y) + (L_{n-1} \otimes L_{n-1}).$$

Now Δ is a k-algebra morphism, and hence for each $i \leqq m$,

$$\Delta y^i = (\Delta y)^i \in \left(\sum_{j=0}^{i} L_{n-1} y^j \right) \otimes \left(\sum_{j=0}^{i} L_{n-1} y^j \right).$$

On the other hand, since $a_i \in L_{n-1}$, we have $\Delta a_i \in L_{n-1} \otimes L_{n-1}$, so that

$$\Delta(a_i y^i) = \Delta a_i (\Delta y)^i \in \left(\sum_{j=0}^{i} L_{n-1} y^j \right) \otimes \left(\sum_{j=0}^{i} L_{n-1} y^j \right),$$

$$\begin{aligned}\Delta(a_m y^m) &= a_m y^m \otimes 1 + a_m \otimes y^m + y^m \otimes a_m + 1 \otimes a_m y^m \\ &\quad + (y^m \otimes 1)s + (1 \otimes y^m)s + (a_m \otimes 1)t_1 + (1 \otimes a_m)t_1 + t_1 s,\end{aligned}$$

$$\text{where} \quad t_1 \in \left(\sum_{j=0}^{m-1} L_{n-1} y^j\right) \otimes \left(\sum_{j=0}^{m-1} L_{n-1} y^j\right).$$

If we choose $f \in H^*$ satisfying $f(C_{j_0-1}) = 0$ and $f(a_m) = 1$, then

$$\begin{aligned} f \rightharpoonup a_m y^m &= (1 \otimes f)\Delta(a_m y^m) \\ &\equiv a_m f(y^m) + y^m + f(a_m y^m) \quad \left(\operatorname{mod} \sum_{j=0}^{m-1} L_{n-1} y^j\right), \end{aligned}$$

and hence

$$f \rightharpoonup (a_m y^m + \cdots + a_1 y + a_0) = y^m + v, \quad v \in \sum_{j=0}^{m-1} L_{n-1} y^j.$$

This implies that y is integral over L_{n-1}.

LEMMA 4.5.3 Let k be a field of characteristic $p > 0$. If the polynomial ring $k[x]$ over k admits a k-Hopf algebra structure and x is its primitive element, then

$$P(k[x]) = \left\{ \sum_{i=0}^{m} a_i x^{p^i} ; a_i \in k \right\}.$$

Proof Note that $\{x^i \otimes x^j\}_{i,j \geqq 1}$ is a basis for $k[x]^+ \otimes k[x]^+$ over k. For $g = \sum_{i,j \geqq 0} c_{ij} x^i \otimes x^j \in k[x] \otimes k[x]$, let $\operatorname{ht}(g) = \max\{i + j; c_{ij} \neq 0\}$. Suppose $z = \sum_{i=0}^{m} a_i x^i \in P(k[x])$ where $a_m \neq 0$ and m is not a power of p. Then we have $q(z) = \Delta z - z \otimes 1 - 1 \otimes z = 0 \Leftrightarrow z \in P(k[x])$. In turn, $q(z^p) = \Delta z^p - z^p \otimes 1 - 1 \otimes z^p = (\Delta z)^p - (z \otimes 1)^p - (1 \otimes z)^p = q(z)^p$. Hence $q(z^{p^i}) = q(z)^{p^i}$. Since $x \in P(k[x])$, we have $x^{p^i} \in P(k[x])$. Therefore if we eliminate terms of degree a power of p from z, then the resulting element is also primitive. Thus we may assume that $a_i = 0$ if i is a power of p. Moreover, $\varepsilon(z) = 0$ implies $a_0 = 0$. Now we will show that $\operatorname{ht}(q(z)) = m$. If this is verified, we obtain $m = 0$, namely that $z = 0$ since $z \in P(k[x]) = \ker q$.

Since $\operatorname{ht}(q(z - a_m x^m)) < m$, we have $\operatorname{ht}(q(z)) = m \Leftrightarrow \operatorname{ht}(q(x^m)) = m$. Setting $m = p^u v$ where $(p, v) = 1$, the hypothesis implies $v \neq 1$, so that we may write $v = ps + r$ $(0 < r < p)$. Hence $q(x^m) = q((x^v)^{p^u})$

$=(x^{sp})^{p^u} \otimes (x^r)^{p^u} + t$, where $t \in k[x]^+ \otimes k[x]^+$ and t is a linear combination of elements of the basis differing from the first term. Therefore we conclude that $\mathrm{ht}(q(x^m)) = (sp)p^u + rp^u = m$.

Let k be a field of characteristic $p > 0$ and H a unipotent k-Hopf algebra which is an integral domain and let $x \in H$ be a transcendental element over a k-sub-Hopf algebra K of H such that $q(x) \in K^+ \otimes K^+$. If $\{b_i\}_{i \in I}$ is a basis for K over k, then $\{b_i x^j \otimes b_{i'} x^{j'}\}$ is a basis for $K[x] \otimes K[x]$ over k. For $g = \sum c_{ii'\,jj'} b_i x^j \otimes b_{i'} x^{j'} \in K[x] \otimes K[x] (c_{ii'\,jj'} \in k)$, if we set $\mathrm{ht}(g) = \max\{j + j'; c_{ii'\,jj'} \neq 0]$, then we have the next lemma.

LEMMA 4.5.4 If $z = \sum_{i=0}^{m} c_i x^i \in K[x]$, $c_i \in K$, and $c_m \neq 0$, then

(i) $\mathrm{ht}(q(z)) \leqq m$,

(ii) $\mathrm{ht}(q(z)) < m \Leftrightarrow m = p^j$ for some j, $c_m \in k$.

Proof (i) Since $q(x^i) \in \left(\sum_{j=0}^{i} Kx^{i-j} \otimes Kx^j \right) + \sum_{j+j'<i} Kx^j \otimes Kx^{j'}$, we have $\mathrm{ht}(q(x^i)) \leqq i$ and $\mathrm{ht}(q(z)) \leqq m$.

(ii) If $c_m \in k$, we see as in the proof of Lemma 4.5.3 that $\mathrm{ht}(q(z)) = m$ if m is not a power of p. In the case $c_m \notin k$, let $\{R_n\}_{n \geqq 0}$ be the coradical filtration on H and assume $c_m \in R_j - R_{j-1}$. Then we have

$$q(c_m x^m) = c_m \otimes x^m + \sum e_i \otimes e_i' x^m = t, \quad e_i \in R_{j-1},$$

where t does not include terms such as $b_i \otimes x^m$. On the other hand, since $c_m \in R_j - R_{j-1}$ and $e_i \in R_{j-1}$, we get $c_m \otimes x^m + \sum e_i \otimes e_i' x^m \neq 0$. Hence $\mathrm{ht}(q(c_m x^m)) = m$.

LEMMA 4.5.5 Let H be a unipotent k-Hopf algebra which is an integral domain. If $w \in H$, $q(w) \in K[x]^+ \otimes K[x]^+$ and $z = \sum_{i=0}^{n} c_i w^{p^i}$ for $c_i \in k$, $c_n \neq 0$, then $\mathrm{ht}(q(z)) = \mathrm{ht}(q(w))p^n$.

Proof Let

$$q(w) = \sum_{j+j' = ht(q(w))} a_{i\,ji'\,j'} b_i x^j \otimes b_{i'} x^{j'} + t,$$

where $\mathrm{ht}(t) < \mathrm{ht}(q(w))$. Then

$$q(z) = \sum_{l=0}^{n} c_l q(w)^{p^l} = \sum_{l=0}^{n} c_l \sum (a_{i\,ji'\,j'} b_i x^j \otimes b_{i'} x^{j'}) p^l + c_l p^l$$

$$= c_n \left(\sum_{j+j'=ht(q(w))} a_{i\,ji'\,j'} b_i x^j \otimes b_{i'} x^{j'} \right)^{p^n} + s,$$

where $\mathrm{ht}(s) < \mathrm{ht}(q(w))p^n$. Now H has no nilpotent elements except 0, and k is an algebraically closed field (and hence a perfect field). Therefore we have $\mathrm{ht}(q(z)) = \mathrm{ht}(q(w))p^n$.

THEOREM 4.5.6 Let $H = k[x_1, \cdots, x_n]$ be a k-Hopf algebra which is an integral domain. If $x_i \in P(H)$ $(1 \leqq i \leqq n)$ and if Kdim $H = 1$, then we can choose $x \in P(H)$ appropriately so that $H = k[x]$.

Proof It is harmless to assume that $H = k[x_1, x_2]$. Moreover, since Kdim $H = 1$, we may assume x_1 is transcendental over k and x_2 is algebraic over $k[x_1]$. By Lemma 4.5.2, x_2 is integral over $k[x_1]$ and $H/k[x_1]^+H$ is a finite dimensional k-linear space. Thus there exist $a_i \in k$ $(0 \leqq i \leqq m-1)$ satisfying

$$z = x_2^{p^m} + a_{m-1} x_2^{p^{m-1}} + \cdots + a_1 x_2^p + a_0 x_2 \in k[x_1]^+ H.$$

Since $q(z) = \sum_{i=0}^{m} a_i q(x_2)^{p^i}$ $(a_m = 1)$, we get $q(z) \in k[x_1]^+ \otimes k[x_1]^+$. Therefore $k[x_1, z]$ is a k-sub-Hopf algebra of H and $k[x_1, z]^+ H = k[x_1]^+ H$. From Theorem 4.4.7, we obtain $k[x_1, z] = k[x_1]$. By Lemma 4.5.3, we may write

$$z = \sum_{i=0}^{n} b_i x_1^{p^i} \quad \text{for} \quad b_1 \in k \ (0 \leqq i \leqq n)$$

because $z \in P(k[x_1])$. Hence

$$x_2^{p^m} + \sum_{j=0}^{m-1} a_j x_2^{p^j} = \sum_{i=0}^{n} b_i x_1^{p^i}, \quad a_j, b_i \in k.$$

Now we choose $x, y \in P(H)$ $(x, y \neq 0)$ such that $H = k[x, y]$ and such that if x, y satisfy

$$\sum_{i=0}^{s} c_i y^{p^i} = \sum_{j=0}^{r} d_j x^{p^j}, \quad (*)$$

then $r+s$ is the smallest possible. If $s=r=0$, then we have $H=k[x]$. Suppose $0\neq s\geqq r$. Not all the d_j are zero since H is an integral domain. Adding

$$-\sum_{i=0}^{r} d_i\left((c_s/d_r)^{p^{-r}}y^{p^{(s-r)}}\right)^{p^i}$$

to both sides of (*), we obtain an equality as in (*) with $r+s-1$ terms in $H=k[x-(c_s/d_r)^{p^{-r}}y^{p^{(s-r)}}, y]$. This contradicts the choice of x and y, forcing $x=(c_s/d_r)^{p^{-r}}y^{p^{(s-r)}}$. Therefore $H=k[y]$.

***Proof of the first part of Theorem* 4.5.1** Suppose that H is a unipotent k-Hopf algebra which is a finitely generated integral domain. We will show that $H=k[x], x\in P(H)$ if $\mathrm{K}\dim H=1$.

Let K be the k-sub-Hopf algebra of H generated by $P(H)$. Since H is a finitely generated k-algebra, by Theorem 4.2.7, K is a finitely generated k-algebra. Thus by Theorem 4.5.6, $K=k[x]$, $x\in P(K)$. Now we prove that $k[x]=H$. Suppose $\{R_n\}_{n\geqq 0}$ is the coradical filtration on H and let A_n be the k-subalgebra of H generated by R_n and set $E_n=k[x]A_n$. To show $E_1=k[x]$ and $H=E_1$, it is enough to verify that $E_n=E_1$ for each E_n. By induction on n, we will prove $E_j=E_1$ if $E_{j-1}=E_1$. Setting $E_{j-1}=E_1$, we want to show that $y\in E_1$ for $y\in R_j^+-R_1$. We get $E_1^+\cap P(E_1)\neq\{0\}$ since $E_1\neq k$. In fact, if $E_1^+\cap P(E_1)=\{0\}$, the restriction of the canonical projection $\pi:E_1\to E_1/E_1^+$ to $P(E_1)$ is injective. Now by Theorem 2.4.11, π is also injective and this contradicts the fact that $E_1\neq k$. Hence $\mathrm{K}\dim E_1=1$ and H is integral over E_1. Since $H/k[x]^+H$ is a finite dimensional k-linear space, there exist $a_i\in k\ (0\leqq i\leqq s)$ such that $z=\sum_{i=0}^{s} a_i y^{p^i}\in k[x]^+H$. Since $y\in R_j$, $R_{j-1}\subset k[x]$ and $q(y)\in k[x]^+\otimes k[x]^+$, if we set $K=k[x]+ky+ky^p+\cdots+ky^{p^s}$, then we have $q(K)\subset k[x]^+\otimes k[x]^+$. If we define $\mathrm{ht}(g)=\max\{i+j; c_{ij}\neq 0\}$ for $g=\sum c_{ij}x^i\otimes x^j\in k[x]\otimes k[x]$ where $c_{ij}\in k$, then, for any $z\in K$, we get $\mathrm{ht}(q(z))=0\Leftrightarrow z\in P(H)$. Moreover, Theorem 4.4.7 forces $k[x]=k[x,z]$ since $k[x]^+H=k[x,z]^+H$.

Now choose $w\in K$ satisfying $k[x,w]=k[x,y]$ and

$$z=\sum_{i=0}^{n} c_i w^{p^i}=\sum_{j=0}^{m} d_j x^j,\quad c_i, d_j\in k \qquad (**)$$

such that $n+m$ is the smallest possible. When $n=0$ or $w\in P(H)$, we have $w\in k[x]$. Now assume $n\neq 0$ and $w\notin P(H)$. Then $\mathrm{ht}(q(w))\neq 0$. By Lemma 4.5.5, we have $\mathrm{ht}(q(z))=\mathrm{ht}(q(w))p^n$. In turn, $\mathrm{ht}(q(z))\leqq m$ implies $\mathrm{ht}(q(w))p^n\leqq m$. When $\mathrm{ht}(q(z))<m$, Lemma 4.5.4 says that $m=p^j(j\geqq n)$, and, if $\mathrm{ht}(q(z))=m$, m is divisible by p^n. Namely, in either case, m is divisible by p^n. Adding

$$-\sum_{i=0}^{n} c_i((d_m/c_n)^{p^{-n}}x^{mp^{-n}})^{p^i}$$

to both sides of $(**)$, we get

$$\sum_{i=0}^{n} c_i(w-(d_m/c_n)^{p^{-n}}x^{mp^{-n}})^{p^i}=\sum_{i=0}^{m-1} d_i'x^i,\quad d_i'\in k\,(0\leqq i\leqq m-1).$$

Therefore if we take $w'=w-(d_m/c_n)^{p^{-n}}x^{mp^{-n}}$ instead of w, we get $k[x,w']=k[x,y]$ and $w'\in K$ satisfies an equality similar to $(**)$ consisting of $m+n-1$ terms. This contradicts the choice of w. Thus $w\in P(H)$ or $n=0$. This forces $w\in k[x]$, so that $k[x,y]=k[x]$.

***Proof of the latter part of Theorem* 4.5.1** We will show that if H is a unipotent k-Hopf algebra which is a finitely generated integral domain and if $\mathrm{Kdim}\, H\geqq 1$, then there exist a k-sub-Hopf algebra K of H and an element x of H satisfying the following properties.

(i) $\mathrm{Kdim}\, H=\mathrm{Kdim}\, K+1$.

(ii) x is transcendental over K and $q(x)\in K[x]^+\otimes K[x]^+$.

(iii) H is integral over $K[x]$.

(iv) H/K^+H is an integral domain of Krull dimension one.

Let $\{R_n\}_{n\geqq 0}$ be the coradical filtration on H and A_n the k-sub-Hopf algebra of H generated by R_n. Pick j so that $\mathrm{Kdim}\, A_{j-1}<\mathrm{Kdim}\, H$ and $\mathrm{Kdim}\, A_j=\mathrm{Kdim}\, H$. We have that H is integral over A_j. Now set $m_j=\mathrm{Kdim}\, A_j-\mathrm{Kdim}\, A_{j-1}$ and choose $y_1,\ldots,y_{m_j}\in R_j$ in such a way that A_j is integral over $A_{j-1}[y_1,\ldots,y_{m_j}]$ (cf. Theorem 1.5.3). Then $B_1=A_{j-1}[y_1,\ldots,y_{m_j-1}]$ is a k-sub-Hopf algebra of H, H is integral over $B_1[y_{m_j}]$, and we have $q(y_{m_j})\in B_1\otimes B_1$. We note that in general $\bar{H}=H/B_1{}^+H$ is not necessarily an integral domain. Now we consider an extension of B_1 which satisfies the conditions of the theorem. Here $\bar{\bar{H}}=\bar{H}/\pi_0(\bar{H})^+\bar{H}$ is an integral domain. Let $\rho:H\to\bar{\bar{H}}$ be the canonical projection and set $\mathrm{Ker}\,\rho=\mathfrak{a}$. Take a k-sub-Hopf algebra K of H satisfying $\mathfrak{a}=K^+H$ (cf. Theorem 4.4.7). Now we show that K and $x = y_{m_j}$ satisfy the conditions of the theorem. Clearly H/K^+H is an integral domain. Since H is integral over $K[x]$, H/K^+H is integral over $k[\rho(x)]$. Thus Kdim $H/K^+H \leq 1$. If Kdim $H/K^+H = 0$, then H/K^+H is a finite dimensional k-linear space. Hence $z = \sum_{i=0}^{n} a_i x^{p^i} \in K^+H$. We then obtain $z \in K^+$ in the same way as in the proof of Theorem 4.5.6. Thus there exists a positive integer s such that $z^{p^s} \in B_1$. This contradicts the fact that x is transcendental over B_1. Therefore we get Kdim $H/K^+H = 1$. Since H is integral over $K[x]$, if x is algebraic over K, we have Kdim $H/K^+H = 0$. Hence x is transcendental over K and Kdim $K <$ Kdim H. On the other hand, $B_1 \subset K$ implies Kdim $K =$ Kdim $H - 1$. This completes the proof of Theorem 4.5.1.

5.2 The decomposition theorem for solvable groups

When H is a commutative pointed k-Hopf algebra, every irreducible representation of the affine k-group $G = \mathbf{M}_k(H, k)$ is of dimension one. We call such an affine k-group a **linearly solvable group**. This subsection will be devoted to proving the following theorem.

THEOREM 4.5.7 Let H be a commutative pointed k-Hopf algebra. Then there exists a Hopf ideal $\mathfrak{a}$ such that $H = \mathfrak{a} \oplus \operatorname{corad} H$.

COROLLARY 4.5.8 Let H be a commutative pointed k-Hopf algebra and let $R = \operatorname{corad} H$. Then R is a diagonalizable k-sub-Hopf algebra of H and $L = H/R^+H$ is a unipotent k-Hopf algebra. If $i: R \to H$ and $p: H \to L$ are respectively the canonical embedding and the projection, then the extension (H, i, p) of R by L is split.

Corollary 4.5.8. shows that a connected solvable algebraic k-group admits a decomposition into a semi-direct product of a diagonalizable k-group and a unipotent k-group. First, we proceed to show that Corollary 4.5.8 is obtained from Theorem 4.5.7. We see that R is k-sub-Hopf algebra of H and is diagonalizable since $R = kG(H) = kG(R)$. It thus suffices to show that L is irreducible. If we set $S = \operatorname{corad} L$, we get a commutative diagram of k-Hopf algebras.

$$\begin{array}{ccc} H & \xrightarrow{p} & L \\ {\scriptstyle r}\downarrow & & \downarrow{\scriptstyle s} \\ R & \xrightarrow{q} & S \end{array}$$

By Theorem 4.5.7, there exist injective k-Hopf algebra morphism $i:R \to H$ and $j:S \to L$ satisfying $r \circ i = 1_R$ and $s \circ j = 1_S$. Since $R = k1 \oplus R^+$, Ker $p = R^+H$ and $s \circ p$ is surjective, we get $S = s \circ p(H) = k1$. Therefore L is irreducible. In this situation, we obtain $H \cong R \flat L$ (co-semi-direct product) as was observed in the preceding section.

***Proof of Theorem* 4.5.7** Let $\mathscr{I}$ be the family of all Hopf ideals $\mathfrak{a}$ of H satisfying: (1) $R \cap \mathfrak{a} = \{0\}$; (2) $\mathfrak{a} + R$ is a k-sub-Hopf algebra of H; (3) $\mathfrak{a}$ is a Hopf ideal of $\mathfrak{a} + R$. Then $\mathscr{I}$ is an ordered set with respect to inclusion, and $\mathscr{I} \neq \varnothing$ since $\{0\} \in \mathscr{I}$. Now Zorn's lemma can be applied to $\mathscr{I}$, so there exists a maximal element in $\mathscr{I}$, say $\mathfrak{a}$. Set $K = \mathfrak{a} + R$. We contend that $K = H$. Let $\{R_n\}_{n \geqq 0}$ be the coradical filtration on H. Assume $K \subsetneqq H$. Then there exists a positive integer j such that $R_{j-1} \subset K$ and $R_j \not\subset K$. Theorem 2.3.11 gives us a decomposition of H as a k-coalgebra, that is, there exists a coideal $\mathfrak{b} \supset \mathfrak{a}$ of H such that $H = \mathfrak{b} \oplus R$. Let $G = G(H)$, $\mathfrak{b}_i = \mathfrak{b} \cap R_i$ and $\mathfrak{a}_i = \mathfrak{a} \cap R_i$. Moreover, define elements $e_g \in H^*$ $(g \in G)$ by $e_g|_{\mathfrak{b}} = 0$ and $e_g(h) = \delta_{g,h}$ for $h \in G$. Then we have

$$R_i = \sum_{(g,h) \in G \times G} e_g \rightharpoonup R_i \leftharpoonup e_h$$

$$= \sum_{(g,h) \in G \times G} (e_g \rightharpoonup R \leftharpoonup e_h) \oplus (e_g \rightharpoonup \mathfrak{b}_i \leftharpoonup e_h).$$

Due to the choice of j, we may take an element x of H such that $x \in (e_g \rightharpoonup \mathfrak{b}_j \leftharpoonup e_h)$ for suitable $(g, h) \in G \times G$ and also so that $x \notin K$. Then we obtain

$$\Delta x = x \otimes g + h \otimes x + t, \quad t \in \mathfrak{a}_{j-1} \otimes \mathfrak{a}_{j-1}.$$

Indeed we have $\sum_{i=0}^{j} R_{j-i} \otimes R_i = R_j \otimes R + R \otimes \mathfrak{b}_j + \sum_{i=1}^{j-1} \mathfrak{b}_{j-i} \otimes \mathfrak{b}_i \subseteq R_j \otimes R + R \otimes \mathfrak{b}_j + \mathfrak{b}_{j-1} \otimes \mathfrak{b}_{j-1}$. Thus we can write $\Delta x = \sum x_i \otimes g_i + \sum h_i \otimes x_i' + t$, where $t \in \mathfrak{b}_{j-1} \otimes \mathfrak{b}_{j-1}$, $g_i, h_i \in G$, $x_i \in R_j$, $x_i' \in \mathfrak{b}_j$. On the other hand, $R_{j-1} \subset K$ implies $\mathfrak{b}_{j-1} \subset \mathfrak{b}_{j-1} \cap K$

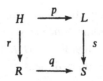

$=\mathfrak{a}_{j-1}$. Hence $t\in\mathfrak{a}_{j-1}\otimes\mathfrak{a}_{j-1}$. Moreover, $e_{g'}e_{h'}=\delta_{g',h'}e_{g'}$ for $g',h'\in G$ implies that if $g_i\neq g$, then $0=e_{g_i}e_g\rightharpoonup x=e_{g_i}\rightharpoonup(e_g\rightharpoonup x)=e_{g_i}\rightharpoonup x=x_i$. Similarly, if $x_i'\neq x$, then $x_i'=0$. Therefore $\Delta x=x\otimes g+h\otimes x+t$. Moreover, since $\mathfrak{a}$ is an ideal of K and $h^{-1}\in K$, we get $h^{-1}\mathfrak{a}_{j-1}\subset R_{j-1}\cap\mathfrak{a}=\mathfrak{a}_{j-1}$. Accordingly, $\Delta h^{-1}x=h^{-1}x\otimes h^{-1}g+1\otimes h^{-1}x+(h^{-1}\otimes h^{-1})t$. Therefore we may substitute $h^{-1}x$ for x and write

$$\Delta x=x\otimes g+1\otimes x+t,\quad t\in\mathfrak{a}_{j-1}\otimes\mathfrak{a}_{j-1}.$$

LEMMA 4.5.9 When $\Delta x=x\otimes g+1\otimes x+t$, $t\in\mathfrak{a}_{j-1}\otimes\mathfrak{a}_{j-1}$, we have

(i) $K[x]$ is a k-sub-Hopf algebra of H,
(ii) if x is algebraic over K, then x is integral over K.

Proof (i) We see that $K+kx$ is a k-subcoalgebra and $\Delta K[x]\subset K[x]\otimes K[x]$. Moreover, $0=\varepsilon(x)=\mu(S\otimes 1)\Delta x=S(x)g+x+\mu(S\otimes 1)t$. Hence $S(x)=-g^{-1}x-g^{-1}\mu(S\otimes 1)t\in K[x]$. Thus $K[x]$ is stable under Δ and S. In other words, $K[x]$ is a k-sub-Hopf algebra.

(ii) Let $f(X)=\sum_{i=0}^{n}a_iX^i\in K[X]$ be the minimal polynomial of x. If $\{R_n'\}_{n\geq 0}$ is the coradical filtration on K, there exists a positive integer s satisfying $a_n\in R_s'-R'_{s-1}$. Since $\Delta a_n\in\sum_{i=0}^{s}R'_{s-i}\otimes R'_i$, we may write $\Delta a_n=\sum_{i=1}^{m}b_i\otimes c_i+r$ where $r\in R_s'\otimes R'_{s-1}$, $\{c_i\}_{1\leq i\leq m}$ is linearly independent over R'_{s-1}, and for $b_i\in R$, $b_i\neq 0$. For any $h\in G$ and any i, we have $hR'_i=R'_i$, so that $\{g^nc_i\}_{1\leq i\leq m}$ is linearly independent over R'_{s-1}. Meanwhile,

$$\Delta(a_nx^n)=\Delta a_n(\Delta x)^n\in\left(\sum_{i=0}^{s}R'_{s-i}\otimes R'_i\right)(x^n\otimes g^n+1\otimes x^n+\cdots).$$

Hence, if we choose $f_1\in H^*$ satisfying $f_1(R'_{s-1})=0$ and $f_1(g^nc_i)=\delta_{1i}$, we get

$$0=f_1\rightharpoonup 0=f_1\rightharpoonup\left(\sum_{i=0}^{n}a_ix^i\right)=b_1x^n+t_1,\quad t_1\in\sum_{i=0}^{n-1}Kx^i.$$

Moreover, if we set $b_1 = \sum d_i g_i$ where $d_i \in k$, $g_i \in G$ and pick $f_2 \in H^*$ such that $f_2(d_i g_i g^n) = \delta_{1i}$, then we have

$$0 = f_2 \rightharpoonup 0 = f_2 \rightharpoonup (b_1 x^n + t_1) = x^n + t_2, \quad t_2 \in \sum_{i=0}^{n-1} Kx^i.$$

Therefore x is integral over K.

Now we will prove $H = K$ by considering the two possible cases separately as follows.

(1) When x is transcendental over K, we have $K[x] = R \oplus (\mathfrak{a} + \sum_{i=1}^{\infty} Kx^i)$. Set $\mathfrak{a} + \sum_{i=1}^{\infty} Kx^i = \mathfrak{c}$. Then $\mathfrak{c}$ is an ideal of $K[x]$ and is generated by the coideal $\mathfrak{a} + kx$, and hence $\mathfrak{c}$ is also a coideal. Since $S(x) = -g^{-1}x - g^{-1}\mu(S \otimes 1)t \in \mathfrak{a}$, we obtain $S(\mathfrak{c}) \subset \mathfrak{c}$. Moreover, we have $\varepsilon(\mathfrak{c}) = 0$. Hence $\mathfrak{c}$ is a Hopf ideal. Meanwhile, $\mathfrak{a} \subsetneqq \mathfrak{c}$. This contradicts the maximality of $\mathfrak{a}$. Therefore $K = H$.

(2) When x is algebraic over K, we set $K[x] = K \oplus Kx \oplus \cdots \oplus Kx^{n-1}$. Then we have $\mathfrak{a}K[x] = \mathfrak{a} \oplus \mathfrak{a}x \oplus \cdots \oplus \mathfrak{a}x^{n-1}$, and, since $\mathfrak{a}$ is a Hopf ideal of K, $\mathfrak{a}[x]$ is a Hopf ideal of $K[x]$. Since x is integral over K, we can write

$$x^n + \sum_{i=0}^{n-1} a_i x^i + \sum_{i=0}^{n-1} a_i' x^i = 0, \quad a_i \in R, \quad a'_i \in \mathfrak{a}.$$

Let $\bar{x}$ be the residue class of $K[x]/\mathfrak{a}[x]$ containing x. Then

$$K[x]/\mathfrak{a}[x] = K/\mathfrak{a} \oplus (K/\mathfrak{a})\bar{x} \oplus \cdots \oplus (K/\mathfrak{a})\bar{x}^{n-1} \cong R[\bar{x}].$$

This implies that

$$\Delta\bar{x} = \bar{x} \otimes g + 1 \otimes \bar{x}, \quad \bar{x}^n + \sum_{i=0}^{n-1} a_i \bar{x}^i = 0.$$

Choose $f \in (R[\bar{x}])^*$ satisfying $f\left(\sum_{i=1}^{n-1} R\bar{x}^i\right) = 0$ and $f(h) = \delta_{1h}$ for $h \in G$. Then

$$0 = \left(\bar{x}^n + \sum_{i=0}^{n-1} a_i \bar{x}^i\right) \leftharpoonup f = \bar{x}^n + f(\bar{x}^n)g^n + \sum_{i<n} f(a_i)\bar{x}^i + f(a_0).$$

Applying ε to the right side, we get $f(\bar{x}^n) = -f(a_0)$ since $\varepsilon(\bar{x}) = 0$ and

$\varepsilon(g^i) = 1$. Setting $a = f(a_0)$, we get

$$\bar{x}^n + \sum_{i=0}^{n-1} f(a_i)\bar{x}^i = a(1 - g^n). \qquad (***)$$

When the characteristic of k is 0, the equality obtained by applying Δ to both sides of (***) shows that $\{\bar{x}^i \otimes \bar{x}^j\}_{1 \leq i, j \leq n-1}$ is linearly dependent over R. This contradicts the linear independence of $\bar{x}, \ldots, \bar{x}^{n-1}$ over R. Therefore x must be transcendental over K, and this case reduces to case (1). When the characteristic of k is $p > 0$, by the same reason as above, i must be a power of p if $f(a_i) \neq 0$. Thus, if we set $n = p^m$, (***) becomes

$$\bar{x}^{p^m} + \sum_{i=0}^{m-1} f(a_i)\bar{x}^{p^i} = a(1 - g^{p^m}).$$

Take $f_1 \in (R[\bar{x}])^*$ satisfying $f_1\left(\sum_{i=1}^{n-1} R\bar{x}^i\right) = 0$ and $f_1(h) = \delta_{g^r, h}(r = p^m)$ for $h \in G$. Then we have

$$\begin{aligned} 0 = f_1 \rightharpoonup 0 &= f_1 \rightharpoonup \left(\bar{x}^{p^m} + \sum_{i=0}^{m-1} f(a_i)\bar{x}^{p^i} - a(1 - g^{p^m})\right) \\ &= \bar{x}^{p^m} + \sum_{i=0}^{m-1} f_1(g^{p^i}) f(a_i)\bar{x}^{p^i} - a + ag^{p^m}. \end{aligned}$$

The choice of f_1 allows us to assume that $g^{p^i} = g^{p^m}$ if $f(a_i) \neq 0$. Since k is algebraically closed, there is an element $b \in k$ such that $b^{p^m} + \sum_{i=0}^{m-1} f(a_i) b^{p^i} = a$. Here, setting $u = b - bg$, we have

$$(\bar{x} - u + u)^{p^m} + \sum_{i=0}^{m-1} f(a_i)(\bar{x} - u + u)^{p^i} = a(1 - g^{p^m}).$$

Hence

$$\begin{aligned} &(\bar{x} - u)^{p^m} + \sum_{i=0}^{m-1} f(a_i)(\bar{x} - u)^{p^i} = 0 \\ &\Leftrightarrow u^{p^m} + \sum_{i=0}^{m-1} f(a_i) u^{p^i} = a(1 - g^{p^m}) \\ &\Leftrightarrow (b - bg)^{p^m} + \sum_{i=0}^{m-1} f(a_i)(b^{p^i} - b^{p^i} g^{p^i}) = a(1 - g^{p^m}) \\ &\Leftrightarrow b^{p^m} + \sum_{i=0}^{m-1} f(a_i) b^{p^i} = a. \end{aligned}$$

Therefore we get

$$(x-u)^{p^m} + \sum_{i=0}^{m-1} f(a_i)(x-u)^{p^i} \in \mathfrak{a}[x-u].$$

In turn,

$$\begin{aligned}\Delta(x-u) &= x\otimes g + 1\otimes x + t - 1\otimes b + bg\otimes g \\ &= (x-b+bg)\otimes g + 1\otimes(x-b+bg)+t \\ &= (x-u)\otimes g + 1\otimes(x-u)+t.\end{aligned}$$

Thus, if we replace x by $x-u$, $\mathfrak{a}+kx$ is a coideal of K. Moreover $\mathfrak{b}=\mathfrak{a}\oplus Kx\oplus Kx^2\oplus\cdots\oplus Kx^{p^m-1}$ is a coideal of K and we have $K[x]=R\oplus Kx\oplus Kx^2\oplus\cdots\oplus Kx^{p^m-1}=R+\mathfrak{b}$. Since we also have $x^{p^m}+\sum_{i=0}^{m-1} f(a_i)x^{p^i}\in\mathfrak{b}$, it follows that $x^{p^m}\in\mathfrak{a}\mathfrak{b}\subset\mathfrak{b}$ and $\mathfrak{b}$ turns out to be a Hopf ideal. This contradicts the maximality of $\mathfrak{a}$. Therefore $K=H$.

6 Completely reducible groups

When H is a co-semi-simple affine k-Hopf algebra, the affine algebraic k-group $G=\mathbf{M}_k(H,k)$ is called a **completely reducible group**. A completely reducible group is an affine algebraic k-group such that all of its representations are completely reducible.

When the characteristic of the field k is 0, a completely reducible group admits a decomposition into a semi-direct product of a torus and a semi-simple group. In general, a connected affine algebraic k-group admits a decomposition into a semi-direct product of a unipotent group and a completely reducible group. When k has characteristic $p>0$, the situation changes drastically and a connected completely reducible group turns out to be a torus (Nagata's theorem). In this section, we will give Sweedler's version of the proof of this theorem.

THEOREM 4.6.1 (i) Assume the characteristic of k is 0. The coradical R of an affine k-Hopf algebra H is a k-sub-Hopf algebra of H.

(ii) With the same notation as in (i), $L=H/R^+H$ is an irreducible

k-Hopf algebra and the exact sequence of k-Hopf algebras $0 \to R \xrightarrow{i} H \xrightarrow{p} L \to 0$ is split.

We omit a detailed proof of this theorem. By Lemma 4.2.10, $G = \mathbf{M}_k(H, k)$ is dense in H^*, so we may regard a right H-comodule – namely, a rational left H^*-module – as a G-module. Since the characteristic of k is 0, if M_1 and M_2 are completely reducible G-modules, so is $M_1 \otimes M_2$ (cf. for example, Hochschild [4], Theorem 12.2). If D_1 and D_2 are simple k-subcoalgebras of H, Theorem 3.1.4 implies that there exist simple G-modules M_1 and M_2 such that $D_i = C(M_i)$ $(i = 1, 2)$. Therefore $D_1 D_2 = C(M_1)C(M_2) = C(M_1 \otimes M_2) \subseteq R$, that is, R is closed under multiplication. Clearly R is stable under S. Therefore R is a k-sub-Hopf algebra of H.

As for (ii), refer to Hochschild [4], Theorem 14.2, and Takeuchi [12].

In general, the maximal connected solvable normal subgroup of a connected affine algebraic k-group G is called the **radical** of G, denoted by rad G. The solvable group rad G admits a decomposition into a semi-direct product of a closed normal unipotent group U and a diagonalizable group T (cf. Theorem 4.5.1). When the characteristic of k is 0, we have $U = \mathbf{M}_k(H/R^+H, k)$. We will denote U by $\mathrm{rad}_u\, G$ and call it the **unipotent radical** of G. Then U is a closed normal subgroup of G. If $\mathrm{rad}_u\, G = \{e\}$, we call G a **reductive group** and if rad $G = \{e\}$, we call G a **semi-simple group**. We can completely classify connected semi-simple groups and the types correspond one-to-one to the types of complex semi-simple Lie groups (cf. Chevalley [2] and Borel [3]).

When the characteristic of k is 0, a reductive group is a completely reducible group. However, in case k has characteristic $p > 0$, the following theorem holds.

THEOREM 4.6.2 A co-semi-simple affine k-Hopf algebra H over a field k of characteristic $p > 0$ is cocommutative if $\pi_0(H) = k$, and, in such a case, $H = kG(H)$.

Hence, if a connected affine algebraic k-group $G = \mathbf{M}_k(H, k)$ is a completely reducible group, then G is a diagonalizable group (namely, a torus).

Before proving this theorem, we will provide some lemmas. For an ideal $\mathfrak{a}$ of a k-algebra A, set $\mathfrak{a}^0 = A$ and $\mathfrak{a}^\infty = \bigcap_{n=0}^{\infty} \mathfrak{a}^n$.

LEMMA 4.6.3 Let $\mathfrak{a}_1, \mathfrak{a}_2, \mathfrak{a}_3$ be two-sided ideals of a k-Hopf algebra H and let $\mathfrak{m} = \operatorname{Ker} \varepsilon$.

(i) If $\Delta\mathfrak{a}_1 \subset \mathfrak{a}_2 \otimes H + H \otimes \mathfrak{a}_3$ and $\mathfrak{a}_2{}^\infty \subset \mathfrak{m}$, then $\mathfrak{a}_1{}^\infty \subset \mathfrak{a}_3{}^\infty$.

(ii) $\mathfrak{a}_1{}^\infty \subset \mathfrak{m} \Leftrightarrow \mathfrak{a}_1{}^\infty \subset \mathfrak{m}^\infty$.

Proof (i) From the hypothesis,

$$\Delta\mathfrak{a}_1{}^{2n} \subset \sum_{i=0}^{2n} \mathfrak{a}_2{}^{2n-i} \otimes \mathfrak{a}_3{}^i \subset \mathfrak{a}_2{}^n \otimes H + H \otimes \mathfrak{a}_3{}^n.$$

Hence $\Delta\mathfrak{a}_1{}^\infty \subset \mathfrak{a}_2{}^\infty \otimes H + H \otimes \mathfrak{a}_3{}^\infty$. In turn, we have $\varepsilon(\mathfrak{a}_2{}^\infty) = 0$ by hypothesis, so that $\mathfrak{a}_1{}^\infty = (\varepsilon \otimes 1)\Delta\mathfrak{a}_1{}^\infty \subset \mathfrak{a}_3{}^\infty$.

(ii) $(1 \otimes \varepsilon)\Delta\mathfrak{a}_1 = \mathfrak{a}_1$ implies $\Delta\mathfrak{a}_1 \subset \mathfrak{a}_1 \otimes H + H \otimes \mathfrak{m}$. Thus from (i), we ahve $\mathfrak{a}_1{}^\infty \subset \mathfrak{m}^\infty$ if $\mathfrak{a}_1{}^\infty \subset \mathfrak{m}$. The converse is clear.

LEMMA 4.6.4 Let H be a finitely generated commutative k-Hopf algebra and assume $\pi_0(H) = k$. Then for any proper ideal $\mathfrak{a}$ of H, $\mathfrak{a}^\infty = \{0\}$.

Proof We may assume $\mathfrak{a}$ is a maximal ideal. Since $H/\text{nil}\, H$ is an integral domain and $\text{nil}\, H \subset \mathfrak{a}$, letting $p : H \to H/\text{nil}\, H$ be the canonical projection, we see that $p(\mathfrak{a})$ is a proper ideal of $H/\text{nil}\, H$. With the aid of Krull's intersection theorem (Theorem 1.5.9), we obtain $p(\mathfrak{a})^\infty = p(\mathfrak{a}^\infty) = \{0\}$. Hence $\mathfrak{a}^\infty \subset \text{nil}\, H \subset \mathfrak{m} = \operatorname{Ker} \varepsilon$. By Lemma 4.6.3 (ii), we have $\mathfrak{a}^\infty \subset \mathfrak{m}^\infty$. On the other hand, since H is a finitely generated k-algebra, Hilbert's *Nullstellensatz* (Theorem 1.5.5) guarantees the existence of an element $x \in \mathbf{M}_k(H, k) = G$ such that $\operatorname{Ker} x = \mathfrak{a}$. Now setting $\mathfrak{b} = \operatorname{Ker} x \circ S$, since $\varepsilon = (x \circ S \otimes x)\Delta$, we see that $\Delta\mathfrak{m} \subset \mathfrak{b} \otimes H + H \otimes \mathfrak{a}$. Moreover $\mathfrak{b}^\infty \subset \mathfrak{m}$. Thanks to Lemma 4.6.3 (i), we have $\mathfrak{m}^\infty \subset \mathfrak{a}^\infty$. Therefore $\mathfrak{a}^\infty = \mathfrak{m}^\infty$. Meanwhile, we see that

$$\bigcap_{\mathfrak{n} \in \text{Spm}\, H} \mathfrak{n}^\infty = \{0\},$$

where Spm H is the family of all maximal ideals of H. Indeed, for

$a \in H$, $a \neq 0$, let $\mathfrak{n}$ be a maximal ideal of H containing the ideal $\mathfrak{c} = \{b \in H; ba = 0\}$. If $a \in \mathfrak{n}^\infty$, then again by Krull's intersection theorem, there exists $x \in \mathfrak{n}$ such that $xa = a$. Thus $x - 1 \in \mathfrak{c} \subset \mathfrak{n}$. This contradiction forces $a \notin \mathfrak{n}^\infty$. Namely, $\bigcap_{\mathfrak{n} \in \mathrm{Spm}\, H} \mathfrak{n}^\infty = \{0\}$. Hence we get $\mathfrak{a}^\infty = \mathfrak{m}^\infty = \{0\}$.

COROLLARY 4.6.5 If a finitely generated commutative k-Hopf algebra H satisfies $\pi_0(H) = k$, then $\bigcap_n H(H^+)^{(p^n)} = \{0\}$, where $(H^+)^{(p^n)} = \{x^{p^n}; x \in H^+\}$.

Proof Now $(H^+)^{(p^n)} \subset (H^+)^{p^n} \subset \mathfrak{m}^{p^n}$. By Lemma 4.6.3, $\mathfrak{m}^\infty = \{0\}$, hence the lemma clearly holds.

***Proof of Theorem* 4.6.2** (1) If H is a finite dimensional commutative local co-semi-simple k-Hopf algebra, the dual k-Hopf algebra $A = H^*$ of H is a finite dimensional cocommutative irreducible k-coalgebra as well as a semi-simple k-algebra. Thus Ker $\varepsilon = \mathfrak{m}$ is the only maximal ideal of H and $k = \mathfrak{m}^\perp$ is the only simple k-subcoalgebra of A. Then it is enough to show that A is a commutative k-algebra. Now $\mathfrak{a}_m = H(H^{(p^m)})^+$ is a Hopf ideal of H and $A_{[m]} = \mathfrak{a}_m{}^\perp$ is a k-sub-Hopf algebra of A, so we obtain the following exact sequences of k-Hopf algebras,

$$0 \to H^{(p^m)} \to H \to H/\mathfrak{a}_m \to 0,$$
$$0 \leftarrow A/A(A_{[m]})^+ \overset{\pi}{\leftarrow} A \leftarrow A_{[m]} \leftarrow 0,$$

where $A_{[m]} \cong (H/\mathfrak{a}_m)^* = \{a \in A; (1 \otimes \pi)\Delta a = a \otimes 1\}$.

LEMMA 4.6.6 Let U be an irreducible k-Hopf algebra. If a commutative U-module k-algebra A satisfies $\pi_0(A) = A$, then for any $u \in U$ and $a \in A$, we have $ua = \varepsilon(u)a$.

Proof If $\pi_0(A) = A$, then $\mathrm{Der}_k(A) = \{0\}$. Indeed we may write $A = ke_1 \oplus \cdots \oplus ke_r$ $(e_ie_j = \delta_{ij}e_i, 1 \leqq i, j \leqq r, e_1 + \cdots + e_r = 1)$. For $D \in \mathrm{Der}_k(A)$, $D(e_i) = D(e_i{}^2) = 2e_iD(e_i)$. Hence $e_iD_i(e_i) = 0$. Meanwhile, if $i \neq j$, we have $0 = e_jD(e_ie_j) = e_jD(e_i)$. Therefore $D(e_i) = (e_1 + \cdots + e_r)D(e_i) = 0$, so that $D = 0$. Let $\{U_n\}_{n \in I}$ be the coradical filtration on

U. Then $U_1 = k \oplus P(U)$. Since $P(U)$ acts on A as k-derivations, the action of $P(U)$ on A is trivial. Thus, for $u \in U_1$, we have $ua = \varepsilon(u)a$ $(a \in A)$. By induction on n, we see that if U_{n-1} acts on A trivially, so do the U_n. Hence $U = \bigcup_n U_n$ acts trivially on A.

For $a, b \in A$, define an action of b on a by ${}^b a = \sum_{(b)} b_{(1)} a S(b_{(2)})$. Then A becomes a left A-module, and $A_{[m]}$ is an A-submodule of A. In fact, since A is cocommutative, we get

$$\begin{aligned}\Delta({}^b a) &= \sum (b_{(1)} \otimes b_{(2)})(a_{(1)} \otimes a_{(2)})(S(b_{(4)}) \otimes S(b_{(3)})) \\ &= \sum b_{(1)} a_{(1)} S(b_{(4)}) \otimes b_{(2)} a_{(2)} S(b_{(3)}) \\ &= \sum {}^{b_{(1)}} a_{(1)} \otimes {}^{b_{(2)}} a_{(2)}.\end{aligned}$$

Hence, for $a \in A_{[m]}$ and $b \in A$, we get

$$\begin{aligned}(1 \otimes \pi)\Delta({}^b a) &= \sum {}^{b_{(1)}} a_{(1)} \otimes \pi({}^{b_{(2)}} a_{(2)}) \\ &= \sum {}^{b_{(1)}} a_{(1)} \otimes {}^{\pi(b_{(2)})} \pi(a_{(2)}) \\ &= \sum b_{(1)} a_{(1)} S(b_{(2)}) \otimes \pi(b_{(3)}) \varepsilon(a_{(2)}) \pi(S(b_{(4)})) \\ &= \sum b_{(1)} a S(b_{(2)}) \otimes \varepsilon(b_{(3)}) 1 \\ &= \sum b_{(1)} a S(b_{(2)}) \otimes 1 = {}^b a \otimes 1.\end{aligned}$$

Thus ${}^b a \in A_{[m]}$. Moreover, since H^* is a semi-simple k-algebra, so is $A_{[m]}$. Now we will prove that A is commutative by induction on $\dim A = n$.

Case (i) $A = A_{[1]}$. Now A is the p-universal enveloping k-bialgebra of $P(A)$. By Theorem 1.3.5, $P(A)$ is commutative if A is semi-simple. Hence A is commutative. Henceforth we assume $A_{[1]} \subsetneqq A$.

Case (ii) $\dim P(A) > 1$. Since $A_{[1]} \subsetneqq A$, the inductive hypothesis implies that $A_{[1]}$ is commutative. Thus $\bar{H} = H/H(H^{(p)})^+$ is cocommutative and $\bar{H}$ is pointed (cf. Theorem 2.3.3). But $\bar{H}$ is co-semi-simple, so that $\bar{H} = kG(\bar{H})$ where $G(\bar{H})$ is an abelian p-group. Therefore we have $\bar{H} \cong \overset{r}{\otimes} k(\mathbb{Z}/p\mathbb{Z})$ (r-fold tensor product) and $A_{[1]} \cong \bigotimes_{i=1}^{r} V_i$, where the V_i are k-Hopf algebras generated by $P(V_i) = kv_i$ and $\{v_i\}_{1 \leqq i \leqq r}$ is a basis for $P(A)$. Now $\mathfrak{b}_i = AV_i^+$ is a Hopf ideal of A. Set $W_i = A/\mathfrak{b}_i$. Then W_i is a semi-simple irreducible cocommutative

k-Hopf algebra and by the inductive hypothesis, W_i is commutative. Let $p: A \to W_i$ be the canonical projection and define a map $\psi: A \otimes A \to A$ by

$$\psi: u \otimes v \mapsto \sum u_{(1)} v_{(1)} S(u_{(2)}) S(v_{(2)}).$$

Then ψ is a k-coalgebra morphism and $\operatorname{Im} \psi$ is a k-subcoalgebra of A since A is cocommutative. In turn, since W_i is commutative, for $x \in \operatorname{Im} \psi$, we have $p(x) = \varepsilon(x)$. Hence we get $(p \otimes 1)\Delta x = 1 \otimes x$. Thus $V_i \supset \operatorname{Im} \psi$. Since $\dim P(A) = \dim P(A_{[1]}) > 1$, we have $A_{[1]} = \bigotimes_{i=1}^{r} V_i (r > 1)$. Therefore $\operatorname{Im} \psi \subset V_1 \cap V_2 = k$ and $\varepsilon \circ \psi = \psi$. For $a, b \in A$, we see that

$$\begin{aligned} ab &= \sum \psi(a_{(1)} \otimes b_{(1)}) b_{(2)} a_{(2)} \\ &= \sum \varepsilon(a_{(1)}) \varepsilon(b_{(1)}) b_{(2)} a_{(2)} \\ &= ba. \end{aligned}$$

This shows that A is commutative.

Case (iii) $\dim P(A) \leqq 1$. We first show that, if $A_{[m]} \subsetneqq A$, $A_{[m]}$ is a central k-subalgebra of A. By the inductive hypothesis, $A_{[m]}$ is a commutative semi-simple k-algebra, and hence is a finite k-algebra direct product of copies of k, and we have $\pi_0(A_{[m]}) = A_{[m]}$. Now $A_{[m]}$ is an A-module k-algebra and by Lemma 4.6.6, A acts on $A_{[m]}$ trivially. Thus, for $a \in A_{[m]}$, $b \in A$, we obtain

$$ba = \sum {}^{b_{(1)}} a b_{(2)} = \sum \varepsilon(b_{(1)}) a b_{(2)} = ab.$$

In other words, $A_{[m]}$ is central in A.

When $\dim P(A) = 0$, $A = k$ is commutative. Let $\dim P(A) = 1$. Then A has a basis consisting of the elements of a sequence of divided powers $\{d_i\}_{0 \leqq i \leqq p^s - 1}$. Let $\{x^i\}_{0 \leqq i \leqq p^s - 1}$ be the dual basis for H with respect to the former basis. Then

$$\varepsilon(x^i) = \delta_{i0}, \quad x^i x^j = x^{i+j}, \quad 1 = x^0.$$

Thus A is generated by x^1 and $(x^1)^{p^s} = 0$. The ideal $H(H^+)^{p^{s-1}}$ of H has $\{x^{p^{s-1}}, \ldots, x^{p^s - 1}\}$ as a basis, and $A_{[p^{s-1}]}$ has $\{d_i\}_{0 \leqq i \leqq p^{s-1} - 1}$ as a basis. Thus $A_{[p^{s-1}]} \subsetneqq A$ and $\dim A_{[p^{s-1}]} = p^{s-1}$. Since $A_{[p^{s-1}]}$ is a central k-subalgebra of A, the k-subalgebra C of A generated by $B = A_{[p^{s-1}]} \oplus k d_{p^{s-1}}$ is commutative and, since B is a k-coalgebra, C is moreover a k-sub-bialgebra. In turn, the irreducibility of C implies that C is a k-sub-Hopf algebra of A (cf. Theorem 2.4.27) and is commutative. Furthermore we have $\dim C = p^s$. Since

$$p^{s-1} = \dim A_{[p^{s-1}]} < \dim C \leqq \dim A = p^s,$$

we see that $C = A$ and that A is commutative. Thus the proof of case (1) is complete.

(2) In the general case, letting C be the k-subcoalgebra of H generated by a given $h \in H$, we get $\dim C < \infty$. On the other hand, from Lemma 4.6.4, we have $\bigcap_m \mathfrak{a}^m = \{0\}$. Hence there exists a positive integer m satisfying $C \cap \mathfrak{a}^m = \{0\}$. Now $H/\mathfrak{a}^m$ is a co-semi-simple commutative finite dimensional local k-Hopf algebra and is cocommutative from (1). If we let $p : H \to H/\mathfrak{a}^m$ be the canonical projection, then the fact that the restriction of p to C is injective implies that C is cocommutative. Therefore H is cocommutative.

5

Applications to field theory

Let K be a finite Galois extension field of a field k and let G be its Galois group $\mathrm{Gal}(K/k)$ over k. Given a subfield M of K containing k (such a field is said to be an **intermediate field** of K/k),

$$F = \{\sigma \in G;\ \sigma(x) = x \quad \forall x \in M\}$$

becomes a subgroup of G. Conversely, given a subgroup F of G,

$$M = \{x \in K;\ \sigma(x) = x \quad \forall \sigma \in F\}$$

is an intermediate field of K/k. By such correspondences, the family of all intermediate fields of K/k is in one-to-one correspondence with the family of all subgroups of G (the fundamental theorem of Galois). However, we cannot construct such correspondences when K is a purely inseparable extension field of k. Nonetheless, when K is a finite purely inseparable extension field of k of exponent 1 (namely, if k is a field of characteristic $p > 0$ and if, for any $x \in K$, $x^p \in k$ holds), the family of all p-k-sub-Lie algebras of the p-k-Lie algebra $\mathrm{Der}_k(K)$ corresponds one-to-one with the family of all intermediate fields of K/k (Jacobson's theorem).

Recall that a group or a p-k-Lie algebra is associated with the group k-bialgebra or the p-universal enveloping k-bialgebra. Here we present how, given an arbitrary field extension K/k, one can construct a Galois correspondence as above by associating an extension field K of k with a suitable k-bialgebra. In this chapter, k is not assumed to be an algebraically closed field. We will denote the tensor product over k by the unadorned $\otimes$ and in all other cases the base field will be specified, as for instance $\otimes_K$.

1 K/k-bialgebras

Given an extension field K of a field k, we define an algebraic system which is not only a K-coalgebra but also a k-algebra and deduce its properties which are analogous to those of k-bialgebras.

1.1 K/k-bialgebras

A set A satisfying the following properties is called a K/k-**algebra**.

(1) A is a K-linear space and admits a k-algebra structure when regarded as a k-linear space.

(2) If we regard the k-linear space $A \otimes A$ as a K-linear space by defining an action of K on $A \otimes A$ by

$$c(a \otimes b) = ca \otimes b, \quad c \in K, \quad a \otimes b \in A \otimes A,$$

then the multiplication map $\mu : A \otimes A \to A$ of the k-algebra A turns out to be K-linear.

If A, B are K/k-algebras and $f : A \to B$ is a k-algebra morphism, then f is called a K/k-**algebra morphism** when f is a K-linear map. A K-linear subspace B of a K/k-algebra A is a K/k-algebra if B is a k-subalgebra of A, and, in that event, B is said to be a K/k-**subalgebra** of A. We note that a k/k-algebra is precisely a k-algebra.

EXAMPLE 5.1 Suppose that K is an extension field of a field k. Let V be a K-linear space and let $\mathrm{End}_k(V)$ be the set of all k-linear endomorphisms of V, where V is regarded as a k-linear space. Then $\mathrm{End}_k(V)$ is a K-linear space defined by

$$(f \pm g)(x) = f(x) \pm g(x), \quad (cf)(x) = cf(x), \quad x \in V,$$

for $f, g \in \mathrm{End}_k(V)$ and $c \in K$. Moreover, $\mathrm{End}_k(V)$ is a k-algebra with composition of maps as a multiplication and is also a K/k-algebra since $c(f \circ g) = (cf) \circ g$ for $c \in K$ and $f, g \in \mathrm{End}_k(V)$.

EXAMPLE 5.2 Regard an extension field K of a field k as a k-algebra and let $\mathrm{Aut}_k(K)$ be the group of all k-algebra automorphisms of K. Then $\mathrm{Aut}_k(K) \subset \mathrm{End}_k(K)$. If G is a subgroup of $\mathrm{Aut}_k(K)$, the K-linear subspace $A = K[G]$ of $\mathrm{End}_k(K)$ generated by G is a k-subalgebra of $\mathrm{End}_k(K)$ and A turns out to be a K/k-subalgebra of $\mathrm{End}_k(K)$. Similarly let $\mathrm{Der}_k(K)$ be the K-Lie algebra of all k-derivations from K to K. Then the k-subalgebra of $\mathrm{End}_k(K)$ generated by $\mathrm{Der}_k(K)$ is a K/k-subalgebra of $\mathrm{End}_k(K)$.

A K/k-algebra H is a K/k-**bialgebra** if H is also a K-coalgebra and satisfies the following conditions.

(1) $\Delta(1)=1\otimes_K 1,\quad \Delta(xy)=\sum_{(x)(y)} x_{(1)}y_{(1)}\otimes_K x_{(2)}y_{(2)},$

(2) $\varepsilon(1)=1,\quad \varepsilon(xy)=\varepsilon(x)\varepsilon(y)$ for $x, y\in H$ such that $\varepsilon(y)\in k$.

A K-linear subspace B of a K/k-bialgebra H is a K/k-bialgebra if B is a k-subalgebra and a K-subcoalgebra, in which case B is called a K/k-**sub-bialgebra** of H. A k/k-bialgebra is simply a k-bialgebra.

Let H, H' be K/k-bialgebras. If a map $f: H\to H'$ is a k-algebra morphism as well as a K-coalgebra morphism, then f is called a K/k-**bialgebra morphism**. If B is a K/k-sub-bialgebra of H, the canonical embedding $i: B\to H$ is a K/k-bialgebra morphism. A K-linear subspace I of a K/k-bialgebra H is said to be a K/k-**bi-ideal** if I is an ideal of the underlying k-algebra of H and is a coideal of the underlying K-coalgebra of H. In these circumstances, H/I is a K/k-bialgebra and the canonical projection $p: H\to H/I$ is a K/k-bialgebra morphism. If H, H' are K/k-bialgebras and $f: H\to H'$ is a K/k-bialgebra morphism, then

$$\operatorname{Ker} f=\{x\in H;\ f(x)=0\},\quad \operatorname{Im} f=\{f(x)\in H';\ x\in H\}$$

are a K/k-bi-ideal of H and a K/k-sub-bialgebra of H' respectively. Here we have $H/\operatorname{Ker} f\cong \operatorname{Im} f$ as K/k-bialgebras.

Let C be a K-coalgebra (or a K-algebra). A k-coalgebra (or a k-algebra) D satisfying $D\otimes K=C$ is called a k-**form** of C. A k-linear subspace B of a K/k-bialgebra H is said to be a k-**form** of the K/k-bialgebra H if B is a k-subalgebra of the underlying k-algebra of H and is a k-form of the underlying K-coalgebra of H. In this case, B is a k-bialgebra. Let H be a k-bialgebra and suppose k_0 is a subfield of k. If B is a k_0-form of the underlying k-algebra of H and also of the underlying k-coalgebra of H, then B is called a k_0-**form** of H. A k_0-form of a k-bialgebra is a k_0-bialgebra.

EXAMPLE 5.3 Let K be an extension field of k, G a group, and KG the group K-coalgebra of G. Given a group morphism $\rho: G\to \operatorname{Aut}_k(K)$, define a multiplication on KG by

$$(ax)(by)=a\rho(x)(b)xy,\quad a, b\in K,\quad x, y\in G.$$

Then KG is a k-algebra as well as a K/k-bialgebra. Here the group k-bialgebra kG of G is a k-form of the K/k-bialgebra KG.

EXAMPLE 5.4 Let K be an extension field of a field k of characteristic $p > 0$. Given a p-k-Lie algebra L, by extending the base field to K, construct a K-coalgebra $\bar{U}(L)_K = K \otimes \bar{U}(L)$ from the underlying k-coalgebra of the p-universal enveloping k-bialgebra $\bar{U}(L)$. Given a p-k-Lie algebra morphism $\rho : L \to \mathrm{Der}_k(K)$, define a multiplication on $\bar{U}(L)_K$ by

$$(ax)(by) = a\rho(x)(b)y + abxy, \quad a, b \in K, \quad x, y \in L.$$

Then $\bar{U}(L)_K$ is a k-algebra and also a K/k-bialgebra. In this situation, the k-bialgebra $\bar{U}(L)$ is a k-form of $\bar{U}(L)_K$.

EXAMPLE 5.5 Let k be a field of characteristic $p > 0$, and let K be a finite purely inseparable extension field of k such that $K^p \subset k$. If we choose a p-basis $S = \{x_1, \ldots, x_n\}$ for K/k, Lemma 4.3.7 shows that $|S| = \dim_K \mathrm{Der}_k(K)$ and thus a set $\{T_1, \ldots, T_n\}$ of elements of $\mathrm{Der}_k(K)$ satisfying $T_i(x_j) = \delta_{ij} x_j$ $(1 \leqq i, j \leqq n)$ is a basis for $\mathrm{Der}_k(K)$. In these circumstances, we have $[T_i, T_j] = 0$ and $T_i^p = T_i$ $(1 \leqq i, j \leqq n)$. If P is the prime field of k, the P-linear space M spanned by $\{T_1, \ldots, T_n\}$ over P is a P-form of $\mathrm{Der}_k(K)$. Now T_i satisfies a separable polynomial $X^p - X = \prod_{a \in P} (X - a)$ over P.

For $\alpha \in G = \mathbf{Mod}_P(M, P)$, set

$$K_\alpha = \{ s \in K;\ T(s) = \alpha(T)s \quad \forall\, T \in M \}.$$

Then $K_0 = \{s \in K;\ T(s) = 0 \quad \forall\, T \in M\}$ is a subfield of K and we have

$$K = \coprod_{\alpha \in G} K_\alpha, \quad K_\alpha K_\beta \subset K_{\alpha+\beta}, \alpha, \beta \in G.$$

Moreover, for a non-zero element $s_\beta \in K_\beta$, since we have $Ks_\beta = K = \sum_{\alpha \in G} K_\alpha s_\beta$ and $K_\alpha s_\beta \subset K_{\alpha+\beta}$, it follows that $K_\alpha s_\beta = K_{\alpha+\beta}$ for any $\alpha \in G$. A set $\{\alpha_1, \ldots, \alpha_n\}$ of elements of G defined by $\alpha_i(T_j) = \delta_{ij}$ is a basis for G over P and, if $\alpha = \sum_{i=1}^n n_i \alpha_i (0 \leqq n_i \leqq p-1)$, we see that $s = \prod_{i=1}^n s_i^{n_i} \in K_\alpha$. Thus, for any $\alpha \in G$, we have $K_\alpha \neq 0$, so that $[K : K_0] = |G|$ (the cardinality of G). Regarding G as an abelian group with respect to addition, let kG be the group k-bialgebra of G and let B be the dual k-bialgebra $(kG)^* = \mathrm{Map}(G, k)$ of kG. Then $K \otimes B$ has a K-coalgebra structure and moreover, if we define a multiplication by

$$(a \otimes x)(b \otimes y) = \sum_{(x)} a x_{(1)}(\alpha) b \otimes x_{(2)} y$$

for $a, b \in K$ such that $b \in K_\alpha$ and $x, y \in B$, $K \otimes B$ becomes a k-algebra and also a K/k-bialgebra. Here $1 \otimes B$ is a k-form of the K/k-bialgebra $K \otimes B$.

EXAMPLE 5.6 Let K be an extension field of a field k. For the underlying semi-group K_m of K defined via multiplication, as was defined in Chapter 2, § 2.2, we consider the set $R_K(K_m)$ of all representative functions defined on K_m with values in K as a K-coalgebra. As defined in Example 5.1, $\mathrm{End}_k(K)$ is a K/k-algebra when we view K as a k-linear space. Let $H(K/k) = R_K(K_m) \cap \mathrm{End}_k(K)$. Then $H(K/k)$ is a K-linear subspace of $\mathrm{End}_k(K)$ and, for $f, g \in H(K/k)$ and $x, y \in K$, we see that

$$\begin{aligned}(f \circ g)(xy) &= \sum_{(g)} f(g_{(1)}(x) g_{(2)}(y)) \\ &= \sum_{(f)(g)} f_{(1)}(g_{(1)}(x)) f_{(2)}(g_{(2)}(y)).\end{aligned}$$

Hence $f \circ g \in H(K/k)$. Thus $H(K/k)$ is a k-subalgebra of $\mathrm{End}_k(K)$. Furthermore, $H(K/k)$ is a K-subcoalgebra of $R_K(K_m)$. Indeed, for $f \in H(K/k)$, let $\Delta(f) = \sum_{i=1}^{n} f_i \otimes_K g_i$, where $\{g_1, \ldots, g_n\}$ is taken to be linearly independent over K. Then

$$\begin{aligned}\Delta(f)((ax + by) \otimes_K z) &= \sum_{i=1}^{n} f_i(ax + by) g_i(z) = f((ax + by)z) \\ &= f(axz + byz) = af(xz) + bf(yz) \\ &= \sum_{i=1}^{n} af_i(x) g_i(z) + \sum_{i=1}^{n} bf_i(y) g_i(z) \\ &= \sum_{i=1}^{n} (af_i(x) + bf_i(y)) g_i(z), \\ & \qquad x, y, z \in K, \quad a, b \in k.\end{aligned}$$

Thus we have

$$\sum_{i=1}^{n} f_i(ax+by)g_i = \sum_{i=1}^{n} (af_i(x)+bf_i(y))g_i.$$

Since $\{g_1, \ldots, g_n\}$ is linearly independent over K, for any $x, y \in K$, $a, b \in k$, it follows that

$$f_i(ax+by) = af_i(x) + bf_i(y).$$

In other words, $f_i \in \mathrm{End}_k(K)$ $(1 \leqq i \leqq n)$. Similarly we get $g_i \in \mathrm{End}_k(K)$ $(1 \leqq i \leqq n)$. Hence $H(K/k) = H$ is a k-algebra and also a K-coalgebra. In this situation, we see that

(1) $\Delta(1_H)(x \otimes_K y) = 1_H(xy) = xy = 1_H(x)1_H(y)$, hence $\Delta(1) = 1 \otimes_K 1$.

(2) $\Delta(f \circ g) = \sum_{(f)(g)} f_{(1)} \circ g_{(1)} \otimes_K f_{(2)} \circ g_{(2)}, \quad f, g \in H(K/k)$.

(3) $\varepsilon(1_H)(1) = 1_H(1) = 1$.

(4) For $f, g \in H(K/k)$ such that $\varepsilon(g) \in k$,

$$\varepsilon(f \circ g) = f \circ g(1) = f(g(1)) = f(\varepsilon(g)1) = \varepsilon(g)\varepsilon(f).$$

Therefore $H(K/k)$ is a K/k-bialgebra.

Remark In Chapter 2, § 2.2, we consider $R_K(K_m)$ as a K-bialgebra with the multiplication given by $(fg)(x) = f(x)g(x)$ for $f, g \in R_K(K_m)$. However, the multiplication in $H(K/k)$ is defined by the composition of maps in contrast to the former case.

1.2 Tensor products and semi-direct products

When H is a K/k-bialgebra and B is a k-bialgebra $H \otimes B$ is a k-algebra and a K-linear space. Moreover, the definition

$$\Delta(x \otimes y) = \sum (x_{(1)} \otimes y_{(1)}) \otimes_K (x_{(2)} \otimes y_{(2)}),$$
$$\varepsilon(x \otimes y) = \varepsilon(x)\varepsilon(y)$$

makes $H \otimes B$ a K-coalgebra as well as a K/k-bialgebra. We call it the **tensor product** of H and B. If H is a k-bialgebra, $H \otimes B$ is just the tensor product of k-bialgebras.

Given a K-linear space V, $\mathrm{End}_k(V)$ is a K/k-algebra as seen in Example 5.1. A K/k-algebra morphism $\rho: B \to \mathrm{End}_k(V)$ from a K/k-algebra B to $\mathrm{End}_k(V)$ is called a **representation** of a K/k-algebra B on V. In this situation, with respect to a K-linear space V and a K/k-bialgebra B, we can say the following.

(1) V is a B-module when we regard V and B respectively as a k-linear space and a k-algebra.

(2) Regard $B \otimes V$ as a K-linear space by defining $a(x \otimes b) = ax \otimes b$ for $a \in K$, $x \in B$, $b \in V$. Then the structure map $\varphi: B \otimes V \to V$ of a B-module V is a K-linear map.

We call such a K-linear space V a K/k-B-**module**. To give a K/k-B-module V is equivalent to giving a representation of the K/k-algebra B on V.

Let A be a K/k-algebra. If a K-linear map $\rho: C \to \mathrm{End}_k(A)$ from a K-coalgebra C to $\mathrm{End}_k(A)$ satisfies the conditions

$$\rho(x)(1_A) = \varepsilon(x)1_A, \quad \rho(x)(ab) = \sum_{(x)} \rho(x_{(1)})(a)\rho(x_{(2)})(b),$$
$$x \in C, \quad a, b \in A,$$

then ρ is said to be a **representation** of a K-coalgebra C on A. If we write $\rho(x)(a) = x \rightharpoonup a$, then

$$x \rightharpoonup 1_A = \varepsilon(x)1_A, \quad x \rightharpoonup (ab) = \sum_{(x)} (x_{(1)} \rightharpoonup a)(x_{(2)} \rightharpoonup b)$$

are satisfied. Now $\mathrm{Im}\,\rho = \rho(C)$ is a K-coalgebra with structure maps given by

$$\varepsilon(\rho(x)) = \varepsilon(x)1_A, \quad \Delta(\rho(x)) = \sum_{(x)} \rho(x_{(1)}) \otimes_K \rho(x_{(2)}), \quad x \in C.$$

Given a K/k-algebra A and a K/k-bialgebra H, if a K-linear map $\rho: H \to \mathrm{End}_k(A)$ is a representation of H on A simultaneously as a K/k-algebra and a K-coalgebra, ρ is called **representation** of a K/k-bialgebra H on A. In these circumstances, the K/k-algebra A has the following properties.

(1) A is a K/k-H-module. Here let $\varphi: H \otimes A \to A$ be the structure map.

(2) $A \otimes A$ becomes an H-module with the K-linear map $\psi = (\varphi \otimes \varphi)(1 \otimes \tau \otimes 1)(\Delta_H \otimes 1 \otimes 1): H \otimes A \otimes A \to A \otimes A$ as the structure

map. Then the multiplication map $\mu_A : A \otimes A \to A$ of A is an H-module morphism.

(3) $\varepsilon \otimes \eta_A = \varphi(1 \otimes \eta_A)$, namely $\varepsilon(x)1 = \varphi(x \otimes 1)$, $x \in H$.

We call such A an **H-module K/k-algebra**. In particular, if H is a k-bialgebra and A is a k-algebra, these conditions amount to stating that A is an H-moduel k-algebra. Giving a representation of H on A is equivalent to giving an H-module K/k-algebra A.

Let H be a k-bialgebra and A an H-module k-algebra. Then $H_A = A \otimes H$ becomes a k-algebra with structure maps $\mu_{H_A} : H_A \otimes H_A \to H_A$ and $\eta_{H_A} : k \to H_A$ defined by

$$\mu_{H_A}((a \otimes x) \otimes (b \otimes y)) = \sum_{(x)} a(x_{(1)} \rightharpoonup b) \otimes x_{(2)} y,$$

$$\eta_{H_A}(1) = \eta_A(1) \otimes \eta_H(1),$$

which we call the **semi-direct product k-algebra** of A and H.

Let H be a k-bialgebra and B be a K/k-bialgebra whose underlying k-algebra is an H-module k-algebra. Setting $H_B = B \otimes H$, H_B becomes the semi-direct product k-algebra of B and H, and moreover a K-coalgebra as well as a K/k-bialgebra with structure maps $\Delta_{H_B} : H_B \to H_B \otimes_K H_B$ and $\varepsilon_{H_B} : H_B \to K$ defined by

$$\Delta_{H_B}(b \otimes x) = \sum_{(b)(x)} (b_{(1)} \otimes x_{(1)}) \otimes_K (b_{(2)} \otimes x_{(2)}),$$

$$\varepsilon_{H_B}(b \otimes x) = \varepsilon_B(b)\varepsilon_H(x).$$

What we defined above is called the **semi-direct product K/k-bialgebra** of B and H.

Suppose H is a K/k-bialgebra. The irreducible component $H_1 = C$ of H containing the identity element 1 is an irreducible K-subcoalgebra of H, and, letting $C_i = \sqcap^{i+1} K1$, we obtain $C = \bigcup_{i=0}^{\infty} C_i$ and $C_i C_j = C_{i+j}$ in the same manner as in the proof of Theorem 2.4.24. Therefore H_1 is a k-subalgebra and a K/k-subbialgebra of H.

We call a K/k-bialgebra H a **normal K/k-bialgebra** if H is cocommutative pointed as a K-coalgebra and further if the semigroup $G(H)$ of all group-like elements of H forms a group. If H is a normal K/k-bialgebra, thanks to Corollary 2.4.28, H can be written as a K-coalgebra in the form $H = \coprod_{x \in G(H)} H_x$ as a direct sum of K-subcoalgebras. Now $kG(H)$ is a k-sub-bialgebra of H and the k-linear map

$$\rho : kG(H) \to \mathrm{End}_k(H_1),\ \rho(g)(x) = gxg^{-1},\quad g \in G(H),\quad x \in H_1$$

turns out to be a representation of the k-bialgebra $kG(H)$ on H_1. In fact, we see that

$$\begin{aligned}\rho(g)(xy) &= g(xy)g^{-1} = \rho(g)(x)\rho(g)(y)\\ &= \sum_{(g)} \rho(g_{(1)})(x)\rho(g_{(2)})(y),\\ \rho(g)(1) &= g1g^{-1} = 1,\quad g \in G(H),\quad x, y \in H_1.\end{aligned}$$

Hence H_1 is a $kG(H)$-module k-algebra. Let $H_1 \otimes kG(H)$ be the semi-direct product K/k-bialgebra of H_1 and $kG(H)$. Then the K-linear map

$$\sigma : H_1 \otimes kG(H) \to H,\quad a \otimes x \mapsto x \rightharpoonup a$$

is a K/k-bialgebra isomorphism (cf. Example 3.1).

If a normal K/k-bialgebra H is co-semi-simple as a K-coalgebra, we have $H = KG(H)$, and *if* H is irreducible as a K-coalgebra, we have $H = H_1$.

1.3 K-measuring K/k-bialgebras

A pair (H, ρ_H) of a K/k-bialgebra H and a representation $\rho_H : H \to \mathrm{End}_k(K)$ of H on K is called a **K-measuring K/k-bialgebra**. For $x \in H, a \in K$, we denote $\rho_H(x)(a)$ by $x \rightharpoonup a$ and call it the action of x on a. Given K-measuring K/k-bialgebras (H, ρ_H), $(H', \rho_{H'})$, if a K/k-bialgebra morphism $f : H \to H'$ satisfies $\rho_{H'} \circ f = \rho_H$, f is called a **K-measuring K/k-bialgebra morphism**. For KG in Example 5.3, $\bar{U}(L)_K$ in Example 5.4, and $H(K/k)$ in Example 5.6, we may define representations on K canonically which turn out to be K-measuring K/k-bialgebras. If (H, ρ_H) is a K-measuring K/k-bialgebra, Ker $\rho_H = \{x \in H;\ \rho_H(x) = 0\}$ is a K/k-bi-ideal of H and $H/\mathrm{Ker}\ \rho_H$ is isomorphic to a K/k-sub-bialgebra of $H(K/k)$. A K-measuring K/k-bialgebra with $\operatorname{Ker}\rho_H = \{0\}$ is said to be **faithful**. As we shall see in the sections to follow, $H(K/k)$ will play a role tantamount to the Galois group of an extension K/k.

For a subset C of a K-measuring K/k-bialgebra H,

$$K^C = \{a \in K;\ x \rightharpoonup (ab) = a(x \rightharpoonup b) \quad \forall x \in C \quad \forall b \in K\}$$

is a subfield of K. In particular, if C is a K-subcoalgebra of H, we see that

$$K^C = \{a \in K;\ x \rightharpoonup a = \varepsilon(x)a \quad \forall x \in C\}.$$

Indeed, for $a \in K$, if $x \rightharpoonup (ab) = a(x \rightharpoonup b)$ for all $x \in C$ and $b \in K$, setting $b = 1$, we get $x \rightharpoonup a = a(x \rightharpoonup 1) = \varepsilon(x)a$.

Conversely, if $x \rightharpoonup a = \varepsilon(x)a$ for all $x \in C$, then we obtain

$$\begin{aligned} x \rightharpoonup (ab) &= \sum_{(x)} (x_{(1)} \rightharpoonup a)(x_{(2)} \rightharpoonup b) \\ &= \sum_{(x)} a(\varepsilon(x_{(1)})x_{(2)} \rightharpoonup b) = a(x \rightharpoonup b). \end{aligned}$$

A K-measuring K/k-bialgebra H is said to be **semi-linear** if it satisfies

$$x(by) = \sum_{(x)} (x_{(1)} \rightharpoonup b)x_{(2)}y, \quad b \in K, \quad x, y \in H.$$

THEOREM 5.1.1 Let H be a k-bialgebra. Assume that an extension field K of k is an H-module k-algebra associated with a representation $\rho_H : H \to \operatorname{End}_k(K)$ of H on K. Let $H_K = K \otimes H$ be the semi-direct product K/k-algebra of K and H and define a K-linear map $\rho : H_K \to \operatorname{End}_k(K)$ by $\rho(a \otimes x) = a\rho_H(x)$ $(a \in K, x \in H)$. Then (H_K, ρ) is a semi-linear K-measuring K/k-bialgebra.

Proof First, we will show that H is a K/k-bialgebra. For $a, b \in K$, $x, y \in H$, identifying $H_K \otimes_K H_K$ with $K \otimes H \otimes H$, we have

$$\begin{aligned} \Delta((a \otimes x)(b \otimes y)) &= \Delta\left(\sum_{(x)} a(x_{(1)} \rightharpoonup b) \otimes x_{(2)}y\right) \\ &= \sum_{(x)(y)} a(x_{(1)} \rightharpoonup b) \otimes x_{(2)}y_{(1)} \otimes x_{(3)}y_{(2)} \end{aligned}$$

$$= \sum_{(x)} (a \otimes x_{(1)} \otimes x_{(2)}) \sum_{(y)} (b \otimes y_{(1)} \otimes y_{(2)})$$
$$= \Delta(a \otimes x)\, \Delta(b \otimes y),$$

and, for $x, y \in H_K$ such that $\varepsilon(y) \in k$, since $x \rightharpoonup 1 = \varepsilon(x)1$ (to be shown later), we get

$$\varepsilon(xy) = xy \rightharpoonup 1 = x \rightharpoonup (\varepsilon(y)1) = \varepsilon(y)\varepsilon(x) = \varepsilon(x)\varepsilon(y).$$

Hence H is a K/k-bialgebra. Moreover, for $b \in K$, we see that

$$(a \otimes x)(b(c \otimes y)) = (a \otimes x)(bc \otimes y)$$
$$= \sum_{(x)} a(x_{(1)} \rightharpoonup bc) \otimes x_{(2)} y$$
$$= \sum_{(x)} a(x_{(1)} \rightharpoonup b)(x_{(2)} \rightharpoonup c) \otimes x_{(3)} y$$
$$= \sum_{(x)} (a(x_{(1)} \rightharpoonup b) \otimes x_{(2)})(c \otimes y).$$

Here, since $\Delta(a \otimes x) = \sum_{(x)} a \otimes x_{(1)} \otimes x_{(2)}$, H is semi-linear.

Next, we proceed to show that ρ is a representation of the K/k-bialgebra H_K. Since we have

$$\rho((a \otimes x)(b \otimes y))(c) = \rho\left(\sum_{(x)} a(x_{(1)} \rightharpoonup b) \otimes x_{(2)} y \right)(c)$$
$$= \sum_{(x)} a(x_{(1)} \rightharpoonup b)((x_{(2)} y) \rightharpoonup c)$$
$$= \rho(a \otimes x)\rho(b \otimes y)(c),$$

ρ is a representation as a k-algebra. We also see that

$$(\Delta(a \otimes x))(b \otimes c) = \left(\sum_{(x)} a \otimes x_{(1)} \otimes x_{(2)} \right)(b \otimes c)$$
$$= a \sum_{(x)} (x_{(1)} \rightharpoonup b)(x_{(2)} \rightharpoonup c)$$
$$= a(x \rightharpoonup bc) = \rho(a \otimes x)(bc),$$
$$\varepsilon(a \otimes x) = a\varepsilon(x) = a(x \rightharpoonup 1) = (a \otimes x) \rightharpoonup 1.$$

Therefore ρ is a representation as a K-coalgebra. Thus (H_K, ρ) is a semi-linear K-measuring K/k-bialgebra.

Remark Now $1 \otimes H$ is a k-form of the K/k-bialgebra H_K and conversely, if B is a semi-linear K-measuring K/k-bialgebra of which H is a k-form, then K is an H-module k-algebra and $K \otimes H$ is a K-measuring K/k-bialgebra. In this situation, $K \otimes H$ is isomorphic to B via the map $a \otimes x \mapsto x \rightharpoonup a$, where $a \in K$, $x \in H$.

1.4 Toral k-bialgebras

Let k be a field of characteristic $p > 0$ and V a k-linear space. A subset T of $\mathrm{End}_k(V)$ is said to be **diagonalizable** if there is a basis for V such that all elements of T can be represented by diagonal matrices with respect to the basis. A k-linear subspace T of $\mathrm{End}_k(V)$ consisting of **semi-simple elements** (elements whose minimal polynomials are separable) is **toral** if T satisfies the condition that

$$s, t \in T \Rightarrow st = ts \quad \text{and} \quad t^p \in T.$$

Given a subset S of $\mathrm{End}_k(V)$ consisting of semi-simple elements, the k-linear subspace T spanned by 1 and $\{s^{p^e}; e \geqq 0, s \in S\}$ over k is toral. Now, T is diagonalizable $\Leftrightarrow S$ is diagonalizable.

Let T be a toral k-linear subspace of $\mathrm{End}_k(V)$. A finite Galois extension field L of k with the following property is called a **splitting field** for T. Namely, when we let $V_L = L \otimes V$, the L-linear subspace T_L of $\mathrm{End}_L(V_L)$ spanned by T is a diagonalizable toral L-linear subspace.

Let T be a subset of $\mathrm{End}_k(V)$ and, for $\alpha \in \mathrm{Map}\,(T, k)$, set

$$V_\alpha(T) = \{v \in V;\, t(v) = \alpha(t)v \quad \forall t \in T\}.$$

If T is a toral k-linear space, letting $T^* = \mathbf{Mod}_k(T, k)$, it follows that T is diagonalizable $\Leftrightarrow V = \prod_{\alpha \in T^*} V_\alpha(T)$. Let P be the prime field of k and let $T_p = \{t \in T;\ t^p = t\}$. Then T_p is a P-linear subspace of T. The minimal polynomial of $t \in T_p$ is $X^p - X = \prod_{a \in P} (X - a)$ and all eigenvalues of an element of T lie in P and T_p is diagonalizable. Thus we may write

$$V = \prod_{\alpha \in (T_p)^*} V_\alpha(T).$$

LEMMA 5.1.2 For a toral k-linear subspace T of $\mathrm{End}_k(V)$, T is diagonalizable $\Leftrightarrow T = kT_p$. Here kT_p is the k-linear space spanned by T_p over k.

Proof Suppose T is diagonalizable. Then we see that $V = \coprod_{\alpha \in R} V_\alpha(T)$, where $R = \{\alpha \in T^*;\ V_\alpha(T) \neq 0\}$. Since R is dense in T^*, R contains a basis $\{\alpha_1, \ldots, \alpha_n\}$ for T^*. Let $\{t_1, \ldots, t_n\}$ be the dual basis for T. Then we have $\alpha_i(t_j) = \delta_{ij}$ and $\alpha_i(t_j{}^p) = \alpha_i(t_j) = \delta_{ij}$. The fact that R is dense in T^* implies $t_i{}^p = t_i$ $(1 \leq i \leq n)$, so that $T = kT_p$. Conversely, if $T = kT_p$, it is clear that a toral k-linear subspace T is diagonalizable.

LEMMA 5.1.3 Let T be a toral k-linear subspace of $\mathrm{End}_k(K)$. Then a basis for T over k is also a basis for KT over K. Here KT is the K-linear subspace of $\mathrm{End}_k(K)$ spanned by T.

Proof Let $T^* = \mathbf{Mod}_k(T, k)$ and $R = \{\alpha \in T^*;\ K_\alpha(T) \neq 0\}$. Then we have $K = \coprod_{\alpha \in (T_p)^*} K_\alpha(T)$. Since R is dense in T^*, R contains a basis $\{\alpha_1, \ldots, \alpha_n\}$ for T^*. Let $\{t_1, \ldots, t_n\}$ be the dual basis for T. Pick $x_i \in K_{\alpha_i}(T)$ such that $x_i \neq 0$. Then we see that $t_i(x_j) = \delta_{ij} x_j$, and $\{t_1, \ldots, t_n\}$ is linearly independent over K and hence becomes a basis for KT over K.

A k-sub-bialgebra (or k-subalgebra, k-subcoalgebra) of $H(K/k)$ which is a toral k-linear subspace of $\mathrm{End}_k(K)$ is said to be **toral**. Given a diagonalizable toral k-coalgebra T, if we set

$$G = G(T) = \{\alpha \in T^*;\ K_\alpha(T) \neq 0\},$$

then, for $x \in K_\alpha(T)$ and $y \in K_\beta(T)$ such that $x \neq 0$, $y \neq 0$, we have

$$t(xy) = \sum_{(t)} t_{(1)}(x) t_{(2)}(y) = \sum_{(t)} \alpha(t_{(1)}) \beta(t_{(2)}) xy, \quad t \in T.$$

Thus if we set $\alpha * \beta = \mu \circ (\alpha \otimes \beta) \circ \Delta$, then $xy \in K_{\alpha * \beta}(T)$. Hence G is closed under convolution. In other words, G is an abelian semigroup with respect to convolution $*$. For $x_\beta \in K_\beta(T)$ such that $x_\beta \neq 0$, we have $K = Kx_\beta = \sum_{\alpha \in G} K_\alpha(T) x_\beta$ and $K_\alpha(T) x_\beta \subset K_{\alpha * \beta}(T)$, so that $K_\alpha(T) x_\beta = K_{\alpha * \beta}(T)$. Since G is a finite semigroup and right translation is a surjection, it follows that $\alpha=\alpha'$ if $\alpha*\beta=\alpha'*\beta$. Moreover, among the $\alpha^n=\alpha*\ldots*\alpha$ (n-fold product of α, where $n=0, 1, \ldots, n$), there are some that agree, and hence there exists a positive integer d such that $\alpha=\alpha^{d+1}=\alpha*\alpha^d$. Therefore $\alpha*\beta=\alpha*(\alpha^d*\beta)$ and we get $\alpha^d*\beta=\beta$ for $\beta\in G$. Hence $\alpha^d=e$ is the identity element of G. Moreover, α^{d-1} is the inverse element of α because $\alpha^{d-1}*\alpha=e$. Thus G is a finite abelian group with respect to convolution. Now $K_eK_e\subset K_{e*e}=K_e$, and $1\in K_e$ since for $u\in K_e$, $u\neq 0$, we have $K_e=K_eu$ and $u\in K_eu$. Therefore K_e is a subfield of K. In addition, for $\alpha\in G$, taking $x_\alpha\in K_\alpha(T)$, $x_\alpha\neq 0$, it follows that $K_\alpha=K_ex_\alpha$, so that we get $[K:K_e]=|G|$ (the order of G).

In general, if K is a finite extension field of k, a finite abelian group G is called a **splitting group** for K/k when, to each element α of G, there corresponds a k-linear subspace K_α of K satisfying $K_\alpha K_\beta\subset K_{\alpha*\beta}$ (where $\alpha*\beta$ is the product of α and β) and such that K admits a direct sum decomposition $K=\coprod_{\alpha\in G}K_\alpha$ as a k-linear space. The above observation gives us the next theorem.

THEOREM 5.1.4 Assume K is a finite extension field of k and $T\subset H(K/k)$ is a diagonalizable toral k-coalgebra. Then

$$G=G(T)=\{\alpha\in T^*;\ K_\alpha(T)\neq 0\}$$

is a finite abelian group with regard to convolution and G is a splitting group for K/k. Moreover, if e is the identity element of G, K_e is a subfield of K and $[K:K_e]=|G|$.

Conversely, if a finite abelian group G is a splitting group for a finite field extension K/k, letting T be the dual k-bialgebra $(kG)^*\cong\mathrm{Map}(G, k)$ of the group k-bialgebra kG of G, we may construct a diagonalizable toral k-bialgebra $T(G)\subset H(K/k)$ in the following manner. For $t\in T$, if we define $\bar{t}(x)=t(\alpha)x\ (x\in K_\alpha)$, then $\bar{t}\in H(K/k)$. Also define a k-linear map $\rho:T\to H(K/k)$ by $\rho(t)=\bar{t}$. Then ρ is a representation of T on K. In fact, for $x\in K_\alpha$ and $y\in K_\beta$,

$$\Delta(\bar{t})(x\otimes y)=\bar{t}(xy)=t(\alpha*\beta)(xy)$$

$$=\sum_{(t)}t_{(1)}(\alpha)t_{(2)}(\beta)(xy)$$

$$= \sum_{(t)} \bar{t}_{(1)}(x)\bar{t}_{(2)}(y),$$

$$\overline{st}(x) = s(\alpha)t(\alpha)x = \bar{s}(\bar{t}(x)).$$

Thus $\rho(T)$ is a k-sub-bialgebra of $H(K/k)$ which we denote by $T(G)$. If we define $\hat{\alpha} \in T(G)^*$ by $\hat{\alpha}(\bar{t}) = t(\alpha)$ $(\bar{t} \in T(G))$, then we have

$$\begin{aligned} K_{\hat{\alpha}}(T(G)) &= \{x \in K;\ \bar{t}(x) = \hat{\alpha}(\bar{t})(x) = t(\alpha)x \quad \forall\, t \in T\} \\ &= K_{\hat{\alpha}}(T) \end{aligned}$$

and $K = \coprod_{\alpha \in G} K_{\hat{\alpha}}(T(G))$. Hence $T(G) \subset H(K/k)$ is a diagonalizable toral k-bialgebra and $G(T(G)) = G$ holds. Thus we have proved

THEOREM 5.1.5 If a finite abelian group G is a splitting group for a finite extension field K of k, then the map

$$\rho : (kG)^* \to H(K/k)$$

given by $\rho(t)(x) = \bar{t}(x) = t(\alpha)x$, where $t \in T$, $x \in K_\alpha$, $\alpha \in G$, is a representation of the k-bialgebra $(kG)^* = T$ on K and $\rho(T) = T(G)$ is a diagonalizable toral k-bialgebra. Moreover, $G(T(G)) = G$ holds.

Remark In the next section, we will verify that $T(G(T)) = T$ if $T \subset H(K/k)$ is a diagonalizable toral k-bialgebra and that the correspondences $G \mapsto T(G)$ and $T \mapsto G(T)$ give a one-to-one correspondence between the family of all splitting groups for K/k and the family of all diagonalizable toral k-sub-bialgebras of $H(K/k)$.

2 Jacobson's theorem

In this section, we will prove Jacobson's theorem which gives a Galois correspondence between purely inseparable extension fields of exponent 1 and p-Lie algebras.

2.1 Jacobson–Bourbaki's theorem

Assume K is an extension field of a field k. Given a K/k-subalgebra A of the K/k-algebra $\mathrm{End}_k(K)$,

$$K^A = \{y \in K;\ f(xy) = f(x)y \quad \forall\, f \in A \quad \forall\, x \in K\}$$

is a subfield of K and we get $A \subset \mathrm{End}_{K^A}(K)$.

EXAMPLE 5.7 Let G be a subgroup of $\mathrm{Aut}_k(K)$. If $A = K[G]$ is the K-linear subspace generated by G in $\mathrm{End}_k(K)$, then we see that

$$K^A = K^G = \{y \in K;\ g(y) = y \quad \forall g \in G\}.$$

In this situation, we may identify A with the group K-algebra KG of G. In other words, elements of G are linearly independent over K. Suppose not and assume that $\sum_{i=1}^{n} c_i g_i = 0$ for $c_i \in K$, $c_i \neq 0$ such that $g_i \in G$ $(1 \leqq i \leqq n)$ are distinct elements and n $(n > 1)$ is the smallest possible number for this to hold. For $a, b \in K$, we have $0 = \sum_{i=1}^{n} c_i g_i(ab)$ $= \sum_{i=1}^{n} c_i g_i(a) g_i(b)$. In turn, $0 = g_n(a) \sum_{i=1}^{n} c_i g_i(b) = \sum_{i=1}^{n} c_i g_n(a) g_i(b)$. Hence we have $\sum_{i=1}^{n-1} (g_i(a) - g_n(a)) g_i(b) c_i = 0 \ \forall\, b \in K$, so that the choice of n implies that $(g_i(a) - g_n(a))c_i = 0$ $(1 \leqq i \leqq n-1)$, $c_i \neq 0$. Hence $g_i(a)$ $= g_n(a)$ $\forall\, a \in K$. Thus we obtain $g_i = g_n$ $(1 \leqq i \leqq n-1)$, which contradicts the assumption that the elements g_i $(1 \leqq i \leqq n)$ are distinct.

Given a K/k-subalgebra A of $\mathrm{End}_k(K)$, for $x \in K$, we define a K-linear map $\hat{x} : A \to K$ from A to K by $\hat{x}(f) = f(x)$, $f \in A$, and let $\hat{S} = \{\hat{x};\, x \in S\}$ for a subset S of K. It is easily seen that $\hat{K} \subset$ $\mathbf{Mod}_K(A, K) = A^*$ is dense in A^*. If we set

$$A^A = \{g \in A;\ f(xg) = f(x)g \quad \forall f \in A \quad \forall x \in K\},$$

then A^A is a K^A-linear subspace of A and, if $\hat{S}$ is dense in A^*, we have

$$A^A = \{g \in A;\, g(S) \subset K^A\}.$$

Indeed, $\hat{y}(f(xg)) = f(xg)(y) = f(xg(y))$. On the other hand, $\hat{y}(f(x)g) = f(x)g(y)$, so we see that

$$\begin{aligned} & f(x)g = f(xg) \quad \forall f \in A \ \forall x \in K \\ \Leftrightarrow\ & f(xg(y)) = f(x)g(y) \ \forall f \in A \ \forall x \in K. \end{aligned}$$

In particular, we have $A^A = \{g \in A;\, g(K) \subset K^A\}$.

THEOREM 5.2.1 If a K/k-subalgebra A of $\mathrm{End}_k(K)$ is finite dimensional over K, A^A is a K^A-form of A and $\dim_K A = [K : K^A]$.

Proof The density of $\hat{K}$ in A^* allows us to take $\hat{S} = \{\hat{x}_1, \ldots, \hat{x}_n\}$ as a basis for A^* over K. Let $\{g_1, \ldots, g_n\}$ be the dual basis for A over K. Since $g_j(x_i) = \hat{x}_i(g_j) = \delta_{ij}$ $(1 \leqq i, j \leqq n)$, we have $g_j(S) \subset K^A$ and thus $g_j \in A^A$. If $g = \sum_{j=1}^{n} y_j g_j \in A^A$ for $y_j \in K$ $(1 \leqq j \leqq n)$, then $y_i = \sum_{j=1}^{n} y_j g_j(x_i) = g(x_i) \in K^A$. This shows that $\{g_1, \ldots, g_n\}$ is a basis for A^A over K^A. Therefore A^A is a K^A-form of A. To prove $\dim_K A = [K : K^A]$, it suffices to show that $\{x_1, \ldots, x_n\}$ is a basis for K over K^A. If $x = \sum_{i=1}^{n} x_i y_i$ $(y_i \in K^A)$, then $g_j(x) = \sum_{i=1}^{n} g_j(x_i) y_i = y_j$. In turn, $g_j\left(\sum_{i=1}^{n} x_i g_i(x)\right) = \sum_{i=1}^{n} g_j(x_i) g_i(x) = g_j(x)$. So we have $x = \sum_{i=1}^{n} x_i g_i(x)$ with $g_i(x) \in K^A$, and any element of K can be uniquely represented as a linear combination of $x_1, \ldots, x_n$ over K^A. In other words, $\{x_1, \ldots, x_n\}$ is a basis for K over K^A.

THEOREM 5.2.2 (Jacobson–Bourbaki.) Given a K/k-subalgebra A of $\mathrm{End}_k(K)$, if A is of finite dimension over K, then $A = \mathrm{End}_{K^A}(K)$.

Proof We clearly have $A \subset \mathrm{End}_{K^A}(K)$. It follows from Theorem 5.2.1 that

$$\dim_K \mathrm{End}_{K^A}(K) = [K : K^A] = \dim_K A.$$

This forces $A = \mathrm{End}_{K^A}(K)$.

COROLLARY 5.2.3 Let P be the prime field of a field K. Let $\mathscr{K}$ be the family of all subfields k of K such that K is finite dimensional over k, and let $\mathscr{A}$ be the family of all K/P-subalgebras A of $\mathrm{End}_P(K)$ of finite dimension over K. Then

$$k \mapsto \mathrm{End}_k(K), \quad A \mapsto K^A$$

give a one-to-one correspondence between $\mathscr{K}$ and $\mathscr{A}$ and they are each other's inverses.

Proof We see that $\mathrm{End}_k(K)$ is a K/P-subalgebra of $\mathrm{End}_P(K)$. Trivially, $k \subset K^{\mathrm{End}_k(K)} = k'$. Thanks to Theorem 5.2.1, we have $[K : k] = \dim_K \mathrm{End}_k(K) = [K : k']$. Hence $k = k'$. Moreover, by

Theorem 5.2.2, $A = \mathrm{End}_{K^A}(K)$. Thus, indeed, the two correspondences above are inverses.

We will apply the above theorem in order to verify that the correspondences $G \mapsto T(G)$, $T \mapsto G(T)$ given in the last section between splitting groups for K/k and diagonalizable toral k-bialgebras are one-to-one. Denote by $\langle T \rangle$ the k-subalgebra of $\mathrm{End}_k(K)$ generated by a subset T of $\mathrm{End}_k(K)$.

THEOREM 5.2.4 For a diagonalizable toral k-subcoalgebra T of $H(K/k)$, we have $T(G(T)) = \langle T \rangle$. If we let $\mathscr{T}$ be the family of all diagonalizable toral k-sub-bialgebras of $H(K/k)$ and $\mathscr{G}$ the family of all splitting groups for K/k, then

$$G \mapsto T(G), \quad T \mapsto G(T)$$

give a one-to-one correspondence between $\mathscr{T}$ and $\mathscr{G}$, and they are inverses.

Proof Suppose T is a diagonalizable toral k-subcoalgebra of $H(K/k)$ and let $t \in T$. Define $\hat{t} \in (kG(T))^* = \mathrm{Map}(G(T), k)$ by $\hat{t}(\alpha) = \alpha(t)\ (\alpha \in G(T))$. Then $\rho(\hat{t})(x) = \hat{t}(\alpha)x = \alpha(t)x = t(x)\ (x \in K_\alpha, \alpha \in G(T))$. Hence $\rho(\hat{t}) = t \in T(G(T))$, namely, $T \subset T(G(T))$. Since $T(G(T))$ is a k-bialgebra, we have $\langle T \rangle \subset T(G(T))$. Moreover, since $G(\langle T \rangle) = G(T) = G(T(G(T)))$, we get $K^T = K^{\langle T \rangle} = K^{T(G(T))}$. Thus, by Corollary 5.2.3, we obtain $K\langle T \rangle = KT(G(T))$. Lemma 5.13 show that

$$\dim_k \langle T \rangle = \dim_K K\langle T \rangle = \dim_K KT(G(T)) = \dim_k T(G(T)).$$

Hence $\langle T \rangle = T(G(T))$. In particular, if T is a k-bialgebra, we get $T = T(G(T))$. Since $G = G(T(G))$ by Theorem 5.1.5, the assertion of the theorem is verified.

THEOREM 5.2.5 Assume that K is a finite extension field of a field k of characteristic $p > 0$ and P is the prime field of k. For a diagonalizable toral k-subcoalgebra (resp. k-sub-bialgebra) T of $H(K/k)$, let $G = G(T)$. Then $T_P = \{t \in (kG)^*;\ t(PG) \subseteq P\} \cong (PG)^*$ is a P-subcoalgebra (resp. P-sub-bialgebra) of $H(K/P)$ and we have

$$H(K/K^T) = K\langle T \rangle = K\langle T_P \rangle \text{ (resp. } H(K/K^T) = KT = KT_P).$$

Proof Given a diagonalizable toral k-subcoalgebra (resp. k-sub-bialgebra) T of $H(K/k)$, Lemma 5.1.2 shows that a basis for T_P over P is also a basis for T over K and that $\varepsilon(T_P) = P$. Let $\{t_1, \ldots, t_n\}$ be a basis for T_P over P and $t \in T_P$. Then from the fact that $t^p = t$, we obtain

$$\Delta(t) = \sum_{i=1}^{n} u_i \otimes t_i = \Delta(t^p) = \sum_{i=1}^{n} u_i^p \otimes t_i^p = \sum_{i=1}^{n} u_i^p \otimes t_i,$$

so that $u_i^p = u_i (1 \leqq i \leqq n)$. Hence $u_i \in T_P$. This is to say that T_P is a P-coalgebra. Since $K^T = K^{T_P} = K^{KT} = K^{\langle T \rangle} = K^{\langle T_P \rangle} = K^{K\langle T \rangle}$, Theorem 5.2.2 implies that $H(K/K^T) = K\langle T \rangle = K\langle T_P \rangle$. If T is closed under multiplication, so is T_P. Hence, when T is a k-bialgebra, we have $T = \langle T \rangle$, $T_P = \langle T_P \rangle$.

Suppose A is a K/k-subalgebra of $\mathrm{End}_k(K)$ and V is a K-linear space as well as an A-module. Set

$$V^A = \{v \in V;\ f(xv) = f(x)v \quad \forall f \in A \quad \forall x \in K\}.$$

Then V^A is a K^A-linear subspace of V.

EXAMPLE 5.8 For $V = K^n, x = (x_1, \ldots, x_n) \in V, f \in A$, the definition $f(x_1, \ldots, x_n) = (f(x_1), \ldots, f(x_n))$ makes V a K/k-A-module. Now we have $A^A = (K^A)^n$. In general, given a subfield k of K^A and a k-linear space W, if we let $V = K \otimes W$ and define

$$f\left(\sum_{i=1}^{n} x_i \otimes w_i\right) = \sum_{i=1}^{n} f(x_i) \otimes w_i, \qquad f \in A, \quad x_i \in K, \quad w_i \in W,$$

then V is a K/k-A-module and, when $k = K^A$, we obtain $V^A = 1 \otimes W$.

THEOREM 5.2.6 (Jacobson.) Let A be a K/k-subalgebra of $\mathrm{End}_k(K)$. If A^A is a K^A-form of A, then V^A is a K^A-form of V. In particular, when A is of finite dimension over K, V^A is a K^A-form of V.

Proof For $g \in A^A$, $v \in V$, since $f(xg) = f(x)g$, we get

$$f(xg(v)) = f(xg)(v) = f(x)g(v), \quad f \in A, \quad x \in K.$$

Hence $g(v) \in V^A$. Accordingly, $A^A(V) \subset V^A$. Since A^A is a K^A-form of A by hypothesis, we may write $1 = \sum_{i=1}^{n} x_i g_i$ for $x_i \in K, g_i \in A^A$ $(1 \leqq i \leqq n)$.

Thus we have $v = 1(v) = \sum_{i=1}^{n} x_i g_i(v)$, and $A^A(V)$ spans V. On the other hand, since $A^A(V) \subseteq V^A$, V^A spans V over K. We now proceed to show that if $\{v_\lambda\}_{\lambda \in \Lambda}$ is a basis for V^A over K^A, then $\{v_\lambda\}_{\lambda \in \Lambda}$ is also a basis for V over K. If $\sum y_\lambda v_\lambda = 0$, $y_\lambda \in K$, then

$$0 = g_i\left(\sum_\lambda y_\lambda v_\lambda\right) = \sum_\lambda g_i(y_\lambda) v_\lambda, \quad g_i(y_\lambda) \in A^A(K) \subset K^A.$$

Hence $g_i(y_\lambda) = 0$ $(1 \leqq i \leqq n)$. Consequently $y_\lambda = 1(y_\lambda) = \sum_{i=1}^{n} x_i g_i(y_\lambda) = 0$, that is, $\{v_\lambda\}_{\lambda \in \Lambda}$ is a basis for V over K and V^A is a K^A-form of V. If A is finite dimensional over K, it follows from Theorem 5.2.1 that V^A is a K^A-form of V.

2.2 Jacobson's theorem

We go on to prove Jacobson's theorem which gives a Galois correspondence between intermediate fields of a purely inseparable extension field K of a field k of exponent 1 whose characteristic is $p > 0$ and p-k-sub-Lie algebras of the p-k-Lie algebra $\mathrm{Der}_k(K)$.

Assume K is an extension field of a field k of characteristic $p > 0$. A K-linear space L is called a **p-K/k-Lie algebra** when L has a p-k-Lie algebra structure, when regarded as a k-linear space. For a p-K/k-sub-Lie algebra L of the p-k-Lie algebra $\mathrm{Der}_k(K)$ consisting of all k-derivations from K to K,

$$\begin{aligned} K^L &= \{y \in K;\ D(xy) = D(x)y \quad \forall D \in L \quad \forall x \in K\} \\ &= \{y \in K;\ D(y) = 0 \quad \forall D \in L\} \end{aligned}$$

is a subfield of K. Moreover, for $x \in K$, $D \in L$, we have $D(x^p) = px^{p-1}D(x) = 0$, so $K^p \subset K^L$.

LEMMA 5.2.7 If k is a subfield of K such that $K^p \subset k$, then $K^{\mathrm{Der}_k(K)} = k$.

Proof Set $M = \mathrm{Der}_k(K)$. If we choose an element $x \in K$ such that $x \notin k$ and let k' be the largest subfield of K not containing x, then $k'(x) = K$. In fact, if we let $k'(x) \subsetneqq K$, there exists $y \in K - k'(x)$. In this situation, we have $k' \subsetneqq k'(x) \subset k'(y)$ because $x \in k'(y)$. Since x^p, $y^p \in k'$, we see that

$[k'(y):k'] = [k'(x):k'] = p$. Thus $k'(x) = k'(y)$, which contradicts the fact that $y \notin k'(x)$. Hence $k'(x) = K$. Now if we define a k'-linear map $D: K \to K$ by $D(x^i) = ix^i$ $(0 \leqq i \leqq p-1)$, then $D \in \mathrm{Der}_k(K)$, so that $x \notin K^M$. Thus we get $K^M \subset k$, and hence $K^M = k$ since $K^M \supset k$ is clear.

Let L be a p-K/k-sub-Lie algebra of $\mathrm{Der}_k(K)$. For $x \in K$, define a K-linear map $\hat{x}: L \to K$ by $\hat{x}(D) = D(x)$. Now $\hat{K}$ is dense in $L^* = \mathbf{Mod}_K(L, K)$.

THEOREM 5.2.8 If L is a p-K/k-sub-Lie algebra of $\mathrm{Der}_k(K)$ of finite dimension over K, then there exists a P-form L_P of L consisting of derivations diagonalizable over P which commute with each other. Moreover, we have $[K:K^L] = p^{\dim_K L}$.

Proof Since L is finite dimensional over K and $\hat{K}$ is dense in L^*, $\hat{K}$ contains a basis $\{\hat{x}_1, \ldots, \hat{x}_n\}$ for L^* over K. Let $D_i (1 \leqq i \leqq n)$ be elements of L defined by $\hat{x}_j(D_i) = \delta_{ij} x_j = D_i(x_j)$. Then $\{D_1, \ldots, D_n\}$ is a basis for L over K. In this situation, we have $[D_i, D_j] = 0$, $D_i^p(x_r) = \delta_{ir} x_r = D_i(x_r)\,(1 \leqq i, j, r \leqq n)$. The P-linear space L^P spanned by $\{D_1, \ldots, D_n\}$ is a P-form of L and, as in Example 5.5, letting $G = \mathbf{Mod}_P(L_P, P)$, we get $[K:K^L] = |G|$. On the other hand, $|G| = p^{\dim_P L}$, so that $[K:K^L] = p^{\dim_K L}$.

THEOREM 5.2.9 (Jacobson.) Let L be a p-K/k-sub-Lie algebra of $\mathrm{Der}_k(K)$ of finite dimension over K. Then $L = \mathrm{Der}_{K^L}(K)$.

Proof Write M for $\mathrm{Der}_{K^L}(K)$. The verification $L \subset M$ is trivial. Theorem 5.2.8 shows that $p^{\dim_K L} = [K:K^L] = p^{\dim_K M}$. Thus $\dim_K L = \dim_K M$, and hence we conclude that $L = M$.

COROLLARY 5.2.10 Assume K is a field of characteristic $p > 0$. Let $\mathscr{P}$ be the family of all subfields k of K such that $K^p \subset k$ and such that K is finite dimensional over k, and let $\mathscr{L}$ be the family of all p-K/P-sub-Lie algebras of $\mathrm{Der}_P(K)$ of finite dimension over K. Then

$$k \mapsto \mathrm{Der}_k(K), \quad L \mapsto K^L$$

are one-to-one correspondences between $\mathscr{P}$ and $\mathscr{L}$, which are each other's inverses.

Proof For $L \in \mathscr{L}$, we have $k = K^L \in \mathscr{P}$. By Lemma 5.2.7, it follows that $K^{\mathrm{Der}_k(K)} = k$. Further, Theorem 5.2.9 implies $L = \mathrm{Der}_{K^L}(K)$. Therefore it follows that the two correspondences are one-to-one and are each other's inverses.

Let V be a K linear space and L a p-K/P-sub-Lie algebra of $\mathrm{Der}_P(K)$. For $x \in K$, $D \in L$, $v \in V$, if we define $x(D \otimes v) = xD \otimes v$, then $L \otimes V$ becomes a K-linear space. We call V a p-K/k-L-**module** if V is a p-L-module whose structure map $\varphi : L \otimes V \to V$ is a K-linear map. In this situation,

$$V^L = \{v \in V;\, D(xv) = D(x)v \quad \forall x \in K \quad \forall D \in L\}$$

is a K^L-linear subspace of V.

EXAMPLE 5.9 Given a p-K/k-sub-Lie algebra L of $\mathrm{Der}_k(K)$, L is a p-K/k-L-module via the k-Lie algebra structure map $\varphi : L \otimes L \to L$. Further, when $V = K^n$, define a map $\varphi : L \otimes V \to V$ by $\varphi(D \otimes v) = (D(x_1), \ldots, D(x_n)) \in V$ where $D \in L$ and $v = (x_1, \ldots, x_n) \in V$. Then V becomes a p-K/k-L-module and we have $V^L = (K^L)^n$. In general, when W is a k-linear space where k is a subfield of K^L, $V = K \otimes W$ becomes a p-K/k-L-module if we define the structure map $\varphi : L \otimes V \to V$ by $\varphi\left(D \otimes \left(\sum_{i=1}^{n} x_i \otimes w_i\right)\right) = \sum_{i=1}^{n} D(x_i) \otimes w_i$, and we see that $V^L = 1 \otimes W$.

THEOREM 5.2.11 Let L be a p-K/k-sub-Lie algebra of $\mathrm{Der}_k(K)$ which is finite dimensional over K. If V is a p-K/k-L-module, then V^L is a K^L-form of V. Moreover, for a P-form L_P of L and $\alpha \in R = \mathbf{Mod}_P(L_P, P)$, if we set

$$K_\alpha = \{x \in K;\, D(x) = \alpha(D)x \quad \forall D \in L_P\},$$
$$V_\alpha = \{v \in V;\, D(v) = \alpha(D)v \quad \forall D \in L_P\},$$

then we have $K = \coprod_{\alpha \in R} K_\alpha$, $V = \coprod_{\alpha \in R} V_\alpha$, $K_\alpha K_\beta \subset K_{\alpha+\beta}$, and $K_\alpha V_\beta \subset V_{\alpha+\beta}$.

Proof Since elements in L_P are diagonalizable over P by Theorem

5.2.8, we can write $K=\coprod_{\alpha\in R} K_\alpha$ and $V=\coprod_{\alpha\in R} V_\alpha$. For $D\in L_P$, $x\in K_\alpha$, $y\in K_\beta$, $v\in V_\beta$, the calculations

$$D(xy)=D(x)y+xD(y)=\alpha(D)xy+\beta(D)xy=(\alpha+\beta)(D)xy,$$
$$D(xv)=D(x)v+xD(v)=\alpha(D)xv+\beta(D)xv=(\alpha+\beta)(D)xv$$

show that $K_\alpha K_\beta\subset K_{\alpha+\beta}$ and $K_\alpha V_\beta\subset V_{\alpha+\beta}$. On the other hand, $\coprod_{\alpha\in R} K_\alpha = \coprod_{\alpha\in R} K_0 x_\alpha$, where $x_\alpha\in K_\alpha$, $x_\alpha\neq 0$. Here we obtain a similar result for V, namely that $V=\coprod_{\alpha\in R} V_\alpha=\coprod_{\alpha\in R} K_\alpha V_0$. Hence $KV_0=V$. We contend that, if a set $\{v_1,\dots,v_m\}$ of elements of the K_0-linear space V_0 is linearly independent over K_0, then it is moreover linearly independent over K. By proving the contention, we can conclude that V_0 is a K_0-form of V. Now suppose that a set $\{v_1,\dots,v_m\}$ of elements of V_0 is linearly independent over K_0 but linearly dependent over K. Let m be the smallest cardinal number of such sets. By hypothesis, we can write $\sum_{i=1}^{m} x_iv_i=0$, $x_i\in K$, $x_i\neq 0$ $(1\leqq i\leqq m)$. Here we may assume $x_m=1$. For $D\in L_P$, we have $0=D\left(\sum_{i=1}^{m} x_iv_i\right)=\sum_{i=1}^{m-1} D(x_i)v_i$. Since $\{v_1,\dots,v_{m-1}\}$ is linearly independent over K, we have $D(x_i)=0$ $(1\leqq i\leqq m-1)$. Thus, for any $D\in L_P$, $D(x_i)=0$ $(1\leqq i\leqq m)$ holds, and hence $x_i\in K_0 (1\leqq i\leqq m)$. Meanwhile, since $\{v_1,\dots,v_m\}$ is linearly independent over K_0, we get $x_i=0 (1\leqq i\leqq m)$. This contradicts $x_m=1$. Accordingly, V_0 is a K_0-form of V. The fact that $K_0=K^L$ and $V_0=V^L$ implies that V^L is a K^L-form of V.

Let K be a purely inseparable extension field of k. The **exponent** of an extension K/k is the smallest m such that $K^{p^m}\subset k$. When K/k is a purely inseparable field extension of exponent 1, $K^{\mathrm{Der}_k(K)}=k$ (cf. Corollary 5.2.10). If $\{x_1,\dots,x_n\}$ is a p-basis for K/k, the set $\{D_1,\dots,D_n\}$ of elements of $\mathrm{Der}_k(K)$ defined by $D_i(x_j)=\delta_{ij}x_j$ turns out to be a basis for $\mathrm{Der}_k(K)$ over K. In fact, if $\sum_{i=1}^{n} y_iD_i=0$, $y_i\in K (1\leqq i\leqq n)$, then we have $\sum_{i=1}^{n} y_iD_i(x_j)=y_jx_j=0$. Since

$x_j \neq 0$, we get $y_j = 0$ $(1 \leqq j \leqq n)$. Hence $\{D_1, \ldots, D_n\}$ is linearly independent over K. Moreover, any $D \in \mathrm{Der}_k(K)$ can be written in the form $D = \sum_{j=1}^{n} D(x_j)x_j^{-1}D_j$. Thus $\{D_1, \ldots, D_n\}$ becomes a basis for $\mathrm{Der}_k(K)$ over K. We see that K is spanned by $\{x_1^{d_1} \cdots x_n^{d_n};\, 0 \leqq d_i \leqq p-1\}$ over k, and $[K:k] = p^{\dim_K \mathrm{Der}_k(K)} = p^n$. In turn, $\alpha \in R = \mathbf{Mod}_P(L_P, P)$ is uniquely determined by $\alpha(D_i) = d_i \in P\,(1 \leqq i \leqq n)$. Thus identifying α with $(d_1, \ldots, d_n)$ $(0 \leqq d_i \leqq p-1)$, we get $K_\alpha = kx_1^{d_1} \cdots x_n^{d_n}$ and $K = \coprod_{\alpha \in R} K_\alpha$. Let A be the K-linear subspace of $\mathrm{End}_k(K)$ spanned by $\{D_1^{d_1} \cdots D_n^{d_n};\, 0 \leqq d_i \leqq p-1\}$ over K. Then A is a K/k-subalgebra of $\mathrm{End}_k(K)$. Now we obtain

$$(xD)(yD') = xD(y)D' + xyDD' \quad \text{and} \quad D_i^p = D_i (1 \leqq i \leqq n),$$
$$x, y \in K, \quad D, D' \in L_P.$$

Moreover, $K^A = K^{\mathrm{Der}_k(K)} = k$ (cf. Theorem 5.2.9), so that we obtain $A = \mathrm{End}_k(K)$ by Theorem 5.2.1. Summing up the above, we have proved

THEOREM 5.2.12 (Jacobson.) Assume K is a finite purely inseparable extension field of a field k of exponent 1. Let $\{x_1, \ldots, x_n\}$ be a p-basis for K/k. Then $\{x_1^{d_1} \cdots x_n^{d_n};\, 0 \leqq d_i \leqq p-1\}$ is a basis for K over k. The set $\{D_1, \ldots, D_n\}$ of elements of $L = \mathrm{Der}_k(K)$ defined by $D_i(x_j) = \delta_{ij}x_j$ $(1 \leqq i, j \leqq n)$ is a basis for L over K and the P-linear space L_P spanned by $\{D_1, \ldots, D_n\}$ over P is a P-form of L, and, furthermore, we have $K = \coprod_{\alpha \in R} K_\alpha$, where $R = \mathbf{Mod}_P(L_P, P)$. Moreover, if $\alpha \in R$ is given by $\alpha(D_i) = d_i$ $(1 \leqq i \leqq n, 0 \leqq d_i \leqq p-1)$, we have $K_\alpha = kx_1^{d_1} \ldots x_n^{d_n}$.

3 Modular extensions

Here we present a Galois correspondence for purely inseparable extensions and the characterization of modular extension fields due to Sweedler.

3.1 Galois correspondence

We assume that K is a field and P is its prime field. We denote by $\mathscr{K}$ the family of all subfields k of K such that K is finite dimensional over k, and by $\mathscr{H}$ the family of all K/P-sub-bialgebras of $\mathrm{End}_P(K)$ of finite dimension over K. For $k \in \mathscr{K}$, we have $\mathrm{End}_k(K) = H(K/k)$. In fact $\mathrm{End}_k(K)$ is a finite dimensional K-linear space and for $f \in \mathrm{End}_k(K)$, $x \in K$, we have xf, $fx \in \mathrm{End}_k(K)$. Therefore $f \in R_K(K_m)$. Namely, we obtain $\mathrm{End}_k(K) = H(K/k)$. Accordingly, the correspondences given in Corollary 5.2.3

$$k \mapsto H(K/k), \quad A \mapsto K^A$$

give a one-to-one correspondence between $\mathscr{K}$ and $\mathscr{H}$ and they are each other's inverses.

THEOREM 5.3.1 Let K be a finite extension field of k.

(1) K/k is a normal extension $\Leftrightarrow H(K/k)$ is a pointed cocommutative normal K/k-bialgebra.

(2) K/k is a Galois extension $\Leftrightarrow H(K/k)$ is pointed cocommutative co-semi-simple.

(3) K/k is a purely inseparable extension $\Leftrightarrow H(K/k)$ is pointed cocommutative irreducible.

Proof Set $H = H(K/k)$.

(2) Suppose K/k is a Galois extension. Then we have $G(H) = \mathrm{Aut}_k(K)$ and $KG(H)$ is a K/P-sub-bialgebra of $\mathrm{End}_P(K)$. Since H is of finite dimension over K, we get $K^H = k = K^{G(H)} = K^{KG(H)}$. Hence $KG(H) = H$. Thus H is pointed cocommutative co-semi-simple. Conversely, if H is pointed cocommutative co-semi-simple, we see that $H = KG(H)$. Thus $k = K^H = K^{G(H)}$. Therefore K/k is a Galois extension.

(3) Assume K/k is a purely inseparable extension. We denote the dual K-algebra of H by $A = H^*$. For $a \in K$, if we define $\hat{a} \in A$ by $\hat{a}(x) = x(a)$, $x \in H = \mathrm{End}_k(K)$, then $A = K\hat{K}$. Now let $M = \left\{ \sum_{i=1}^{n} a_i \hat{b}_i ; n \geqq 1, a_i, b_i \in K, \sum_{i=1}^{n} a_i b_i = 0 \right\}$. Then M is a maximal ideal of A and is nilpotent. In fact, for $f = \sum_{i=1}^{n} a_i \hat{b}_i \in A$, we have $f(1) = 0 \Leftrightarrow \sum_{i=1}^{n} a_i b_i = 0$. Now, if we set $g = f - f(1)1$, then $g \in M$. Thus $A = K1 + M$ and M is a maximal ideal. Moreover, for $f = \sum_{i=1}^{n} a_i \hat{b}_i \in M$, take e

such that $a_i^{p^e} \in k\,(1 \leqq i \leqq n)$. Then we get

$$\left(\sum_{i=1}^{n} a_i \hat{b}_i\right)^{p^e} = \sum_{i=1}^{n} a_i^{p^e} \hat{b}_i^{p^e} = \sum_{i=1}^{n} (a_i^{p^e} b_i^{p^e})\hat{} = 0.$$

Hence M is a nilpotent ideal. Consequently A is a local K-algebra and there exists a K-algebra morphism $\varepsilon : A \to K$ satisfying $\varepsilon \circ \eta_A = 1_K$. Therefore H is irreducible. Conversely, assume H is pointed cocommutative irreducible, and let $H_i = \sqcap^{i+1} K1$. Then $\{H_i\}_{i \in I}$ is the coradical filtration on H and $H = \bigcup_{i=0}^{\infty} H_i$. Because K/k is a finite extension, it is possible to choose a positive integer n such that $H = \bigcup_{i=0}^{n} H_i$. For $x \in (H_{i+1})^+$, we see that

$$\Delta(x) = x \otimes_K 1 + 1 \otimes_K x + \sum_j x_j \otimes_K y_j, \quad x_j, y_j \in (H_i)^+.$$

Hence, for $a \in K^{H_i}$, $b \in K$, we get $x(ab) = x(a)b + ax(b)$. Thus $x(a^p) = 0$. Accordingly $(K^{H_i})^p \subset K^{H_{i+1}}$ holds. From $K = K^{H_0}$, it follows that $K^{p^n} \subset K^{H_n} = K^H = k$. Therefore K/k is a purely inseparable extension.

(1) When K/k is a normal extension, there are subfields K_{gal} and K_{rad} of K satisfying the conditions that K_{gal}/k is a Galois extension, K_{rad}/k is a purely inseparable extension, and that $K = K_{\text{gal}} K_{\text{rad}}$. Letting $H_{\text{gal}} = KG(H)$ and $H_{\text{rad}} = H_1$, we proceed to show that $H_{\text{gal}} = H(K_{\text{gal}}/k)$ and $H_{\text{rad}} = H(K_{\text{rad}}/k)$. For $x \in H(K_{\text{gal}}/k)$ and $y \in H(K_{\text{rad}}/k)$, the element f of $\text{End}_k(K)$ defined by

$$f : ab \mapsto x(a)y(b), \quad a \in K_{\text{gal}}, \quad b \in K_{\text{rad}}$$

will be denoted by $x \otimes y$. Let H' be the k-linear subspace of $\text{End}_k(K)$ spanned by $\{x \otimes y;\ x \in H(K_{\text{gal}}/k),\ y \in H(K_{\text{rad}}/k)\}$ over k. Then H' is a K/k-subalgebra of $\text{End}_k(K)$. In fact, for $a \in K_{\text{gal}}$, $b \in K_{\text{rad}}$, we have $ab(x \otimes y) = ax \otimes by$. Thus H' is a K-linear subspace, and, for $x \in H(K_{\text{gal}}/k)$ and $y \in H(K_{\text{rad}}/k)$, we have

$$\Delta(x \otimes y) = \sum_{(x)(y)} (x_{(1)} \otimes y_{(1)}) \otimes_K (x_{(2)} \otimes y_{(2)}).$$

Hence

$$\Delta(x \otimes y)(a \otimes b \otimes_K a' \otimes b') = \sum_{(x)(y)} x_{(1)}(a) y_{(1)}(b) \otimes x_{(2)}(a') y_{(2)}(b').$$

Therefore H' is a K/k-sub-bialgebra of $\mathrm{End}_k(K)$. On the other hand, since $K^{H(K/k)} = k = K^{H'}$ and the correspondence $k \mapsto H(K/k)$ is one-to-one, it follows that $H' = H(K/k)$. By (2) and (3), we may regard $H(K_{\mathrm{gal}}/k) \subset KG(H)$ and $H(K_{\mathrm{rad}}/k) \subset H_1$, so the equalities hold. Therefore $H = H(K/k)$ is pointed cocommutative normal.

Conversely, if $H = H(K/k)$ is pointed cocommutative normal, H admits a decomposition into a semi-direct product $H = H_1 \otimes kG(H)$ of K/k-bialgebras. Now we get $H_1 = H(K/K^{H_1})$ and $KG(H) = H(K/K^{G(H)})$. Setting $K^{G(H)} = K_{\mathrm{rad}}$ and $K^{H_1} = K_{\mathrm{gal}}$, (2) and (3) imply that K/K_{rad} is a Galois extension and K/K_{gal} is a purely inseparable extension. Thus $K/K_{\mathrm{gal}}K_{\mathrm{rad}}$ is a Galois extension as well as a purely inseparable extension, and hence we have $K = K_{\mathrm{gal}}K_{\mathrm{rad}}$. In turn, we obtain $(K_{\mathrm{gal}})^{KG(H)} \subset K^{G(H)} \cap K^{H_1} = K^H = k$, and similarly, $(K_{\mathrm{rad}})^{H_1} \subset k$. Hence K_{gal}/k is a Galois extension and K_{rad}/k is purely inseparable extension. Therefore K/k is a normal extension.

Remark For a pointed irreducible cocommutative K-Hopf algebra H_1 over a field K of characteristic 0, if $P(H_1) \neq \{0\}$, then $H_1 \cong U(P(H_1))$ is of infinite dimension over K. As for $H = H(K/k)$, since H is finite dimensional over K, we have $H_1 = K$ when the characteristic of k is 0. Thus if H is pointed cocommunative, H is co-semi-simple and we have $H = KG(H)$.

THEOREM 5.3.2 With regard to a finite purely inseparable field extension K/k, the following conditions are equivalent.

(i) K is isomorphic to a tensor product of a finite number of simple extension fields of k.

(ii) There exists a diagonalizable toral k-sub-bialgebra T of $H(K/k)$ satisfying $K^T = k$.

Proof (i)$\Rightarrow$(ii) Suppose $K \cong k(x_1) \otimes \cdots \otimes k(x_n)$. For each i $(1 \leqq i \leqq n)$, take the smallest positive integer e_i for which $x_i^{e_i} \in k$. Let G be the direct product of cyclic groups $\langle g_i \rangle$ $(1 \leqq i \leqq n)$ of order e_i. Then

$$G = \{g_1^{f_1} \cdots g_n^{f_n};\ 0 \leqq f_i < e_i,\quad 1 \leqq i \leqq n\}$$

and $|G| = e_1 e_2 \cdots e_n$. Now G is a finite abelian group and, for $\alpha = g_1^{f_1} \cdots g_n^{f_n} \in G$, if we set $K_\alpha = kx_1^{f_1} \cdots x_n^{f_n}$, then we see that

$K = \coprod_{\alpha \in G} K_\alpha$ and that G is a splitting group for the extension K/k. Letting $T = T(G)$, we obtain $K^T = k$.

(ii) ⇒ (i) Assume that T is a diagonalizable toral k-sub-bialgebra of $H(K/k)$ such that $K^T = k$. Let $\{g_1, \ldots, g_n\}$ be a basis for the splitting group $G = G(T)$ for K/k and let e_i be the order of g_i. Fix elements $x_i \in K_{g_i}$ $(1 \leqq i \leqq n)$ such that $x_i \neq 0$. Then for $\alpha = g_1^{f_1} \cdots g_n^{f_n} \in G$ $(0 \leqq f_i \leqq e_i)$, we get $K_\alpha = kx_1^{f_1} \cdots x_n^{f_n}$ and $K = \coprod_{\alpha \in G} K_\alpha$. On the other hand, we have $K^G = K^{T(G)} = K^{KT} = k$, and hence $K \cong k(x_1) \otimes \cdots \otimes k(x_n)$.

3.2 Higher derivations

Let A and B be k-algebras such that $A \subset B$. Given a sequence $D = \{D_0, D_1, \ldots, D_m\}$ (where m is a positive integer and ∞ is permitted) of k-linear maps from A to B, if D_0 is the canonical embedding $A \to B$ and if

$$D_n(xy) = \sum_{i=0}^{n} D_i(x) D_{n-i}(y), \quad x, y \in A, \quad 0 \leqq n \leqq m$$

is satisfied, D is called a **higher k-derivation** from A to B of **rank** m or simply an **m-k-derivation**. When $B = A$, it is called an m-k-derivation of A.

EXAMPLE 5.10 Let A be the polynomial ring $k[X]$ in one variable over a field k. Define D_i by

$$D_i(X^n) = \binom{n}{i} X^{n-i} \quad (n = 0, 1, 2, \ldots) \quad \text{where} \quad \binom{n}{i} = 0 \quad (i > n).$$

Then we get a sequence $D = \{D_0, D_1, D_2, \ldots\}$ of k-linear maps from A to A. Now D is an ∞-k-derivation of A. Indeed, since

$$\sum_{i=0}^{j} \binom{m}{i}\binom{n}{j-i} = \binom{m+n}{j},$$

we get

$$\sum_{i=0}^{j} D_i(X^m) D_{j-i}(X^n) = \binom{m+n}{j} X^{m+n-j}.$$

Here D_1 is a k-derivation of A. Conversely, if k is of characteristic

0, given $D_1 \in \mathrm{Der}_k(A)$, we obtain an ∞-k-derivation $D = \{D_0, D_1, D_2, \ldots\}$ by setting $D_i = D_1{}^i/i!$ $(i = 0, 1, 2, \ldots)$.

EXAMPLE 5.11 As in Example 5.10, set $A = k[X]$ and define k-linear maps D_i from A to A by

$$D_i(X^n) = \binom{n}{i} X^n \quad (n = 0, 1, 2, \ldots) \quad \text{where} \quad \binom{n}{i} = 0 \quad (i > n).$$

Then $D = \{D_0, D_1, D_2, \ldots\}$ is an ∞-k-derivation of A. Let $\mathfrak{a}$ be the ideal of A generated by X^{m+1}. Since $D_i(\mathfrak{a}) \subset \mathfrak{a}$ (i = 0, 1, 2, . . .), each D_i yields a k-linear transformation $\bar{D}_i$ of $\bar{A} = A/\mathfrak{a}$ and $\{\bar{D}_0, \ldots, \bar{D}_m\}$ becomes an m-k-derivation of $\bar{A}$.

EXAMPLE 5.12 Suppose that $k[t]$ is the factor k-algebra of the polynomial ring $k[X]$ in one variable over k by the ideal (X^{m+1}) where t is the residue class containing X. Given a k-algebra A, set $B = A \otimes k[t] = A[t]$. For each m-k-derivation $D = \{D_0, D_1, \ldots, D_m\}$ of A, define a k-linear map $s = s(D)$ from A to B by

$$s(a) = \sum_{i=0}^{m} D_i(a)t^i, \quad a \in A.$$

Then we have $s(ab) = s(a)s(b)$, $a, b \in A$, and s becomes a k-algebra morphism from A to B. Let $\pi : B \to A$ be the canonical projection, that is, let

$$\pi(a_0 + a_1 t + \cdots + a_m t^m) = a_0, \quad a_i \in A \ (0 \leqq i \leqq m).$$

Then $\pi \circ s = 1_A$ holds. Similarly for each ∞-k-derivation $D = \{D_0, D_1, \ldots\}$ of A, define a k-linear map $s = s(D)$ from A to the power series ring $B = A[[t]]$ in one variable over A by

$$s(a) = \sum_{i=0}^{\infty} D_i(a)t^i, \quad a \in A.$$

Then s is a k-algebra morphism from A to B and satisfies $\pi \circ s = 1_A$ where $\pi : B \to A$ is the canonical projection. Conversely, given a k-algebra morphism $s : A \to B$ $(B = A[t]$ or $A[[t]])$ satisfying $\pi \circ s = 1_A$, if we express $s(a) = \sum_{i=0}^{\infty} D_i(a)t^i$ for $a \in A$, we obtain an m-k-derivation $D = \{D_0, D_1, D_2, \ldots\}$ of A (where m is a positive integer or ∞). We call s the k-algebra morphism associated with an m-k-derivation. Now

$$A^D = \{a \in A; s(a) = a\} = \{a \in A; D_i(a) = 0, i = 1, 2, \ldots\}$$

is a k-subalgebra of A and is called the **k-algebra of constants** for D. In particular, when A is an extension field of k, A^D is a subfield of A and is called the **field of constants** for D.

EXAMPLE 5.13 Let K/k be a finite field extension and C a K-coalgebra. Given an m-sequence of divided powers $\{c_0, c_1, \ldots, c_m\}$ consisting of elements of C, let C' be the K-linear subspace of C spanned by these elements. Then C' is a K-subcoalgebra of C. If $f : C' \to \mathrm{End}_k(K)$ is a representation of the K-coalgebra C' on K, then we see that $\{f(c_0), \ldots, f(c_m)\}$ is an m-k-derivation of K.

As for an m-k-derivation $D = \{D_0, D_1, \ldots, D_m\}$ of a k-algebra A, the number q such that $D_1 = D_2 = \cdots = D_{q-1} = 0$ and $D_q \neq 0$ $(1 \leqq q \leqq m)$ is called the **order** of D and, when $q = 1$, D is said to be **proper**.

THEOREM 5.3.3 Let K be a finite extension field of a field k whose characteristic is $p > 0$. Let D be an m-k-derivation of K of order q where m is not ∞, and let $k' = K^D$ be the field of constants for D. If p^e is the minimal power of p satisfying $p^e > m/q$, K/k' is a purely inseparable field extension of exponent e.

Proof Suppose s is the k-algebra morphism associated with D. Since $s(a) = a + D_q(a)t^q + \cdots$ for $a \in K$, we get

$$\begin{aligned} s(a^{p^e}) = s(a)^{p^e} &= (a + D_q(a)t^q + \cdots + D_m(a)t^m)^{p^e} \\ &= a^{p^e} + D_q(a)^{p^e} t^{p^e q} + \cdots = a^{p^e}. \end{aligned}$$

Thus we have $a^{p^e} \in k'$. Therefore K/k' is a purely inseparable field extension of exponent at most e. In turn, choose $a \in K$ such that $D_q(a) \neq 0$. Then $a^{p^{e-1}} \notin k'$ forces e to be the exponent of K/k'.

Remark Given an m-k-derivation $D = \{D_0, D_1, \ldots\}$ of K, the K-linear space C spanned by $D_0, D_1, \ldots, D_m$ is a K-subcoalgebra of $H(K/k)$ defined by

$$\Delta(D_n) = \sum_{i=0}^{n} D_i \otimes_K D_{n-i}, \quad \varepsilon(D_n) = \delta_{0n} \ (0 \leqq n \leqq m),$$

and it is pointed irreducible cocommutative. This fact also allows us to confirm that K/k' is a purely inseparable extension.

Suppose that $K = k(\alpha)$ is a simple inseparable extension field of k and that $X^{p^e} - a$ is the minimal polynomial of α over k. Define a $(p^e - 1)$-k-derivation $D = \{D_0, D_1, \ldots, D_{p^e-1}\}$ of the polynomial ring $k[X]$ in one variable over k by $D_i(X^n) = \binom{n}{i} X^n (1 \leqq i \leqq p^e - 1)$. Then the fact that $D_i(X^{p^e} - a) = \binom{p^e}{i} X^{p^e} = 0 \ (1 \leqq i \leqq p^e - 1)$ shows that each D_i yields a k-linear transformation $\bar{D}_i$ of $k(\alpha) \cong k[X]/(X^{p^e} - a)$ and hence $\bar{D} = \{\bar{D}_0, \bar{D}_1, \ldots\}$ becomes a $(p^e - 1)$-k-derivation of K. We see that $K^{\bar{D}} = k$. In fact, the verification $k \subset K^{\bar{D}} = k'$ is trivial. If we assume $k \subsetneqq k'$, then the minimal polynomial of α over k' is $X^{p^f} - b$ where $b \in k'$, $f < e$. Now we get $\alpha^{p^f} \in k'$ and $\bar{D}_{p^f}(\alpha^{p^f}) = \alpha^{p^f} \neq 0$, which is a contradition. Thus $k = k'$. Now let T be the k-linear space spanned by $\bar{D}_0, \bar{D}_1, \ldots, \bar{D}_{p^e-1}$ and define

$$\Delta(\bar{D}_n) = \sum_{i=0}^{n} \bar{D}_i \otimes \bar{D}_{n-i}, \quad \varepsilon(\bar{D}_n) = \delta_{0n} \ (0 \leqq i \leqq n).$$

Then T is a diagonalizable toral k-coalgebra and $k = K^T$ holds.

Let $K = k(\alpha_1) \otimes \cdots \otimes k(\alpha_n)$ be a purely inseparable extension field of k expressed as a tensor product of n simple inseparable extension fields of k. Set

$$K_i = k(\alpha_1) \otimes \cdots \otimes k(\alpha_{i-1}) \otimes k(\alpha_{i+1}) \otimes \cdots \otimes k(\alpha_n).$$

Then K/K_i is a simple inseparable extension and the argument above guarantees the existence of an m_i-k-derivation $D^{(i)}$ of K for which K_i is the field of constants (where m_i is the exponent of $k(\alpha_i)/k$). We see that $k = \bigcap_{i=1}^{n} K_i = \bigcap_{i=1}^{n} K^{D^{(i)}}$ and k coincides with the field of constants $K^{\mathscr{D}} = \bigcap_{D^{(i)} \in \mathscr{D}} K^{D^{(i)}}$ for $\mathscr{D}$ where $\mathscr{D} = \{D^{(i)}; 1 \leqq i \leqq n\}$ is a set consisting of a finite number of higher k-derivations of K. In the same manner as in the case of a simple extension field, we obtain a diagonalizable toral k-coalgebra T such that $K^T = k$ from these derivations.

3.3 Modular extensions

Given a finite purely inseparable extension field K of a field k of characteristic $p > 0$, K is called a **modular extension field** of k if K is isomorphic to a tensor product $k(\alpha_1)\otimes\cdots\otimes k(\alpha_n)$ of simple extension fields of k.

THEOREM 5.3.4 For a finite extension field K of a field k, the following conditions are equivalent.

(i) K/k is a modular field extension.

(ii) There exists a diagonalizable toral k-sub-bialgebra T of $H(K/k)$ satisfying $K^T = k$.

(iii) There exists a set $\mathscr{D} = \{D^{(i)};\ 1 \leqq i \leqq n\}$ of higher k-derivations of K such that $k = \bigcap_{i=1}^{n} K^{D^{(i)}}$.

Proof The equivalence (i)$\Leftrightarrow$(ii) has already been proven in Theorem 5.3.2. Since (i)$\Rightarrow$(iii) has been proven in the last subsection, we need only show (iii)$\Rightarrow$(i). First of all, we provide some lemmas.

Let K/k be a finite purely inseparable field extension of exponent r. Set $k^{p^{-i}} \cap K = K_i (0 \leqq i \leqq r)$. Then we get a sequence

$$k = K_0 \subset K_1 \subset \ldots \subset K_r = K$$

of subfields of K. Here K_{i+1} is a purely inseparable extension field of K_i of exponent 1. Choose a p-basis for K over K_{r-1} and denote it by $S_{1,1}$. Then $S_{1,1}{}^p = \{\alpha^p;\ \alpha \in S_{1,1}\}$ is a subset of K_{r-1}. Choose a p-basis for $K_{r-2}(S_{1,1}{}^p)$ over K_{r-2} from $S_{1,1}{}^p$ and denote it by $S_{2,1}$. Next choose $S_{2,2}$ so that $S_{2,1} \cup S_{2,2} (S_{2,1} \cap S_{2,2} = \varnothing)$ is a p-basis for K_{r-1} over K_{r-2}. By repeating the procedure, we obtain subsets $S_{i,1}, \ldots, S_{i,i}$ of $K_{r-(i-1)}$. Choose a p-basis for $K_{r-i-1}(S_{i,1}{}^p \cup \cdots \cup S_{i,i}{}^p)$ from $S_{i,1}{}^p \cup \cdots \cup S_{i,i}{}^p$ and denote it by S. Set $S \cup S_{i,j}{}^p = S_{i+1,j} (1 \leqq j \leqq i)$ and take a subset $S_{i+1,i+1}$ of K_{r-i} such that

$S \cup S_{i+1,i+1}$ is a p-basis for K_{r-i} over K_{r-i-1} and $S \cap S_{i+1,i+1} = \emptyset$. The procedure above gives us a family

$$\begin{array}{llll} S_{1,1} & & & \\ S_{2,1} & S_{2,2} & & \\ \vdots & & & \\ S_{r,1} & S_{r,2} & \ldots & S_{r,r} \end{array}$$

of subsets of K, which is called a **diagram** for the field extension K/k. We have $S_{i,i} \cap S_{j,j} = \emptyset$ if $i \neq j$. Set $N = S_{1,1} \cup S_{2,2} \cup \ldots \cup S_{r,r}$. For $\alpha \in S_{i,i}$, $h(\alpha) = r - i + 1$ is called the **height** of α, and the largest positive integer $j = l(\alpha)$ for which $\alpha^{p^j} \in S_{i+j,i}$ is called the **length** of α. By its construction,

$$\left\{ \prod_{\alpha \in N} \alpha^{e(\alpha)};\ 0 \leqq e(\alpha) < p^{l(\alpha)} \right\}$$

is seen to be a basis for K over k. Moreover, $S_r = S_{r,1} \cup S_{r,2} \cup \ldots \cup S_{r,r}$ is a p-basis for K_1 over k and we have $S_r' = \{\alpha^{p^{h(\alpha)-1}};\ \alpha \in N\} \subset S_r$ and that

$$S_r' = S_r \Leftrightarrow h(\alpha) = l(\alpha) \quad \forall\, \alpha \in N.$$

LEMMA 5.3.5 Let $D = \{D_0, D_1, \ldots, D_m\}$ be an m-k-derivation of K. Then for $a \in K$,

$$\begin{array}{ll} D_i(a^p) = (D_{i/p}(a))^p & \text{if } p \text{ divides } i, \\ D_i(a^p) = 0 & \text{if } i \text{ is relatively prime to } p. \end{array}$$

Proof If we let $s = s(D)$ be the k-algebra morphism from K to $K[x] = K[X]/(X^{m+1})$ associated with D, then $s(a) = \sum_{i=0}^{m} D_i(a)x^i$ $(a \in K)$. Since $s(a^p) = s(a)^p$, we have

$$\sum_{i=0}^{m} D_i(a^p)x^i = \sum_{i=0}^{m} D_i(a)^p x^{p^i}.$$

Thus we obtain the desired equalities of this lemma.

LEMMA 5.3.6 Assume that a finite purely inseparable extension field K of k satisfies condition (iii) in Theorem 5.3.4. If a set

$\{c_1, \ldots, c_r\}$ of elements of k is linearly independent over $k \cap K^{p^n}$, then $\{c_1, \ldots, c_r\}$ is also linearly independent over K^{p^n}.

Proof Assume that there exists a set $\{c_1, \ldots, c_r\}$ of elements of k which is linearly independent over $k \cap K^{p^n}$ but linearly dependent over K^{p^n} and take r to be the smallest possible number for this to hold. By hypothesis, we have $r \geqq 2$ and there exist $a_i \in K^{p^n} (1 \leqq i \leqq r)$ satisfying $c_1 a_1 + \cdots + c_r a_r = 0$. Here we may assume $a_1 = 1$ and $a_2 \notin k \cap K^{p^n}$. Condition (iii) in Theorem 5.3.4 ensures that there is a set $\mathscr{D}$ of higher k-derivations of K such that $k = \bigcap_{D \in \mathscr{D}} K^D$. Since $a_2 \notin k$, there exists $D = \{D_0, D_1, \ldots, D_m\} \in \mathscr{D}$ satisfying $D_q(a_2) \neq 0$ $(q > 1)$. On the other hand, since $D_q(c_1) = 0$, we have

$$c_2 D_q(a_2) + \cdots + c_r D_q(a_r) = 0.$$

From Lemma 5.3.5, $D_q(a_i) \in K^{p^n} (2 \leqq i \leqq r)$ follows. This contradicts the fact that r was taken to be the smallest possible for which such a relation holds.

***Proof of (iii)⇒(i) in Theorem* 5.3.4** It is enough to show $h(\alpha) = l(\alpha)$ for all $\alpha \in N$. Hence it suffices to prove that for the set $S = \left\{ \prod_{\alpha \in S_{i,i}} \alpha^{e(\alpha)};\ 0 \leqq e(\alpha) < p \right\}$ which is linearly independent over K_{r-i}, $S^p = \left\{ \prod_{\alpha \in S_{i,i}} \alpha^{pe(\alpha)};\ 0 \leqq e(\alpha) < p \right\}$ is also linearly independent over K_{r-i-1}. Suppose that S^p is linearly dependent over K_{r-i-1}. Then $S^{p^{r-i}}$ $(\subset k)$ is linearly dependent over $K^{p^{r-i}}$. Hence Lemma 5.3.6 implies that $S^{p^{r-i}}$ is linearly dependent over $k \cap K^{p^{r-i}}$. Therefore S is linearly dependent over $k^{p^{-(r-i)}} \cap K = K_{r-i}$, which is a contradiction.

EXAMPLE 5.14 Let P be the prime field of characteristic $p > 0$, and let x, y, z be indeterminates. If we let $k = P(x^p, y^p, z^{p^2})$ and $K = k(z, xz + y)$, then K is a purely inseparable extension field of k of exponent 2. Now we will show that K/k is not a modular extension. It is sufficient to show that $z^p \in K^D$ for any higher k-derivation $D = \{D_0, D_1, \ldots, D_m\}$ of K (cf. Theorem 5.3.3). Suppose that there is a higher k-derivation $D = \{D_0, D_1, \ldots, D_m\}$ of K satisfying $z^p \notin K^D$ and assume $D_i(z^p) \neq 0$. By Lemma 5.3.5, i is divisible by p and $D_i(z^p) = (D_{i/p}(z))^p$. Moreover, since $x^p \in k$, we have

$$x^p(D_{i/p}(z))^p = D_i(x^p z^p) = D_i(x^p z^p + y^p) = (D_{i/p}(xz + y))^p.$$

Thus

$$x^p = (D_{i/p}(xz + y)/D_{i/p}(z))^p, \text{ that is,}$$
$$x = D_{i/p}(xz + y)/D_{i/p}(z).$$

This contradicts the fact that $x \notin K$. Therefore K/k is not a modular extension. A diagram for K/k is given by $S_{1,1} = \{z, xz + y\}$, $S_{2,1} = \{z^p\}$, $S_{2,2} = \varnothing$. In general, diagrams for purely inseparable extensions of degree p^3 can be categorized into the following four types.

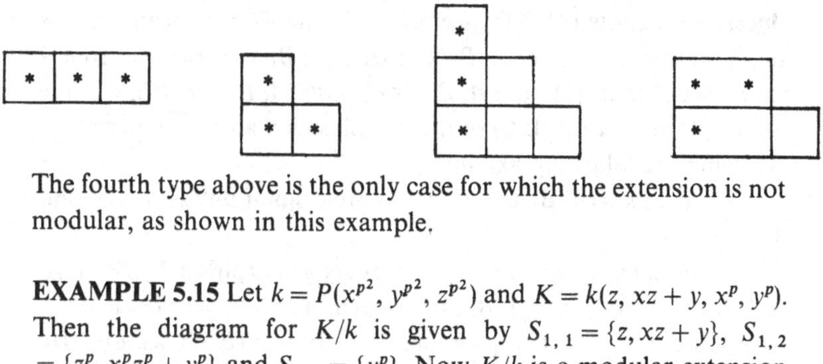

The fourth type above is the only case for which the extension is not modular, as shown in this example.

EXAMPLE 5.15 Let $k = P(x^{p^2}, y^{p^2}, z^{p^2})$ and $K = k(z, xz + y, x^p, y^p)$. Then the diagram for K/k is given by $S_{1,1} = \{z, xz + y\}$, $S_{1,2} = \{z^p, x^p z^p + y^p\}$ and $S_{2,2} = \{y^p\}$. Now K/k is a modular extension and we have $K \cong k(z) \otimes k(xz + y) \otimes k(y^p)$. However, $K/k(x^p, y^p)$ is not a modular extension.

Appendix: Categories and functors

A.1 Categories

In the theory of categories, we consider a 'collection' of objects such as 'all groups' or 'all topological spaces'. Although we cannot deal with an unlimited collecton of objects, it is impossible to talk about such collections if we simply limit ourselves to sets. In set theory, a collection of objects which is larger than a set is called a class when it satisfies the appropriate axioms. A **category** $\mathscr{C}$ consists of (1) a class of **objects** of $\mathscr{C}$ denoted ob $\mathscr{C}$ (henceforth, we write $A \in \mathscr{C}$ to mean that A is an object of $\mathscr{C}$) and (2) for $A, B \in \mathscr{C}$, a set $\mathscr{C}(A, B)$ of **morphisms** from A to B such that (3) for $A, B, C \in \mathscr{C}$, $u \in \mathscr{C}(A, B)$, $v \in \mathscr{C}(B, C)$, the composition $v \circ u \in \mathscr{C}(A, C)$ of the morphisms u and v is determined satisfying the following axioms.

(i) The sets $\mathscr{C}(A, B)$ and $\mathscr{C}(A', B')$ are disjoint unless $A = A'$ and $B = B'$.

(ii) For each object $A \in \mathscr{C}$, there exists a morphism $1_A \in \mathscr{C}(A, A)$ such that for any $u \in \mathscr{C}(A, B)$ and $v \in \mathscr{C}(B, A)$, $u \circ 1_A = u$, $1_A \circ v = v$.

(iii) For $A, B, C, D \in \mathscr{C}$, given $u \in \mathscr{C}(A, B)$, $v \in \mathscr{C}(B, C)$, $w \in \mathscr{C}(C, D)$, we have $w \circ (v \circ u) = (w \circ v) \circ u$.

The morphism 1_A in (ii) is uniquely determined for a given A and is called the **identity map** of A.

For instance, if sets are taken to be the objects, maps as the morphisms, and composition of maps as the composition of morphisms, then the resulting category is called the category of sets, which we denote by **E**. Likewise, we have the category **Mon** of semigroups with identity element, the category **Gr** of groups, and the category **Top** of topological spaces where the objects are topological spaces and the morphisms are continuous maps.

If commutative rings with identity element are taken to be the objects, ring homomorphisms which assign the identity element to the identity element are taken to be the morphisms (such morphisms are called ring morphisms in Chapter 1), and the composition of homomorphisms is taken as the composition of morphisms, then we obtain the category of commutative rings, which we denote by **M**. For $k \in \mathbf{M}$, we have the following categories.

$\mathbf{Mod}_k$ the category of k-modules
$\mathbf{M}_k$ the category of commutative k-algebras
$\mathbf{Cog}_k$ the category of k-coalgebras
$\mathbf{Hopf}_k$ the category of k-Hopf algebras
$\mathbf{Alg}_k$ the category of k-algebras
$\mathbf{Big}_k$ the category of k-bialgebras
$\mathbf{Lie}_k$ the category of k-Lie algebras

For a category $\mathscr{C}$, $u \in \mathscr{C}(A, B)$, X, $Y \in \mathscr{C}$, we define maps

$$h_X(u) : \mathscr{C}(X, A) \to \mathscr{C}(X, B),$$
$$h^Y(u) : \mathscr{C}(B, Y) \to \mathscr{C}(A, Y)$$

by $h_X(u)(v) = u \circ v$ and $h^Y(u)(w) = w \circ u$, respectively. If $h_X(u)$ is injective for each $X \in \mathscr{C}$, then u is called a $\mathscr{C}$**-injection**. Similarly, if $h^Y(u)$ is an injection for each $Y \in \mathscr{C}$, then u is said to be a $\mathscr{C}$**-surjection**. When u is simultaneously a $\mathscr{C}$-injection and a $\mathscr{C}$-surjection, then we say that u is a $\mathscr{C}$**-bijection**. Moreover, given $u \in \mathscr{C}(A, B)$, if there exists $v \in \mathscr{C}(B, A)$ such that $v \circ u = 1_A$ and $u \circ v = 1_B$, then u is called a $\mathscr{C}$**-isomorphism**. In this situation, A is said to be isomorphic to B and we write $A \cong B$. By definition, 1_A is a $\mathscr{C}$-isomorphism. A $\mathscr{C}$-isomorphism is always a $\mathscr{C}$-bijection, but the converse does not necessarily hold. For the category of sets, **E**-injections and **E**-surjections coincide respectively with injections and surjections of sets. Henceforth, **E**-injections and **E**-surjections will be referred to simply as injections and surjections. Furthermore, bijections are **E**-isomorphisms. In general, it does not necessarily follow that injections and surjections are respectively $\mathscr{C}$-injections and $\mathscr{C}$-surjections, nor does the converse necessarily follow. Consider for instance the category **Top**. A **Top**-isomorphism is a homeomorphism of topological spaces and an **E**-isomorphism. However, a bijection is not necessarily a **Top**-isomorphism. Also, the embedding of the field of rational numbers into the field of real numbers is not surjective, yet is a **Top**-surjection in an 'ordinary' topology.

Given a category $\mathscr{C}$, let the objects be the same as those of $\mathscr{C}$ and regard $u \in \mathscr{C}(A, B)$ as a morphism from B to A. Then we obtain the

dual category of $\mathscr{C}$ which we denote $\mathscr{C}^\circ$. By definition, it follows that $\mathscr{C}(A, B) = \mathscr{C}^\circ(B, A)$, and that $u \in \mathscr{C}(A, B)$ is a $\mathscr{C}$-injection if and only if $u \in \mathscr{C}^\circ(B, A)$ is a $\mathscr{C}^\circ$-surjection.

When categories $\mathscr{C}, \mathscr{C}'$ have the following properties, $\mathscr{C}'$ is said to be a **subcategory** of $\mathscr{C}$. (1) $\text{ob}\,\mathscr{C}' \subset \text{ob}\,\mathscr{C}$. (2) $A, B \in \mathscr{C}'$ implies $\mathscr{C}'(A, B) \subset \mathscr{C}(A, B)$. (3) Given $A, B, C \in \mathscr{C}'$, $u \in \mathscr{C}'(A, B)$, $v \in \mathscr{C}'(B, C)$, the composition $v \circ u$ of v and u in $\mathscr{C}'$ coincides with their composition in $\mathscr{C}$. In particular, when $\mathscr{C}'(A, B) = \mathscr{C}(A, B)$, $\mathscr{C}'$ is said to be a **full subcategory** of $\mathscr{C}$. For example, $\mathbf{M}_k$ is a full subcategory of $\mathbf{Alg}_k$.

Let $A \in \mathscr{C}$. If $\mathscr{C}(A, X)$ consists of one element for each $X \in \mathscr{C}$, then A is called an **initial object** of $\mathscr{C}$. Dually, if, for any $X \in \mathscr{C}$, $\mathscr{C}(X, A)$ is a set with a single element, then A is called a **terminal object** of $\mathscr{C}$. In the category **Gr**, the group consisting solely of the identity element is simultaneously an initial and terminal object. In the category **E**, a set with a single element is a terminal object but not an initial object. In the case of categories $\mathbf{M}_k$ or $\mathbf{Alg}_k$, k is an initial object.

Let $A, B \in \mathscr{C}$. A triple (C, u_1, u_2) consisting of an object $C \in \mathscr{C}$ together with $u_1 \in \mathscr{C}(A, C)$ and $u_2 \in \mathscr{C}(B, C)$ or simply C satisfying property (S) below is called the **direct sum** of A and B and is denoted $C = A \coprod B$, where u_1 and u_2 are called the **canonical embeddings**.

(S) Given an object $X \in \mathscr{C}$ and two morphisms $f_1 \in \mathscr{C}(A, X)$, $f_2 \in \mathscr{C}(B, X)$, there exists a unique morphism $v \in \mathscr{C}(C, X)$ such that $f_i = v \circ u_i (i = 1, 2)$.

The morphism v determined as above will be denoted (f_1, f_2). If a direct sum exists, it is unique up to isomorphism.

Dually, for $A, B \in \mathscr{C}$, the triple (C, p_1, p_2) consisting of $C \in \mathscr{C}$, $p_1 \in \mathscr{C}(C, A)$ and $p_2 \in \mathscr{C}(C, B)$ (or simply C) which satisfies property (P) below is called the **direct product** of A and B, and is denoted $C = A \times B$ or $A \Pi B$. Here, p_1, p_2 are called the **canonical projections**.

(P) Given an object $X \in \mathscr{C}$ and morphisms $f_1 \in \mathscr{C}(X, A)$, $f_2 \in \mathscr{C}(X, B)$, there exists a unique morphism $v \in \mathscr{C}(X, C)$ such that $f_i = p_i \circ v \; (i = 1, 2)$.

The morphism v determined in the above manner is written $[f_1, f_2]$. If a direct product exists, it is unique up to isomorphism. Similarly, we can define the direct sum and product of an arbitrary number of objects. For instance, in the category **E** of sets, the direct product is the direct product of sets, the direct sum is the disjoint union of sets. In the category **Gr** of groups, the direct product is the direct product of groups and the direct sum is the free product. In the category $\mathbf{Mod}_k$ of k-modules, the direct sum and product of a finite number of k-modules both turn out to be the direct sum of k-modules. However, in the category $\mathbf{M}_k$ of commutative k-algebras, although the direct product is the direct product of k-algebras, the direct sum is the tensor product over k and hence does not coincide with the direct sum of k-algebras. Dually, the direct product in the category of cocommutative k-coalgebras is the tensor product over k.

A.2 Functors

Let $\mathscr{C}, \mathscr{C}'$ be two categories. A rule F which associates to each object X of $\mathscr{C}$ an object $F(X)$ of $\mathscr{C}'$ and to each morphism $u \in \mathscr{C}(X, Y)$ of $\mathscr{C}$ a morphism $F(u) \in \mathscr{C}'(F(X), F(Y))$ of $\mathscr{C}'$ such that the properties

$$(1)\ F(1_X) = 1_{F(X)},$$
$$(2)\ F(v \circ u) = F(v) \circ F(u)$$

are satisfied is called a **covariant functor** from $\mathscr{C}$ to $\mathscr{C}'$, and is written $F : \mathscr{C} \to \mathscr{C}'$. The collection of all such functors will be denoted $\mathscr{C}\mathscr{C}'$. A covariant functor from the category $\mathscr{C}^{\circ}$ to $\mathscr{C}'$ (or from $\mathscr{C}$ to $\mathscr{C}'^{\circ}$) is called a **contravariant functor** from $\mathscr{C}$ to $\mathscr{C}'$. If F is a covariant functor from $\mathscr{C}$ to $\mathscr{C}'$ such that for each $X, Y \in \mathscr{C}$ the map

$$\mathscr{C}(X, Y) \to \mathscr{C}'(F(X), F(Y))$$

defined by $u \mapsto F(u)$ is injective, then F is said to be a **faithful functor**. On the other hand, if F is surjective, it is called a **full functor**. Now let $\mathscr{C}'$ be a subcategory of $\mathscr{C}$. Let F be the rule which identifies objects and functors of $\mathscr{C}'$ respectively with those of $\mathscr{C}$. Then $F : \mathscr{C}' \to \mathscr{C}$ is a faithful covariant functor. Moreover, if $\mathscr{C}'$ is a full subcategory of $\mathscr{C}$, then F is a **full and faithful functor**. In particular, for $\mathscr{C}' = \mathscr{C}$, F is written $1_{\mathscr{C}}$ and is called the **identity functor**.

Let $F, G \in \mathscr{C}\mathscr{C}'$. If there exists a rule which associates to each $X \in \mathscr{C}$ a morphism $\varphi(X) : F(X) \to G(X)$ of $\mathscr{C}'$ such that for each $u \in \mathscr{C}(X, Y)$ the diagram on p. 268 commutes, the family $\varphi = \{\varphi(X); X \in \mathscr{C}\}$ of morphisms is said to be a **natural transformation** from F to G, and is written $\varphi : F \to G$.

$$\begin{array}{ccc} F(X) & \xrightarrow{\varphi(X)} & G(X) \\ {\scriptstyle F(u)}\downarrow & & \downarrow{\scriptstyle G(u)} \\ F(Y) & \xrightarrow{\varphi(Y)} & G(Y) \end{array}$$

Given $F, G, H \in \mathscr{C}\mathscr{C}'$, if $\varphi : F \to G$, $\psi : G \to H$ are natural transformations, the rule which associates to $X \in \mathscr{C}$ the morphism $\psi \circ \varphi(X) = \psi(X) \circ \varphi(X)$ makes $\psi \circ \varphi = \{\psi \circ \varphi(X); X \in \mathscr{C}\}$ a natural transformation $\psi \circ \varphi : F \to H$.

For $F \in \mathscr{C}\mathscr{C}'$, the natural transformation which associates to each $X \in \mathscr{C}$ the identity map $1_{F(X)} : F(X) \to F(X)$ is denoted 1_F. This natural transformation is such that, given arbitrary natural transformations $\varphi F \to G$ and $\psi : G \to F$, then $\varphi \circ 1_F = \varphi$ and $1_F \circ \psi = \psi$. Moreover, if a natural transformation $\psi : F \to G$ satisfies one of the following equivalent conditions (1) or (2), then we say that φ is a **natural isomorphism** and that F and G are **isomorphic functors**. In this situation, we write $F \cong G$.

(1) There exists a natural transformation ψ from G to F such that $\psi \circ \varphi = 1_F$ and $\varphi \circ \psi = 1_G$.

(2) For any $X \in \mathscr{C}$, $\varphi(X) : F(X) \to G(X)$ is an isomorphism in $\mathscr{C}'$.

EXERCISE Prove that (1) and (2) are equivalent.

For two elements F, G of $\mathscr{F} = \mathscr{C}\mathscr{C}'$, let $\mathscr{F}(F, G)$ be the collection of all natural transformations from F to G. Then $\mathscr{F}$ and $\mathscr{F}(F, G)$ have properties similar to those of a category. Let $\mathscr{C}$ be a category such that $\mathscr{F}(F, G)$ defined as above becomes a set, for instance, a category whose objects form a set (such a category is called a **small category**). Now, a new category can be obtained by taking as the objects the elements of $\mathscr{F}$, and as the morphism from F to G the elements of $\mathscr{F}(F, G)$. We denote this category by the same letter $\mathscr{F} = \mathscr{C}\mathscr{C}'$. Henceforth, when we speak of a category $\mathscr{C}\mathscr{C}'$, we will assume that $\mathscr{C}$ is the type of category mentioned above. Furthermore, natural transformations of functors will be called **functor morphisms**, and natural isomorphisms will be referred to simply as isomorphisms.

Let $\mathscr{C}$, $\mathscr{C}'$ be two categories. If there exist two covariant functors

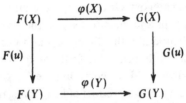

$F:\mathscr{C}\to\mathscr{C}'$ and $G:\mathscr{C}'\to\mathscr{C}$ such that $G\circ F\cong 1_{\mathscr{C}}$ and $F\circ G\cong 1_{\mathscr{C}'}$, then $\mathscr{C}$ and $\mathscr{C}'$ are said to be **categorically equivalent** or simply **equivalent**. In particular, when we have $G\circ F=1_{\mathscr{C}}$ as well as $F\circ G=1_{\mathscr{C}'}$, then we say that $\mathscr{C}$ and $\mathscr{C}'$ are **categorically isomorphic** or simply **isomorphic**. When $\mathscr{C}^\circ$ and $\mathscr{C}'$ are categorically equivalent (resp. categorically isomorphic), then $\mathscr{C}$ and $\mathscr{C}'$ are said to be **categorically anti-equivalent** (resp. **categorically anti-isomorphic**).

Given a category $\mathscr{C}$, we consider the following correspondences from $\mathscr{C}$ to $\mathbf{E}$:

$$h^X : Y\to\mathscr{C}(Y,X)$$
$$h_X : Y\to\mathscr{C}(X,Y).$$

For $u\in\mathscr{C}(Y,Y')$, we define

$$h^X(u):h^X(Y')=\mathscr{C}(Y',X)\to h^X(Y)=\mathscr{C}(Y,X)$$
$$h_X(u):h_X(Y)=\mathscr{C}(X,Y)\to h_X(Y')=\mathscr{C}(X,Y')$$

by $h^X(u)(v)=v\circ u$, $h_Y(u)(w)=u\circ w$. Here h^X, h_X are respectively contravariant and covariant functors from $\mathscr{C}$ to $\mathbf{E}$.

Let $X,X'\in\mathscr{C}$, $w\in\mathscr{C}(X,X')$. For each $Y\in\mathscr{C}$, we let

$$h^w(Y):h^X(Y)\to h^{X'}(Y)$$

be defined by $h^w(Y)(v)=w\circ v$. Since composition of morphisms satisfies the associative law, for $u\in\mathscr{C}(Y,Y')$ the following diagram

$$\begin{array}{ccc} h^X(Y') & \xrightarrow{\;h^X(u)\;} & h^X(Y) \\ {\scriptstyle h^w(Y')}\downarrow & & \downarrow{\scriptstyle h^w(Y)} \\ h^{X'}(Y') & \xrightarrow{\;h^{X'}(u)\;} & h^{X'}(Y) \end{array}$$

commutes. That is to say, $h^w:h^X\to h^{X'}$ is a functor morphism. A covariant functor from the category $\mathscr{C}$ to the category $\mathscr{C}^\circ\mathbf{E}$ is obtained from h^X and h^w. Similarly, for each $Y\in\mathscr{C}$, the map

$$h_w(Y):h_{X'}(Y)\to h_X(Y)$$

defined by $h_w(Y)(v)=v\circ w$ gives the functor morphism $h_w:h_{X'}\to h_X$.

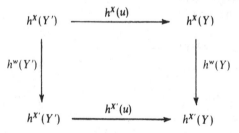

From h_w and h_X we obtain a contravariant functor from the category $\mathscr{C}$ to the category $\mathscr{C}\mathbf{E}$. The following theorem is due to Yoneda.

THEOREM A Let $\mathscr{C}$ be a category. Given $F \in \mathscr{C}^\circ\mathbf{E}$, $X \in \mathscr{C}$, there exists a natural bijection from $\mathscr{C}^\circ\mathbf{E}(h^X, F)$ to $F(X)$. Similarly, for $F \in \mathscr{C}\mathbf{E}$, there exists a natural bijection from $\mathscr{C}\mathbf{E}(h_X, F)$ to $F(X)$. If, in particular, F is h^Y or h_Y, then

$$\mathscr{C}^\circ\mathbf{E}(h^X, h^Y) \cong \mathscr{C}(X, Y), \quad \mathscr{C}\mathbf{E}(h_X, h_Y) \cong \mathscr{C}(Y, X).$$

Therefore the functor $X \mapsto h^X$ (resp. $X \mapsto h_X$) from $\mathscr{C}$ to $\mathscr{C}^\circ\mathbf{E}$(resp. $\mathscr{C}\mathbf{E}$) is full and faithful. Moreover, for X, $Y \in \mathscr{C}$, we have

$$X \cong Y \Leftrightarrow h^X \cong h^Y \quad (\text{resp. } h_X \cong h_Y).$$

Proof We prove the theorem for the case $F \in \mathscr{C}^\circ\mathbf{E}$. The proof is similar for $F \in \mathscr{C}\mathbf{E}$. Given $g \in \mathscr{C}^\circ\mathbf{E}(h^X, F)$, we have $g(X): h^X(X) = \mathscr{C}(X, X) \to F(X)$, so we define the map

$$\alpha : \mathscr{C}^\circ\mathbf{E}(h^X, F) \to F(X)$$

by $\alpha(g) = g(X)(1_X)$. Given $\xi \in F(X)$, we will conversely define a functor morphism $\beta(\xi): h^X \to F$. The morphism $v \in \mathscr{C}(Y, X)$ determines $F(v) \in \mathbf{E}(F(X), F(Y))$, so we define a map

$$\beta(\xi)(Y): h^X(Y) \to F(Y)$$

by $\beta(\xi)(Y)(v) = F(v)(\xi)$. Since $F(v \circ u)(\xi) = (F(u) \circ F(v))(\xi)$ for $u \in \mathscr{C}(Y, Y')$, the diagram

$$\begin{array}{ccc}
h^X(Y) & \xrightarrow{\beta(\xi)(Y)} & F(Y) \\
{\scriptstyle h^X(u)}\uparrow & & \uparrow{\scriptstyle F(u)} \\
h^X(Y') & \xrightarrow{\beta(\xi)(Y')} & F(Y')
\end{array}$$

commutes. Consequently, $\beta(\xi)$ is a functor morphism. Hence we have

$$\alpha(\beta(\xi)) = F(1_X)(\xi) = 1_{F(X)}(\xi) = \xi,$$
$$\beta(\alpha(g))(Y)(v) = F(v)(g(X)(1_X)) = g(Y)(h^X(v)(1_X)) = g(Y)(v).$$

This shows that α and β are inverses, forcing α to be bijective.

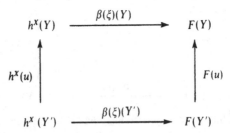

A.3 Adjoint functors

Given categories $\mathscr{C}$, $\mathscr{C}'$, define a new category by taking as objects ordered pairs (X, X') for $X \in \mathscr{C}$, $X' \in \mathscr{C}'$, and by letting the morphisms from (X, X') to (Y, Y') be the elements of $\mathscr{C}(X, Y) \times \mathscr{C}'(X', Y')$. If we define composition of morphisms by $(u, u') \circ (v, v') = (u \circ v, u' \circ v')$, then the resulting category is called the **direct product** of $\mathscr{C}$ and $\mathscr{C}'$, which we denote by $\mathscr{C} \times \mathscr{C}'$. Given covariant functors $F : \mathscr{C} \to \mathscr{C}'$, $G : \mathscr{C}' \to \mathscr{C}$, if the two covariant functors from $\mathscr{C}^\circ \times \mathscr{C}'$ to $\mathbf{E}$ given by

$$(X, X') \mapsto \mathscr{C}'(F(X), X'), \quad (X, X') \mapsto \mathscr{C}(X, G(X'))$$

are isomorphic, then we say that F is a **left adjoint** of G or that G is a **right adjoint** of F. Sometimes, we simply say that F and G are **adjoint**. Similarly, if $F : \mathscr{C} \to \mathscr{C}'$, $G : \mathscr{C}' \to \mathscr{C}$ are contravariant functors which are adjoint when viewed as covariant functors $F : \mathscr{C} \to \mathscr{C}'^\circ$, $G : \mathscr{C}'^\circ \to \mathscr{C}$, namely if

$$\mathscr{C}'(X', F(X)) \cong \mathscr{C}(X, G(X')),$$

then we say that F and G are **adjoint**. (cf. Chapter 1, Exercise 1.7, and §§ 1.3, 1.4, 1.5, 1.6, 1.7.)

A.4 Representable functors

Let $\mathscr{C}$ be a category. Suppose that, given $F \in \mathscr{C}^\circ\mathbf{E}$, $F \cong h^X$ for a suitable choice of $X \in \mathscr{C}$ or that, given $G \in \mathscr{C}\mathbf{E}$, $G \cong h_Y$ for a suitable choice of $Y \in \mathscr{C}$. Then F and G are called **representable functors**, and we say that F and G are 'represented' by X and Y respectively. Thanks to Theorem A, for such functors F and G, the choices of X and Y are unique up to isomorphism. By using representable functors, we can translate many of the properties which can be defined on the category of sets $\mathbf{E}$ to other categories. Here are some examples.

Terminal and initial objects Let $\{a\}$ be a set with a single element, namely, a terminal object of the category $\mathbf{E}$. For a category $\mathscr{C}$, we define a functor $F \in \mathscr{C}^\circ\mathbf{E}$ by assigning to each $X \in \mathscr{C}$ the set $\{a\}$, and by assigning to each morphism $u \in \mathscr{C}(X, X')$ the morphism $1_{\{a\}}$. If F is representable and is represented by $E \in \mathscr{C}$, then, for any $X \in \mathscr{C}$, $h^E(X) = \mathscr{C}(X, E)$ consists of a single element and E is a terminal object of the category $\mathscr{C}$. Therefore the functor F is representable if and only if the category $\mathscr{C}$ has a terminal object. As for initial objects, if $F \in \mathscr{C}\mathbf{E}$ is defined as above, then the functor F is representable if and only if the category $\mathscr{C}$ has an initial object.

Direct products and direct sums Given a category $\mathscr{C}$ and $A, B \in \mathscr{C}$, if the contravariant functor from $\mathscr{C}$ to $\mathbf{E}$ given by

$$F : X \mapsto \mathscr{C}(X, A) \times \mathscr{C}(X, B)$$

is representable and F is represented by $C \in \mathscr{C}$, then C is precisely the direct product of A and B. In this situation, we have

$$\mathscr{C}(X, A \times B) \cong \mathscr{C}(X, A) \times \mathscr{C}(X, B).$$

By substituting $X = A \times B$, and letting (p_1, p_2) be the element corresponding to $1_{A \times B}$, p_1 and p_2 turn out to be the canonical projections from $A \times B$ to A and B respectively, so that the bijection above can be given by $f \mapsto (p_1 \circ f, p_2 \circ f)$. We thus conclude that F is representable if and only if the direct product of A and B exists in the category $\mathscr{C}$. Similarly, if the covariant functor from $\mathscr{C}$ to $\mathbf{E}$ given by

$$G : X \mapsto \mathscr{C}(A, X) \times \mathscr{C}(B, X)$$

is representable, the object of $\mathscr{C}$ which represents G turns out to be the direct sum of A and B. Thus we obtain

$$\mathscr{C}(A \amalg B, X) \cong \mathscr{C}(A, X) \times \mathscr{C}(B, X).$$

Projective limits and inductive limits An ordered set Λ can be regarded as a category when we take as the objects the elements of Λ, and, for $\alpha, \beta \in \Lambda$, we take as the set of morphisms the set $\Lambda(\alpha, \beta)$ which consists of one element when $\alpha \leqq \beta$, and is the empty set when $\alpha \nleqq \beta$. For a category $\mathscr{C}$, a contravariant functor from Λ to $\mathscr{C}$ is called a **projective system** of $\mathscr{C}$, and a covariant functor from Λ to $\mathscr{C}$ is called an **inductive system** of $\mathscr{C}$. More precisely, a projective system of $\mathscr{C}$ is a family $\{A_\alpha\}_{\alpha \in \Lambda}$ of objects of $\mathscr{C}$ together with a set of morphisms $\{u_{\alpha\beta} : A_\beta \to A_\alpha\}_{(\alpha, \beta) \in \Lambda \times \Lambda, \alpha \leqq \beta}$ such that for $\alpha \leqq \beta \leqq \gamma$ the relations $u_{\alpha\gamma} = u_{\alpha\beta} \circ u_{\beta\gamma}$ and $u_{\alpha\alpha} = 1_{A_\alpha}$ hold. Given a projective system $\{A_\alpha, u_{\alpha\beta}\}$ of $\mathscr{C}$, if there exists a system $\{A, u_\alpha, \alpha \in \Lambda\}$ where $A \in \mathscr{C}$ and

$u_\alpha \in \mathscr{C}(A, A_\alpha)$ for each α satisfying properties (P1), (P2) below, then $\{A, u_\alpha, \alpha \in \Lambda\}$ or simply A is called the **projective limit** of the projective system $\{A_\alpha, u_{\alpha\beta}\}$, and is written $A = \varprojlim_\alpha A_\alpha$.

(P1) $u_\alpha = u_{\alpha\beta} \circ u_\beta \ (\alpha \leqq \beta)$,

(P2) Given a system $\{X, v_\alpha, \alpha \in \Lambda\}$ for $X \in \mathscr{C}$ and $v_\alpha \in \mathscr{C}(X, A_\alpha)$ $(\alpha \in \Lambda)$ such that $v_\alpha = u_{\alpha\beta} \circ v_\beta \ (\alpha \leqq \beta)$, there exists a unique $v \in \mathscr{C}(X, A)$ such that $v_\alpha = v \circ u_\alpha$ for each $\alpha \in \Lambda$.

If a projective limit exists, it is determined uniquely up to isomorphism. If $\{A_\alpha, u_{\alpha\beta}\}$ is a projective system in the category of sets **E**, the projective limit exists in **E**. In fact, define

$$A = \{(a_\alpha)_{\alpha \in \Lambda} \in \prod_{\alpha \in \Lambda} A_\alpha;\ u_{\alpha\beta}(a_\beta) = a_\alpha \ (\alpha \leqq \beta)\}$$

and let $u_\alpha : A \to A_\alpha$ be the restriction of the canonical projection $\prod_{\alpha \in \Lambda} A_\alpha \to A_\alpha$ to A. Then $\{A, u_\alpha\}$ becomes the projective limit of the projective system $\{A_\alpha, u_{\alpha\beta}\}$. **Inductive limits** may be defined likewise by changing the direction of the morphisms. We denote the inductive limit of the inductive system $\{A_\alpha, u_{\beta\alpha}\}$ by $A = \varinjlim_\alpha A_\alpha$.

For a projective system $\{A_\alpha, u_{\alpha\beta}\}$ in a category $\mathscr{C}$ and for an object $X \in \mathscr{C}$, $\{h_X(A_\alpha), h_X(u_{\alpha\beta})\}$ is a projective system in the category of sets **E**. Thus we can define the projective limit $\varprojlim_\alpha h_X(A_\alpha)$. The contravariant functor from $\mathscr{C}$ to **E** defined by

$$F : X \mapsto \varprojlim_\alpha h_X(A_\alpha)$$

is representable and is represented by $A \in \mathscr{C}$ if and only if $A = \varprojlim_\alpha A_\alpha$. In this situation, we have

$$\mathscr{C}(X, \varprojlim_\alpha A_\alpha) \cong \varprojlim_\alpha \mathscr{C}(X, A_\alpha).$$

Similarly, given an inductive system $\{A_\alpha, u_{\beta\alpha}\}$ of a category $\mathscr{C}$ and an object $X \in \mathscr{C}$, $\{h^X(A_\alpha), h^X(u_{\beta\alpha})\}$ is a projective system in the category of sets **E**, so we can define a covariant functor from $\mathscr{C}$ to **E** by

$$G : X \mapsto \varprojlim_\alpha h^X(A_\alpha).$$

Then G is representable and is represented by $A \in \mathscr{C}$ if and only if $\varinjlim_{\alpha} A_\alpha = A$. In this situation, we have

$$\mathscr{C}(\varinjlim_{\alpha} A_\alpha, X) \cong \varprojlim_{\alpha} \mathscr{C}(A_\alpha, X).$$

EXERCISE Construct the projective limit and the inductive limit respectively of a projective system and an inductive system in the categories **Gr**, $\mathbf{Mod}_k$, $\mathbf{M}_k$.

Remark (1) Given an ordered set Λ in which $\alpha \leqq \beta$ means $\alpha = \beta$, namely, if for any two distinct elements of Λ there is no order, then the projective limit is simply the direct product and the inductive limit is simply the direct sum.

(2) Let Λ be an ordered set. If every finite subset of Λ has an upper bound, namely, if for each finite subset of elements $\alpha_1, \ldots, \alpha_n$ of Λ, there exists $\beta \in \Lambda$ such that $\alpha_i \leqq \beta$ $(1 \leqq i \leqq n)$, then Λ is called a **directed set**. Let Λ' be a subset of a directed set Λ. If for any $\alpha \in \Lambda$, there exists $\beta \in \Lambda'$ such that $\alpha \leqq \beta$, then Λ' is said to be a **cofinal subset** of Λ. In this situation, Λ' is a directed subset of Λ. The limit of a projective system (resp. an inductive system) $F : \Lambda \to \mathscr{C}$ for a directed set Λ coincides with the limit of the projective system (resp. the inductive system) $F|_{\Lambda'} : \Lambda' \to \mathscr{C}$ which is the restriction of F to the cofinal subset Λ' of Λ provided it exists.

A.5 $\mathscr{C}$-groups and $\mathscr{C}$-cogroups

Let $\mathscr{C}$ be a category with a terminal object such that the direct product of two arbitrary objects exists. Given $G \in \mathscr{C}$, if there exist $\mu \in \mathscr{C}(G \times G, G)$, a terminal object $e \in \mathscr{C}$, and $\eta \in \mathscr{C}(e, G)$ such that the following diagrams

(1) (the associative law)

$$\begin{array}{ccc} G \times G \times G & \xrightarrow{\mu \times 1} & G \times G \\ {\scriptstyle 1 \times \mu}\downarrow & & \downarrow{\scriptstyle \mu} \\ G \times G & \xrightarrow{\mu} & G \end{array}$$

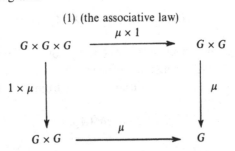

(2) (the unitary property)

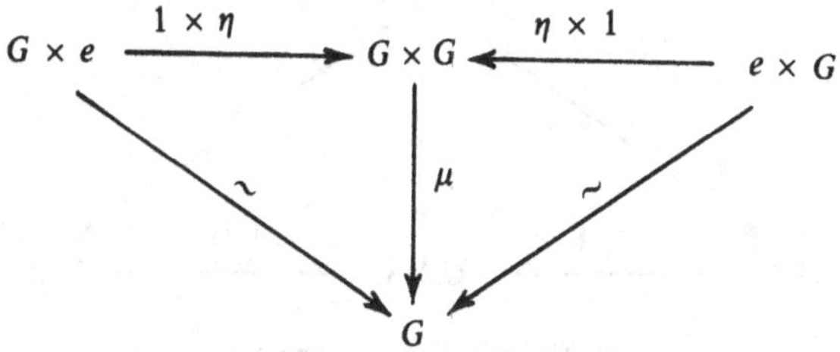

are commutative, then G is called a **$\mathscr{C}$-semigroup**. Furthermore, if there exists $S \in \mathscr{C}(G, G)$ making the diagram

(3) (the inverse property)

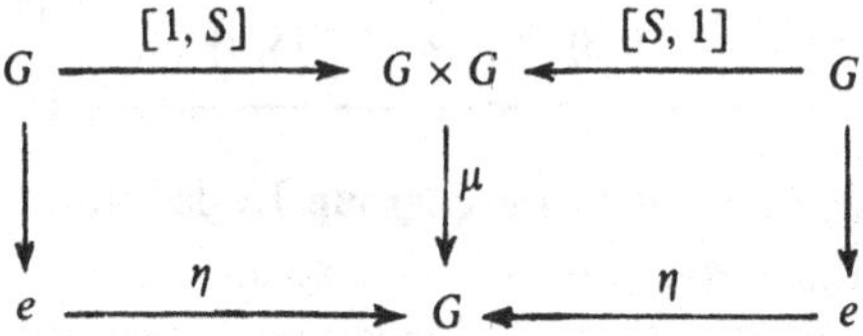

commute, G is called a **$\mathscr{C}$-group**. Here, $[1, S]$ and $[S, 1]$ stand for the morphisms uniquely determined for $1_G = 1 \in \mathscr{C}(G, G)$ and $S \in \mathscr{C}(G, G)$ by the definition of the direct product, and $G \to e$ is the unique morphism to the terminal object e. When $\mathscr{C}$ is the category of sets or the category of topological spaces, then a $\mathscr{C}$-group turns out to be a group or a topological group respectively. When $\mathscr{C}$ is the category of analytic manifolds or the category of affine k-varieties, then a $\mathscr{C}$-group is a Lie group or an affine k-group respectively.

Dually, let $\mathscr{C}$ be a category with an initial object such that the direct sum of any two objects in $\mathscr{C}$ exists. If for $C \in \mathscr{C}$, there exists $\Delta \in \mathscr{C}(C, C \amalg C)$, an initial object $e \in \mathscr{C}$, and $\varepsilon \in \mathscr{C}(C, e)$ such that diagrams (1) and (2) below commute, then C is said to be a **$\mathscr{C}$-cosemigroup**. Moreover, if there exists $S \in \mathscr{C}(C, C)$ making diagram (3)

(1) (the coassociative law)

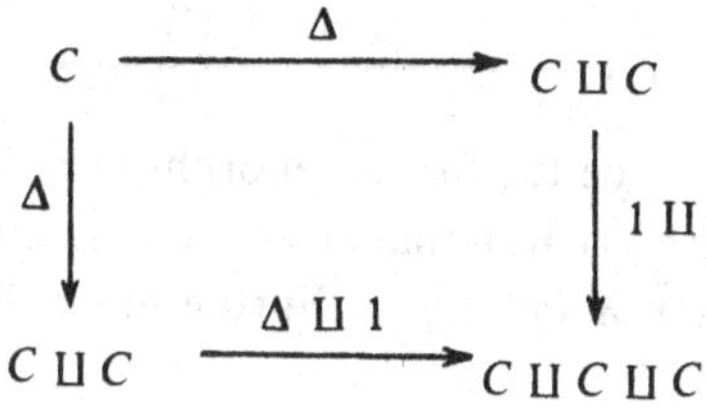

(2) (the counitary property)

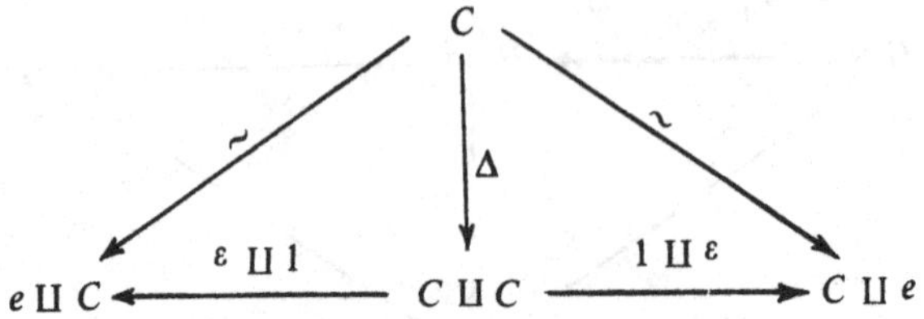

(3) (the coinverse property)

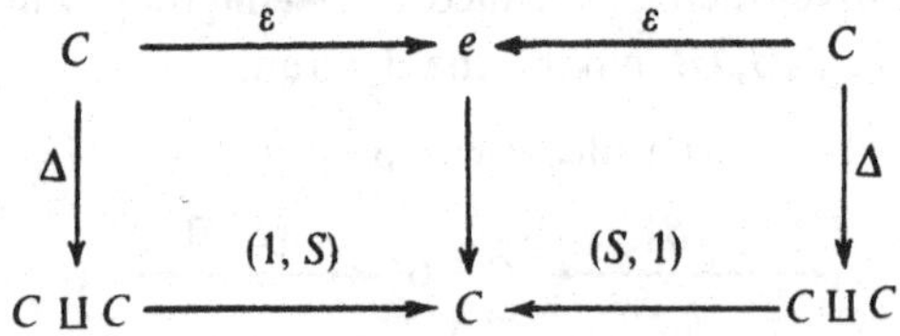

commutative, then C is called a **$\mathscr{C}$-cogroup**. By definition, a $\mathscr{C}$-cogroup is a $\mathscr{C}^\circ$-group and a $\mathscr{C}$-group is a $\mathscr{C}^\circ$-cogroup.

Given a category $\mathscr{C}$, a ($\mathscr{C}^\circ$**E**)-group is a contravariant functor from $\mathscr{C}$ to **Gr**. In particular, suppose $\mathscr{C}$ has a terminal object and direct products. Then, if a representable ($\mathscr{C}^\circ$**E**)-group F can be represented by an object G of $\mathscr{C}$, G is a $\mathscr{C}$-group. In fact, setting $F = h^G$ with $F(X) = \mathscr{C}(X, G)$ and letting $\tilde{\mu} : F \times F \to F$ be the functor morphism which defines the multiplication on F, then for $X \in \mathscr{C}$, we have

$$\tilde{\mu}(X) : \mathscr{C}(X, G) \times \mathscr{C}(X, G) \cong \mathscr{C}(X, G \times G) \to \mathscr{C}(X, G).$$

Now set $\tilde{\mu}(G \times G)(1_{G \times G}) = \mu$. Given $f \in \mathscr{C}(X, G \times G)$, since the diagram below commutes, we can write $\tilde{\mu}(X)(f) = f \circ \mu$.

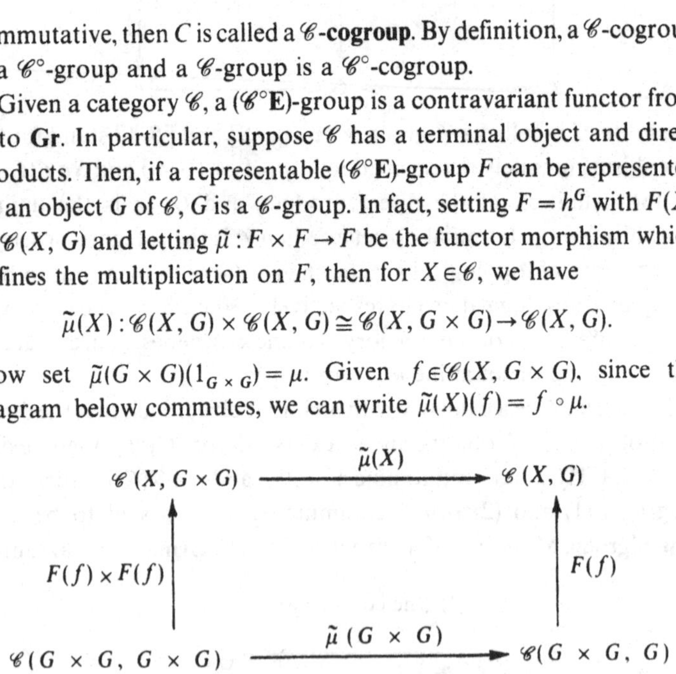

Similarly let $\tilde{\eta} : e \to F$ be the functor morphism defining the identity element and regard e as a terminal object of $\mathscr{C}$. Setting $\tilde{\eta}(e)(1_e) = \eta$, for $f \in \mathscr{C}(X, e)$, we get $\tilde{\eta}(X)(f) = \eta \circ f$. Furthermore, if $\tilde{S} : F + F$ is the functor morphism defining the inverse element, setting $\tilde{S}(G)(1_G) = S$, we have $\tilde{S}(X)(f) = S \circ f$ for $f \in \mathscr{C}(X, G)$. Now μ, η, S satisfy the axioms of a $\mathscr{C}$-group for G.

Dually, a ($\mathscr{C}$**E**)-group is a covariant functor from $\mathscr{C}$ to **Gr**. In particular, for a category $\mathscr{C}$ with an initial object and direct sums, if a representable ($\mathscr{C}$**E**)-group F is represented by an object A of $\mathscr{C}$, then A is a $\mathscr{C}$-cogroup.

For $k \in \mathbf{M}$, $\mathbf{M}_k$-cosemigroups and $\mathbf{M}_k$-cogroups turn out to be k-bialgebras and k-Hopf algebras respectively. An affine k-group is a representable $\mathbf{M}_k\mathbf{E}$-group, and can be represented by an $\mathbf{M}_k$-cogroup, namely, by a k-Hopf algebra.

References

Chapter 1

The following is a list of introductory books which give more detailed discussions on the topics taken up in Chapter 1. Books [1] and [2] are written in Japanese.

[1] Hattori, A. 1968. *Modern algebra* . Asakura shoten.
[2] Nagata, M. 1974. *Commutative algebra.* Kinokuniya shoten.
[3] Bourbaki, N. 1962. *Algèbre*, Chapter 2, Algèbre linéaire. Herman.
[4] Bourbaki, N. 1961. *Algèbre commutative*, Chapter 1. Herman.
[5] Curtis, C. W. & Reiner, I. 1962. *Representation theory of finite groups and associative algebras.* Interscience.
[6] Bourbaki, N. 1960. *Groupes et algèbres de Lie*, Chapter 1. Herman.
[7] Humphreys, J. E. 1972. *Introduction to Lie algebras and representation theory.* Springer.

Chapter 2

[1] gives a readily accessible introduction to the theory of Hopf algebras. [2] to [9] are papers closely related to the material presented in Chapter 2.

[1] Sweedler, M. E. 1969. *Hopf algebras.* Benjamin.
[2] Allen, H. P. 1973. Invariant radical splitting: a Hopf approach. *J. of Pure and App. Alg.* **3**, 1–21.
[3] Hynemann, R. G. & Radford, D. E. 1974. Reflexivity and coalgebras of finite type. *J. Algebra* **28**, 215–46.
[4] Hynemann, R. G. & Sweedler, M. E. 1969, 1970. Affine Hopf algebras I, II. *J. Algebra* **13**, 192–241; **16**, 271–97.
[5] Newmann, K. 1972. Sequences of divided powers in irreducible cocommutative Hopf algebras. *Trans. Amer. Math. Soc.* **163**, 25–34.
[6] Radford, D. E. 1973. Coreflexive coalgebras. *J. Algebra* **26**, 512–35.
[7] Radford, D. E. 1974. On the structure of ideals of the dual algebra of a coalgebra. *Trans. Amer. Math. Soc.* **198**, 123–37.
[8] Sweedler, M. E. 1967. Hopf algebras with one group-like element. *Trans. Amer Math. Soc.* **127**, 515–26.
[9] Taft, E. J. 1972. Reflexivity of algebras and coalgebras. *Amer. J. Math.* **94**, 1111–30.

Chapter 3

[1] to [3] are papers on integrals for Hopf algebras. [4] is on the duality of compact topological groups. Papers [5] to [8] concern the relationship between groups and Hopf algebras.

[1] Larson, R. 1973. Coseparable Hopf algebras. *J. of Pure and App. Alg.* **3**, 261–7.

[2] Sullivan, J. B. 1971. The uniqueness of integrals for Hopf algebras and some existence theorems of integrals for commutative Hopf algebras. *J. Algebra* **19**, 426–40.

[3] Sweedler, M. E. 1969. Integrals for Hopf algebras. *Ann. of Math.* **89**, 323–35.

[4] Hochschild, G. 1965. *Structure of Lie groups.* Holden-Day.

[5] Takahashi, S. 1965. A characterization of group rings as a special class of Hopf algebras. *Canad. Math. Bull.* **8**, 465–75.

[6] Larson, R. 1967. Cocommutative Hopf algebras. *Canad. J. Math.* **19**, 350–60.

[7] Hochschild, G. 1970. Algebraic groups and Hopf algebras. *Illinois J. Math.* **14**, 52–65.

[8] Hochschild, G. & Mostow, G. D. 1969. Complex analytic groups and Hopf algebras. *Amer. J. Math.* **91**, 1141–51.

Chapter 4

[1] to [5] are on affine algebraic groups. [1] includes a treatment of the structure of linear algebraic groups over a field of characteristic 0 and their Lie algebras; this is the first book ever written which deals systematically with the subject. [2], [3] and [5] present the general theory of algebraic groups over a field of arbitrary characteristic and the classification of semi-simple groups. [5] is easy to read. In [4], affine algebraic groups are derived from Hopf algebras which turn out to be their coordinate rings. [6] is one among the many books and papers which discuss group schemes. Papers [7] to [16] deal with applications of Hopf algebras to the theory of algebraic groups which are relevant to this chapter.

[1] Chevalley, C. 1968. *Théorie des groups de Lie.* Hermann.

[2] Séminaire Chevalley. 1956–58. *Classification des groups de Lie algébriques*, I, II.

[3] Borel, A. 1969. *Linear algebraic groups.* Benjamin.

[4] Hochschild, G. 1971. *Introduction to affine algebraic groups.* Holden Day.

[5] Humphreys, J. E. 1975. *Linear algebraic groups.* Springer.

[6] Demazure, M. & Gabriel, P. 1970. *Groupes algébriques*, I. North Holland.

[7] Takeuchi, M. 1972. A correspondence between Hopf ideals and sub-Hopf algebras. *Manuscripta Math.* **7**, 251–70.

[8] Newmann, K. 1975. A correspondence between bi-ideals and sub-Hopf algebras in a cocommutative Hopf algebras. *J. Algebra*, **36**, 1–15.

[9] Sullivan, J. B. 1973. Automorphisms of affine unipotent groups in positive characteristic. *J. Algebra*, **26**, 140–51.

[10] Sullivan, J. B. 1973. A decomposition theorem for solvable pro-affine algebraic groups over algebraically closed fields. *Amer. J. Math.* **95**, 221–8.

[11] Abe, E. & Doi, Y. 1972. Decomposition theorems for Hopf algebras and pro-affine algebraic groups. *J. Math Soc. Japan* **24**, 433–47.

[12] Takeuchi, M. 1972. On a semi-direct product decomposition of affine groups over a field of characteristic 0. *Tohoku Math. J.* **24**, 453–6

[13] Sweedler, M. E. 1971. Connected fully reducible affine group schemes in positive characteristic are abelian. *J. Math. Kyoto Univ.* **11**, 51–70.

[14] Hochschild, G. 1970. Coverings of pro-affine algebraic groups. *Pacific J. Math.* **35**, 399–415.
[15] Takeuchi, M. 1974. Tangent coalgebras and hyperalgebras I. *Jap. J. Math.* **42**, 1–143.
[16] Takeuchi, M. 1975. On coverings and hyperalgebras of affine algebraic groups. *Trans. Amer. Math. Soc.* **211**, 249–75.

Chapter 5

This chapter is based on [1] and [2].

[2] is somewhat difficult to follow since notation for algebraic systems such as bialgebras is used unsystematically.

[3] uses group schemes, and [4] presents the Galois theory of rings.

[1] Sweedler, M. E. 1968. Structure of inseparable extensions. *Ann. of Math.* **87**, 401–10. Correction ibid. **88**, 206–7.
[2] Winter, D. 1974. *The structure of fields*. Springer.
[3] Chase, S. U. 1972. On the automorphism scheme of a purely inseparable field extension. *Proc. of Conf. Ring Theory*. Utah Academic Press.
[4] Chase, S. U. & Sweedler, M. E. 1969. *Hopf algebras and Galois theory*. Lecture Notes in Math. **97**. Springer.

Index

Printed in the United States
By Bookmasters